T0296305

Rugged Embedded Systems

Rugged Embedded Systems

Computing in Harsh Environments

Augusto Vega

Pradip Bose

Alper Buyuktosunoglu

AMSTERDAM • BOSTON • HEIDELBERG • LONDON
NEW YORK • OXFORD • PARIS • SAN DIEGO
SAN FRANCISCO • SINGAPORE • SYDNEY • TOKYO
Morgan Kaufmann is an imprint of Elsevier

Morgan Kaufmann is an imprint of Elsevier
50 Hampshire Street, 5th Floor, Cambridge, MA 02139, United States

Library of Congress Cataloging-in-Publication Data
A catalog record for this book is available from the Library of Congress

British Library Cataloguing-in-Publication Data
A catalogue record for this book is available from the British Library

ISBN: 978-0-12-802459-1

For information on all Morgan Kaufmann publications
visit our website at https://www.elsevier.com/

Acquisition Editor: Todd Green
Editorial Project Manager: Charlotte Kent
Production Project Manager: Mohana Natarajan
Cover Designer: Mathew Limbert

Typeset by SPi Global, India

Dedication

To my wife Chiara and our little boy Niccolò—they teach me every day how to find happiness in the simplest of things. To the amazing group of people I have the privilege to work with at IBM.

Augusto Vega

To all contributing members of the IBM-led project on: "*Efficient resilience in embedded computing*," sponsored by DARPA MTO under its PERFECT program. To my wife (Sharmila) and my mother (Reba) for their incredible love and support.

Pradip Bose

To my wife Agnieszka, to our baby boy John and to my parents. To the wonderful group of colleagues that I work with each day.

Alper Buyuktosunoglu

Contents

Contributors

J. Abella
Barcelona Supercomputing Center, Barcelona, Spain

R.A. Ashraf
University of Central Florida, Orlando, FL, United States

P. Bose
IBM T. J. Watson Research Center, Yorktown Heights, NY, United States

A. Buyuktosunoglu
IBM T. J. Watson Research Center, Yorktown Heights, NY, United States

V.G. Castellana
Pacific Northwest National Laboratory, Richland, WA, United States

F.J. Cazorla
IIIA-CSIC and Barcelona Supercomputing Center, Barcelona, Spain

M. Ceriani
Polytechnic University of Milan, Milano, Italy

E. Cheng
Stanford University, Stanford, CA, United States

R.F. DeMara
University of Central Florida, Orlando, FL, United States

F. Ferrandi
Polytechnic University of Milan, Milano, Italy

R. Gioiosa
Pacific Northwest National Laboratory, Richland, WA, United States

N. Imran
University of Central Florida, Orlando, FL, United States

M. Minutoli
Pacific Northwest National Laboratory, Richland, WA, United States

S. Mitra
Stanford University, Stanford, CA, United States

G. Palermo
Polytechnic University of Milan, Milano, Italy

J. Rosenberg
Draper Laboratory, Cambridge, MA, United States

A. Tumeo
Pacific Northwest National Laboratory, Richland, WA, United States

A. Vega
IBM T. J. Watson Research Center, Yorktown Heights, NY, United States

R. Zalman
Infineon Technologies, Neubiberg, Germany

X. Zhang
Washington University, St. Louis, MO, United States

Preface

The adoption of *rugged* chips that can operate reliably even under extreme conditions has experienced an unprecedented growth. This growth is in tune with the revolutions related to mobile systems and the Internet of Things (IoT), emergence of autonomous and semiautonomous transport systems (such as connected and *driverless* cars), and highly automated factories and the robotics *boom*. The numbers are astonishing—if we consider just a few domains (connected cars, *wearable* and IoT devices, tablets and smartphones), we will end up having around 16 billion embedded devices surrounding us by 2018, as Fig. 1 shows.

A distinctive aspect of embedded systems (probably the most interesting one) is the fact that they allow us to take computing virtually anywhere, from a car's braking system to an interplanetary rover exploring another planet's surface to a computer attached to (or even implanted into!) our body. In other words, there exists a *mobility* aspect—inherent to this type of systems—that gives rise to all sorts of design and operation challenges, high energy efficiency and reliable operation being the most critical ones. In order to meet target energy budgets, one can decide to (1) minimize error detection or error tolerance related overheads and/or (2) enable aggressive power and energy management features, like low- or near-threshold voltage

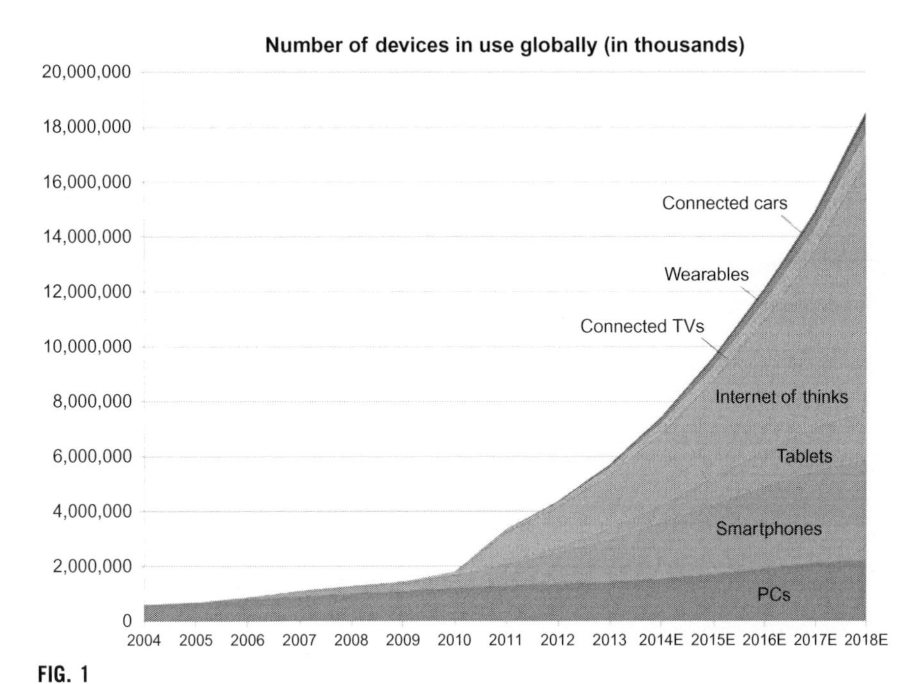

FIG. 1

Embedded devices growth through 2018.

Source: Business Insider Intelligence.

operation. Unfortunately, both approaches have direct impact on error rates. The hardening mechanisms (like hardened latches or error-correcting codes) may not be affordable since they add extra complexity. Soft error rates (SERs) are known to increase sharply as the supply voltage is scaled down. It may appear to be a rather challenging scenario. But looking back at the history of computers, we have overcome similar (or even larger) challenges. Indeed, we have already hit severe power density-related issues in the late 80s using bipolar transistors and here we are, almost 30 years after, still creating increasingly powerful computers and machines.

The challenges discussed above motivated us some years ago to ignite serious discussion and brainstorming in the computer architecture community around the most critical aspects of new-generation harsh-environment-capable embedded processors. Among a variety of activities, we have successfully organized three editions of the workshop on Highly-Reliable Power-Efficient Embedded Designs (HARSH), which have attracted the attention of researchers from academia, industry, and government research labs during the last years. Some of the experts that contributed material to this book had previously participated in different editions of the HARSH workshop. This book is in part the result of such continued efforts to foster the discussion in this domain involving some of the most influential experts in the area of rugged embedded systems.

This book was also inspired by work that the guest editors have been pursuing under DARPA's PERFECT (*Power Efficiency Revolution for Embedded Computing Technologies*) program. The idea was to capture a representative sample of the current state of the art in this field, so that the research challenges, goals, and solution strategies of the PERFECT program can be examined in the right perspective. In this regard, the book editors want to acknowledge DARPA's sponsorship under contract no. HR0011-13-C-0022.

We also express our deep gratitude to all the contributors for their valuable time and exceptional work. Needless to say, this book would not have been possible without them. Finally, we also want to acknowledge the support received from the IBM T. J. Watson Research Center to make this book possible.

Augusto Vega
Pradip Bose
Alper Buyuktosunoglu
Summer 2016

CHAPTER

Introduction

1

A. Vega, P. Bose, A. Buyuktosunoglu
IBM T. J. Watson Research Center, Yorktown Heights, NY, United States

Electronic digital computers are very powerful tools. The so-called *digital revolution* has been fueled mostly by chips for which the number of transistors per unit area on integrated circuits kept doubling approximately every 18 months, following what Gordon Moore observed in 1975. The resulting exponential growth is so remarkable that it fundamentally changed the way we perceive and interact with our surrounding world. It is enough to look around to find traces of this *revolution* almost everywhere. But the dramatic growth exhibited by computers in the last three to four decades has also relied on a fact that goes frequently unnoticed: *they operated in quite predictable environments and with plentiful resources*. Twenty years ago, for example, a desktop personal computer sat on a table and worked without much concern about power or thermal dissipation; security threats also constituted rare episodes (computers were barely connected if connected at all!); and the few mobile devices available did not have to worry much about battery life. At that time, we had to put our eyes on some specific niches to look for truly sophisticated systems—i.e., systems that had to operate on *unfriendly* environments or under significant amount of "stress." One of those niches was (and is) *space exploration*: for example, NASA's Mars Pathfinder planetary rover was equipped with a RAD6000 processor, a radiation-hardened POWER1-based processor that was part of the rover's on-board computer [1]. Released in 1996, the RAD6000 was not particularly impressive because of its computational capacity—it was actually a modest processor compared to some contemporary high-end (or even embedded system) microprocessors. Its cost—in the order of several hundred thousand dollars—is better understood as a function of the chip *ruggedness* to withstand total radiation doses of more than 1,000,000 rads and temperatures between −25°C and +105°C in the thin Martian atmosphere [2].

In the last decade, computers continued growing in terms of performance (still riding on Moore's Law and the multicore era) and chip power consumption became a critical concern. Since the early 2000s, processor design and manufacturing is not driven by *just* performance anymore but it is also determined by strict power budgets—a phenomenon usually referred to as the "power wall." The rationale behind the power wall has its origins in 1974, when Robert Dennard et al. from the IBM T. J. Watson

Rugged Embedded Systems. http://dx.doi.org/10.1016/B978-0-12-802459-1.00001-4

Research Center, postulated the scaling rules of metal-oxide-semiconductor field-effect transistors (MOSFETs) [3]. One key assumption of the Dennard's scaling rule is that operating voltage (V) and current (I) should scale proportionally to the linear dimensions of the transistor in order to keep power consumption ($V \times I$) proportional to the transistor area (A). But manufacturers were not able to lower operating voltages sufficiently over time and power density ($V \times I/A$) kept growing until it reached unsustainable levels. As a result, frequency scaling was knocked down and industry shifted to multicore designs to cope with single-thread performance limitations.

The power wall has fundamentally changed the way modern processors are conceived. Processors became aware of power consumption with additional on-chip "intelligence" for power management—*clock gating* and *dynamic voltage and frequency scaling* (DVFS) are two popular dynamic power reduction techniques in use today. But at the same time, chips turned out to be more susceptible to errors (*transient* and *permanent*) as a consequence of thermal issues derived from high power densities as well as low-voltage operation. In other words, we have hit the *reliability wall* in addition to the power wall. The power and reliability walls are interlinked as shown in Fig. 1. The power wall forces us toward designs that have tighter design margins and "better than worst case" design principles. But that approach eventually degrades reliability ("mean time to failure")—which in turn requires redundancy and hardening techniques that increase power consumption and forces us back against the power wall. This is a vicious "karmic" cycle!

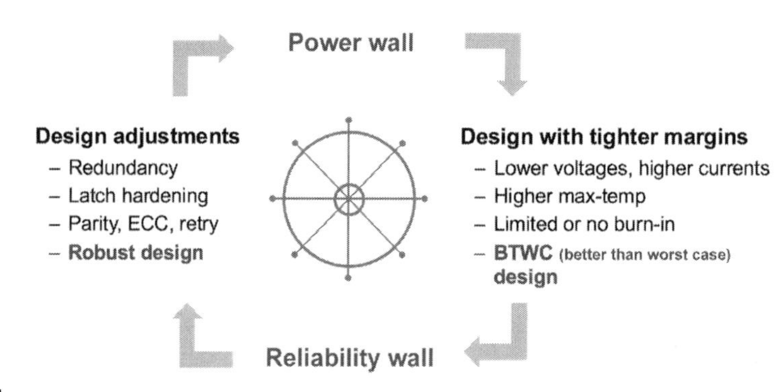

FIG. 1

Relationship and mutual effect between the power and reliability walls.

This already worrying outlook exacerbated in tune with the revolutions related to mobile systems and the Internet of Things (IoT) since the aforementioned constraints (e.g., power consumption and battery life) get more strict and the challenges associated with fault-tolerant and reliable operation become more critical. Embedded computing has become pervasive and, as a result, many of the day-to-day devices that we use and rely on are subject to similar constraints—in some cases, with critical consequences when they are not met. Automobiles are becoming "smarter" and in some cases autonomous (*driverless*), robots can conduct medical surgery as well as other critical roles in the health and medical realm, commercial aviation is heavily automated (modern aircrafts are generally flown by a computer autopilot), just to mention a few examples. In all these cases, highly-reliable, low-power embedded systems are the key enablers and it is not difficult to imagine the safety-related consequences if the system fails or proper operation is not guaranteed. In this context, we refer to a *harsh* environment as a scenario that presents inherent characteristics (like extreme temperature and radiation levels, very low power and energy budgets, strict fault tolerance, and security constraints, among others) that challenge the embedded system in its design and operation. When such a system guarantees proper operation under harsh conditions (eventually with acceptable deviations from its functional specification), we say that it is a *rugged* embedded system.

Interestingly, the mobile systems and IoT boom has also disrupted the scope of the reliability and power optimization efforts. In the past, it was somewhat enough to focus on per-system (underlying hardware + software) optimization. But this is not the case anymore in the context of embedded systems for mobile and IoT applications. In such scenario, systems exhibit much tighter interaction and interdependence with distributed, mobile (*swarm*) computing aspects as well as on-demand support from cloud (server) in some cases (Fig. 2). This interaction takes place mostly over unreliable wireless channels [4] and may require resilient system reconfiguration on node failure or idle rotation. In other words, the architectural vision scope has changed (expanded) and so the resulting optimization opportunities also have.

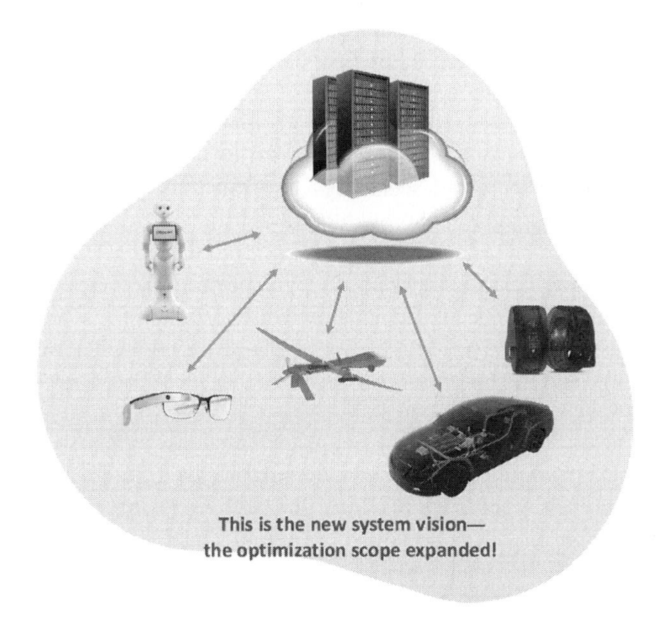

This is the new system vision—
the optimization scope expanded!

FIG. 2

New system architectural vision for the mobile and IoT *eras*.

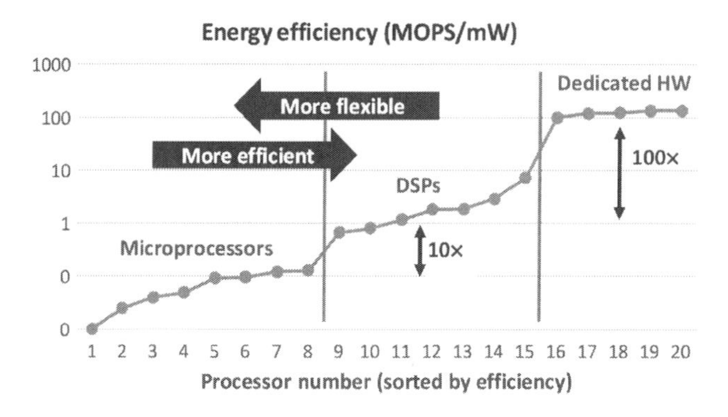

FIG. 3

Energy efficiency via specialization expressed in terms of million operations per second (MOPS) per milliwatts.

Source: Bob Brodersen, Berkeley Wireless Group.

Embedded processors in general (and those targeted to operate in harsh environments in particular) are designed taking into consideration a precise application or a well-defined domain, and only address those requirements (we say they are *domain specific* or *dedicated* or *specialized*). Domain-specific designs are easier to verify since the range of different use cases that the system will face during operation is usually well known in advance. But specialized hardware also means higher power/energy efficiency (compared to a general-purpose design) since the hardware is highly optimized for the specific function(s) that the processor is conceived to support. In general, the advantage in terms of efficiency over general-purpose computation can be huge in the range of 10–100× as shown in Fig. 3.

The aforementioned challenges—i.e., ultra-efficient, fault-tolerant, and reliable operation in highly-constrained scenarios—motivate this edited book. Our main goal is to provide a broad yet thorough treatment of the field through first-hand use cases contributed by experts from industry and academia currently involved in some of the most exciting embedded systems projects. We expect the reader to gain a deep understanding of the comprehensive field of embedded systems for harsh environments, covering the state-of-the-art in unmanned aerial vehicles, autonomous cars, and interplanetary rovers, as well as the inherent security implications. To guarantee robustness and fault tolerance across these diverse scenarios, the design and operation of rugged embedded systems for harsh environments should not be solely confined to the hardware but traverse different layers, involving firmware, operating system, applications, as well as power management units and communication interfaces, as shown in Fig. 4. Therefore, this book addresses the latest ideas, insights, and knowledge related to *all* critical aspects of new-generation harsh environment-capable embedded computers, including architectural approaches, cross-stack hardware/software techniques, and emerging challenges and opportunities.

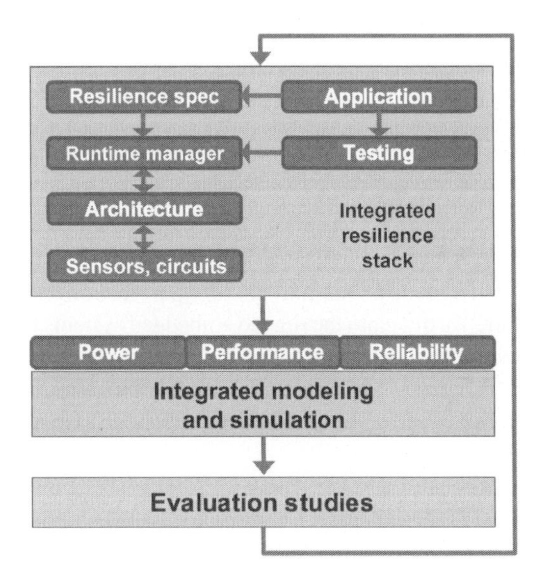

FIG. 4

Cross-layer optimization approach.

Today is a turning point for the embedded computer industry. As it was mentioned before, computers are being deployed almost everywhere and in unimaginable ways having become critical in our daily lives. Therefore, we think that this is the right moment to address the rugged embedded systems field and capture its technological and social challenges in a comprehensive edited book.

1 WHO THIS BOOK IS FOR

The book treats the covered areas in depth and with a great amount of technical details. However, we seek to make the book accessible to a broad set of readers by addressing topics and use cases first from an informational standpoint with a gradual progression to complexity. In addition, the first chapters lead the reader through the fundamental concepts on reliable and power-efficient embedded systems in such a way that people with minimal expertise in this area can still come to grips with the different use cases. In summary, the book is intended for an audience including but not limited to:

- Academics (undergraduates and graduates as well as researchers) in the computer science, electrical engineering, and telecommunications fields. We can expect the book to be adopted as complementary reading in university courses.
- Professionals and researchers in the computer science, electrical engineering, and telecommunications industries.

In spite of our intention to make it accessible to a broad audience, this book is not written with the *newcomer* in mind. Even though we provide an introduction to the field, a minimum amount of familiarity with embedded systems and reliability principles is strongly recommended to get the most out of the book.

2 HOW THIS BOOK IS ORGANIZED

The book is structured as follows: an introductory part that covers fundamental concepts (Chapters 1 and 2), a dive into the rugged embedded systems field (Chapters 3–6) with a detour into the topic of resilience for extreme scale computing (Chapter 5), a set of three case studies (Chapters 7–9), and a final part that provides a cutting-edge vision of cross-layer resilience for next-generation rugged systems (Chapter 10). We briefly describe each Chapter below:

Chapter 2: Reliable and power-aware architectures: Fundamentals and modeling. This Chapter discusses fundamental reliability concepts as well as techniques to deal with reliability issues and their power implications. It also introduces basic concepts related to power-performance modeling and measurement.

Chapter 3: Real-time considerations for rugged embedded systems. This Chapter introduces the characterizing aspects of embedded systems and discusses the specific features that a designer should address to *make* an embedded system "rugged"—i.e., able to operate reliably in harsh environments. The Chapter also presents a case study that focuses on the interaction of the hardware and software layers in reactive real-time embedded systems.

Chapter 4: Emerging resilience techniques for embedded devices. This Chapter presents techniques for highly reliable and survivable Field Programmable Gate Array (FPGA)-based embedded systems operating in harsh environments. The notion of autonomous self-repair is essential for such systems as physical access to such platforms is often limited. In this regard, adaptable reconfiguration-based techniques are presented.

Chapter 5: Resilience for extreme scale computing. This Chapter reviews the intrinsic characteristics of high-performance applications and how faults occurring in hardware propagate to memory. It also summarizes resilience techniques commonly used in current supercomputers and supercomputing applications and explores some resilience challenges expected in the exascale era and possible programming models and resilience solutions.

Chapter 6: Embedded security. Embedded processors can be subject to cyber attacks which constitutes another source of *harshness*. This Chapter discusses the nature of this type of harsh environment, what enables cyber attacks, what are the principles we need to understand to work toward a much higher level of security, as well as new developments that may change the game in our favor.

Chapter 7: Reliable electrical systems for MAVs and insect-scale robots. This Chapter presents the progress made on building an insect-scale microaerial vehicle (MAV) called *RoboBee* and zooms into the critical reliability issues associated with this system. The Chapter focuses on the design context and motivation of customized System-on-Chip for microrobotic application and provides an in-depth investigation of supply resilience in battery-powered microrobotic system using a prototype chip.

Chapter 8: Rugged autonomous vehicles. This Chapter offers an overview of embedded systems and their usage in the automotive domain. It focuses on the constraints particular to embedded systems in the automotive area with emphasis on providing dependable systems in harsh environments specific to this domain. The Chapter also mentions challenges for automotive embedded systems deriving from modern emerging applications like autonomous driving.

Chapter 9: Harsh computing in the space domain. This Chapter discusses the main challenges in spacecraft systems and microcontrollers verification for future missions. It reviews the verification process for spacecraft microcontrollers and introduces a new holistic approach to deal with functional and timing correctness based on the use of randomized probabilistically analyzable hardware designs and appropriate timing analyses—a promising path for future spacecraft systems.

Chapter 10: Resilience in next-generation embedded systems. This final Chapter presents a unique framework which overcomes a major challenge in the design of rugged embedded systems: achieve desired resilience targets at minimal costs (energy, power, execution time, and area) by combining resilience techniques across various layers of the system stack (circuit, logic, architecture, software, and algorithm). This is also referred to as cross-layer resilience.

We sincerely hope that you enjoy this book and find its contents informative and useful!

ACKNOWLEDGMENTS

The book editors acknowledge the input of reliability domain experts within IBM (e.g., Dr. James H. Stathis and his team) in developing the subject matter of relevant sections within Chapter 2.

The work presented in Chapters 1, 2, and 10 is sponsored by Defense Advanced Research Projects Agency, Microsystems Technology Office (MTO), under contract no. HR0011-13-C-0022. The views expressed are those of the authors and do not reflect the official policy or position of the Department of Defense or the U.S. Government.

REFERENCES

[1] Wikipedia, IBM RAD6000—Wikipedia, the free encyclopedia. (2015). https://en.wikipedia.org/w/index.php?title¼IBM_RAD6000&oldid¼684633323 [Online; accessed July 7, 2016].

[2] B.A.E. Systems, RAD6000™ Space Computers, (2004). https://montcs.bloomu.edu/~bobmon/PDFs/RAD6000_Space_Computers.pdf.
[3] R. Dennard, F. Gaensslen, H.N. Yu, L. Rideout, E. Bassous, A. LeBlanc, Design of ion-implanted MOSFET's with very small physical dimensions. IEEE Journal of Solid-State Circuits, vol. SC-9(5), (October 1974), pp. 256–268.
[4] A. Vega, C.C. Lin, K. Swaminathan, A. Buyuktosunoglu, S. Pankanti, P. Bose, Resilient, UAV-embedded real-time computing, in: Proceedings of the 33rd IEEE International Conference on Computer Design (ICCD 2015), 2015, pp. 736–739.

CHAPTER

Reliable and power-aware architectures: Fundamentals and modeling

2

A. Vega*, P. Bose*, A. Buyuktosunoglu*, R.F. DeMara†
*IBM T. J. Watson Research Center, Yorktown Heights, NY, United States**
University of Central Florida, Orlando, FL, United States†

1 INTRODUCTION

Chip power consumption is one of the most challenging and transforming issues that the semiconductor industry has encountered in the past decade, and its sustained growth has resulted in various concerns, especially when it comes to chip reliability. It translates into thermal issues that could harm the chip. It can also determine (i.e., limit) battery life in the mobile arena. Furthermore, attempts to circumvent the power wall through techniques like near-threshold voltage (NTV) computing lead to other serious reliability concerns. For example, chips become more susceptible to soft errors at lower voltages. This scene becomes even more disturbing when we add an extra variable: a hostile (or harsh) surrounding environment. Harsh environmental conditions exacerbate already problematic chip power and thermal issues, and can jeopardize the operation of any conventional (i.e., nonhardened) processor.

This chapter discusses fundamental reliability concepts as well as techniques to deal with reliability issues and their power implications. The first part of the chapter discusses the concepts of *error*, *fault*, and *failure*, the *resolution phases* of resilient systems, and the definition and associated metrics of *hard* and *soft errors*. The second part presents two effective approaches to stress a system from the standpoints of resilience and power-awareness—namely *fault injection* and *microbenchmarking*. Finally, the last part of the chapter briefly introduces basic ideas related to power-performance modeling and measurement.

2 THE NEED FOR RELIABLE COMPUTER SYSTEMS

A *computer system* is a human-designed machine with a sole ultimate purpose: to solve human problems. In practice, this principle usually materializes as a *service* that the system delivers either to a person (the ultimate "consumer" of that service) or to other computer systems. The delivered service can be defined as the system's

Rugged Embedded Systems. http://dx.doi.org/10.1016/B978-0-12-802459-1.00002-6

externally perceived behavior [1] and when it matches what is "expected," then the system is said to operate correctly (i.e., the service is correct). The expected service of a system is described by its *functional specification* which includes the description of the system functionality and performance, as well as the threshold between acceptable versus unacceptable behavior [1]. In spite of the different (and sometimes even incongruous) definitions around system reliability, one idea is unanimously accepted: ideally, a computer system should operate correctly (i.e., stick to its functional specification) all the time; and when its internal behavior experiences anomalies, the impact on the external behavior (i.e., the delivered service) should be concealed or minimized.

In practice, a computer system can face anomalies (faults and errors) during operation which require palliative actions in order to conceal or minimize the impact on the system's externally perceived behavior (failure). The concepts of *error*, *fault*, and *failure* are discussed in Section 2.1. The ultimate goal is to sustain the *quality of the service* (QoS) being delivered in an acceptable level. The range of possible palliative actions is broad and strongly dependent on the system type and use. For example, space-grade computers deployed on earth-orbiting satellites demand more effective (and frequently more complex) fault-handling techniques than computers embedded in mobile phones. But in most cases, these actions usually involve *anomaly detection* (AD), *fault isolation* (FI), *fault diagnosis* (FD), and *fault recovery* (FR). These four *resolution phases* are discussed in detail in Section 2.2.

Today, reliability has become one of the most critical aspects of computer system design. Technology scaling, per Moore's Law has reached a stage where process variability, yield, and in-field aging threaten the economic viability of future scaling. Scaling the supply voltage down per classical Dennard's rules has not been possible lately, because a commensurate reduction in device threshold voltage (to maintain performance targets) would result in a steep increase in leakage power. And, even a smaller rate of reduction in supply voltage needs to be done carefully—because of the soft error sensitivity to voltage. Other device parameters must be adjusted to retain per-device soft error rates at current levels in spite of scaling. Even with that accomplished, the per-chip soft error rate (SER) tends to increase with each generation due to the increased device density. Similarly, the dielectric (oxide) thickness within a transistor device has shrunk at a rate faster than the reduction in supply voltage (because of performance targets). This threatens to increase hard fail rates of processor chips beyond acceptable limits as well. It is uncertain today what will be the future impact of further miniaturization beyond the 7-nm technology node in terms of meeting an acceptable (or affordable) balance across reliability and power consumption metrics related to prospective computing systems. In particular for mission-critical systems, device reliability and system survivability pose increasingly significant challenges [2–5]. Error resiliency and self-adaptability of future electronic systems are subjects of growing interest [3, 6]. In some situations, even survivability in the form of graceful degradation is desired if a full recovery cannot be achieved. Transient, or so-called *soft errors* as well as permanent, *hard errors* in electronic devices caused by aging require autonomous mitigation as manual intervention may not be feasible [7]. In application domains that include harsh operating environments (e.g., high altitude, which exacerbates soft error rates, or extreme temperature swings that exacerbate certain other transient and permanent failure rates), the concerns about future system reliability are of course even more pronounced. The reliability concerns of highly complex VLSI systems in sub-22 nm processes, caused by soft and hard errors, are increasing. Therefore, the importance of addressing reliability issues is on the rise. In general, a system is said to be *resilient* if it is capable of handling failures throughout its lifetime to maintain the desired processing performance within some tolerance.

2.1 SUSTAINING QUALITY OF SERVICE IN THE PRESENCE OF FAULTS, ERRORS, AND FAILURES

To advance beyond static redundancy in the nanoscale era, it is essential to consider innovative resilient techniques which distinguish between *faults*, *errors*, and *failures* in order to handle them in innovative ways. Fig. 1 depicts each of these terms using a layered model of system dependability.

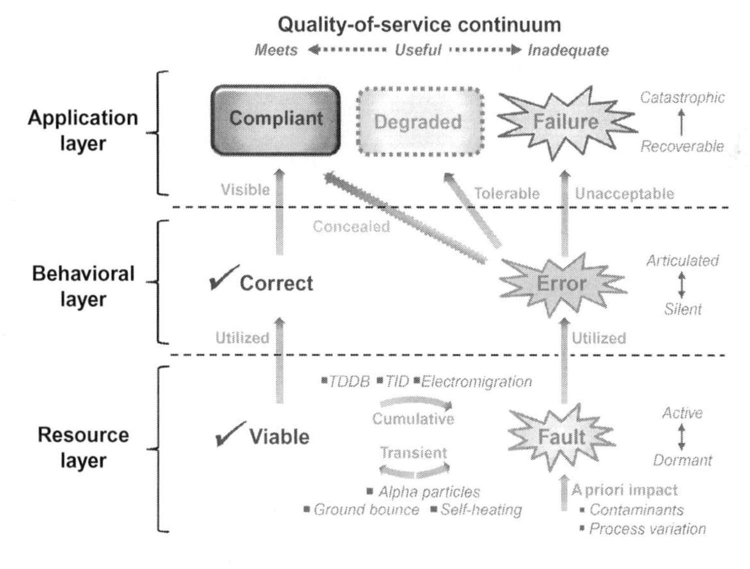

FIG. 1

Layered model of system dependability.

The *resource layer* consists of all of the physical components that underlie all of the computational processes used by an (embedded) application. These physical components span a range of granularities including logic gates, field-programmable gate array (FPGA) look-up tables, circuit functional units, processor cores, and memory chips. Each physical component is considered to be *viable* during the current computation if it operates without exhibiting defective behavior at the time that it is utilized. On the other hand, components which exhibit defective behavior are considered to be *faulty*, either initially or else may become faulty at any time during the mission. Initially faulty resources are a direct result of a priori conditions of manufacturing imperfections such as contaminants or random effects creating process variation beyond allowed design tolerances [8]. As depicted by the *cumulative* arc in Fig. 1, during the mission each component may transition from viable status to faulty status for highly scaled devices. This transition may occur due to cumulative effects of deep submicron devices such as time-dependent dielectric breakdown (TDDB) due to electrical field weakening of the gate oxide layer, total ionizing dose (TID) of cosmic radiation, electromigration within interconnect, and other progressive degradations over the mission lifetime. Meanwhile, transient effects such as incident alpha particles which ionize critical amounts of charge, ground bounce, and dynamic temperature variations may cause either long lasting or intermittent reversible transitions between viable and faulty status. In this sense, faults may lie *dormant* whereby the physical resource is defective, yet currently unused. Later in the computations, dormant faults become *active* when such components are utilized.

The *behavioral layer* shown in Fig. 1 depicts the outcome of utilizing viable and faulty physical components. Viable components result in correct behavior during the interval of observation. Meanwhile, utilization of faulty components manifests *errors* in the behavior according to the input/output and/or timing requirements which define the constituent computation. Still, an error which occurs but does not incur any impact to the result of the computation is termed a *silent* error. Silent errors, such as a flipped bit due to a faulty memory cell at an address which is not referenced by the application, remain isolated at the behavioral layer without propagating to the application. On the other hand, errors which are *articulated* propagate up to the application layer.

The *application layer* shown in Fig. 1 depicts that correct behaviors contribute to sustenance of *compliant* operation. Systems that are compliant throughout the mission at the application layer are deemed to be *reliable*. To remain completely compliant, all articulated errors must be *concealed* from the application to remain within the behavioral layer. For example, error masking techniques which employ voting schemes achieve reliability objectives by insulating articulated errors from the application. Articulated errors which reach the application cause the system to have *degraded* performance if the impact of the error can be tolerated. On the other hand, articulated errors which result in unacceptable conditions to the application incur a *failure* condition. Failures may be catastrophic, but more often are recoverable—e.g., using some of the techniques discussed in Chapter 4. In general, resilience techniques that can provide a *continuum* in QoS (spanning from completing meeting requirements down to inadequate performance from the application perspective) are very desirable. This mapping of the QoS continuum to application states of compliant, degraded, and failure is depicted near the top of Fig. 1.

2.2 PROCESSING PHASES OF COMPUTING SYSTEM RESILIENCY

A four-state model of system resiliency is shown in Fig. 2. For purposes of discussion, the initial and predominant condition is depicted as the lumped representation of the useful operational states of *compliant* or *degraded* performance in the upper center of the figure. To deal with contingencies in an attempt to return to a complaint or degraded state, resilient computing systems typically employ a sequence of resolution phases including AD, FI, FD, and FR using the variety of techniques, some of which are described in this and following chapters. Additionally, methods such as radiation shielding attempt to prevent certain anomalies such as alpha particle-induced soft errors from occurring.

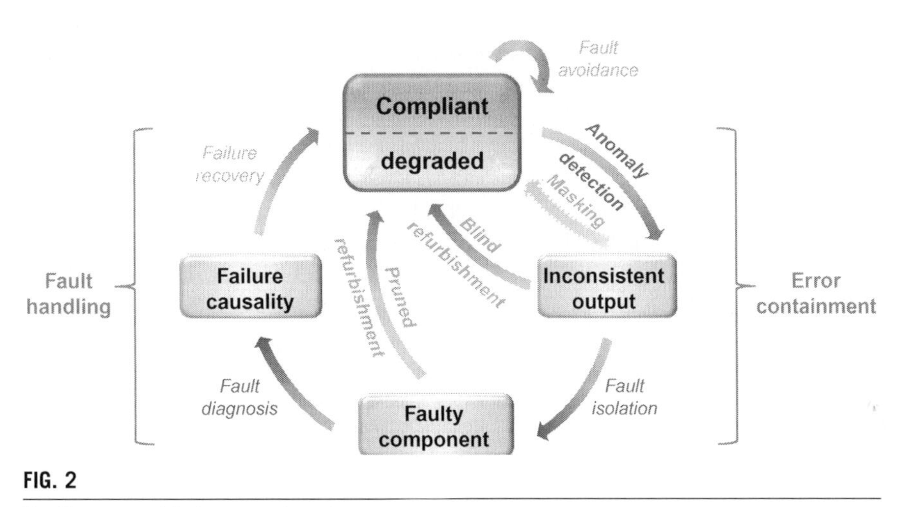

FIG. 2

Resiliency-enabled processing phases.

Redundancy-based AD methods are popular throughout the fault-tolerant systems community, although they incur significant area and energy overhead costs. In the *comparison diagnosis model* [9, 10] units are evaluated in pairs when subjected to identical inputs. Under this AD technique, any discrepancy between the units' outputs indicates occurrence of at least a single failure. However, two or more identical *common-mode failures* (CMF) which occur simultaneously in each module may be undetected. For instance, a *concurrent error detection* (CED) arrangement utilizes either two concurrent replicas of a design [11], or a diverse duplex design to reduce CMFs [12]. This raises the concept of *design diversity* in redundant systems. Namely, *triple modular redundancy* (TMR) systems can be implemented using physically distinct, yet functionally identical designs. Granted, the meaning of physically distinct differs when referring to FPGAs than when referring to application-specific integrated circuits (ASICs). In FPGAs, two modules are said to be physically distinct if the look-up tables in the same relative location on both modules do not implement the same logical function. TMR systems based on diverse designs possess more immunity toward CMF that impact multiple modules at the same time in the same manner, generally due to a common cause.

An additional primary advantage of TMR is its very low fault detection latency. A TMR-based system [13, 14] utilizes three instances of a datapath module. The outputs of these three instances become inputs to a majority voter, which in turn provides the main output of the system. In this way, besides AD capability, the system is able to mask its faults in the output if distinguishable faults occur within one of the three modules. However, this incurs an increased area and power requirement to accommodate three replicated datapaths. It will be shown that these overheads can be significantly reduced by either considering some health metric, such as the instantaneous peak signal-to-noise ratio (PSNR) measure obtained within a video encoder circuit as a precipitating indication of faults, or periodic checking of the logic resources. In contrast, simple *masking* methods act immediately to attempt to conceal each articulated error to return immediately to an operational state of compliant or degraded performance.

As shown in Fig. 2, FI occurs after AD identifies inconsistent output(s). Namely, FI applies functional inputs or additional test vectors in order to locate the faulty component(s) present in the resource layer. The process of FI can vary in granularity to a major module, component, device, or input signal line. FI may be specific with certainty or within some confidence interval. One potential benefit of identifying faulty component(s) is the ability to prune the recovery space to concentrate on resources which are known to be faulty. This can result in more rapid recovery, thus increasing system availability which is defined to be the proportion of the mission which the system is operational. Together, these first two phases of AD and FI are often viewed to constitute *error containment* strategies.

The FD phase consists in distinguishing the characteristics of the faulty components which have been isolated. Traditionally, in many fault-tolerant digital circuits, the components are diagnosed by evaluating their behavior under a set of test inputs. This *test vector strategy* can isolate faults while requiring only a small area overhead. However the cost of evaluating an extensive number of test vectors to diagnose the functional blocks increases exponentially in terms of the number of components and their input domains. The active dynamic redundancy approach presented in Chapter 4 combines the benefits of redundancy with a negligible computational overhead. On the other hand, static redundancy techniques reserve dedicated spare resources for fault handling.

While reconfiguration and redundancy are fundamental components of an FR process, both the reconfiguration scheduling policy and the granularity of recovery affect *availability* during the recovery phase and *quality of recovery* after fault handling. In this case, it is possible to exploit the FR algorithms properties so that the reconfiguration strategy is constructed while taking into account varying priority levels associated with required functions.

A system can be considered to be fault tolerant if it can continue some useful operation in the presence of failures, perhaps in a degraded mode with partially restored functionality [15]. Reliability and availability are desirable qualities of a system, which are measured in terms of service continuity and operational availability in the presence of adverse events, respectively [16]. In recent FPGA-based designs, reliability has been attained by employing the reconfigurable modules in the fault-handling flow, whereas availability is maintained by minimum interruption of the main throughput datapath. These are all considered to constitute fault handling procedures as depicted in Fig. 2.

3 MEASURING RESILIENCE

The design of a resilient system first necessitates an acceptable definition of resilience. Many different definitions exist, and there is no commonly accepted definition or complete set of measurable metrics that allow a system to be definitively classified as "resilient." This lack of a standardized framework or common, complete set of metrics leads to organizations determining their own specific approaches and means of measuring resilience. In general, we refer to resilience as the ability of the system to provide and maintain an acceptable level of service in the face of various faults and challenges to normal operation. This of course begs the question of what is "acceptable" which, in most cases, is determined by the end customer, the application programmer, and/or the system designer. When evaluating resilience technology, there are usually two concerns, namely the *cost* and the *effectiveness*.

3.1 COST METRICS

Cost metrics estimate the impact of providing resilience to the system by measuring the difference in a given property that resilience causes compared to a base system that does not include the provisioning for resilience. These are typically measured as a percentage increase or reduction compared with the base system as a reference. Cost metrics include:

- **Performance Overhead**—This metric is simply the performance loss of the system with implemented resilient techniques measured as the percentage slowdown relative to the system without any resilience features. An interesting but perhaps likely scenario in the 7 nm near-threshold future will be that the system will not function at all without explicitly building in specific resilience features.
- **Energy Overhead**—This corresponds to the increase in energy consumption required to implement varying subsets of resilience features over a baseline system. The trade-off between energy efficiency and resilience is usually a key consideration when it comes to implementing resilience techniques.
- **Area Overhead**—Despite the projected increase in device density for use in future chip designs, factoring in comprehensive resilience techniques will nevertheless take up a measurable quantity of available silicon.

- **Coding Overhead**—In cases where resilience is incorporated across all areas of the system stack (up to and including applications themselves), this metric corresponds to the increase in application program size, development time, and system software size that can be directly attributed to constructs added to improve resilience.

3.2 EFFECTIVENESS METRICS

Effectiveness metrics quantify the benefit in system resilience provided by a given technology or set of resilience techniques. Such metrics tend to be measured as probabilistic figures that predict the expected resilience of a system, or that estimate the average time before an event expected to affect a system's normal operating characteristics is likely to occur. These include:

- **Mean Time to Failure (MTTF)**—Indicates the average amount of time before the system degrades to an unacceptable level, ceases expected operation, and/or fails to produce the expected results.
- **Mean Time to Repair (MTTR)**—When a system degrades to the point at which it has "failed" (this can be in terms of functionality, performance, energy consumption, etc.), the MTTR provides the average time it takes to recover from the failure. Note that a system may have different MTTRs for different failure events as determined by the system operator.
- **Mean Time Between Failures (MTBF)**—The mean time between failures gives an average expected time between consecutive failures in the system. MTBF is related to MTTF as MTBF = MTTF + MTTR.
- **Mean Time Between Application Interrupts (MTBAI)**—This measurement gives the average time between application level interrupts that cause the application to respond to a resilience-related event.
- **Probability of Erroneous Answer**—This metric measures the probability that the final answer is wrong due to an undetected error.

4 METRICS ON POWER-PERFORMANCE IMPACT

Technology scaling and NTV operation are two effective paths to achieve aggressive power/energy efficiency goals. Both are fraught with resiliency challenges in prevalent CMOS logic and storage elements. When resiliency improvements actually enable more energy-efficient techniques (smaller node sizes, lower voltages) metrics to assess the improvements they bring to performance and energy-efficiency need to be also considered. A closely related area is thermal management of systems where system availability and performance can be improved by proactive thermal management solutions and thermal-aware designs versus purely reactive/thermal-emergency management approaches. Thermal-aware design and proactive management techniques focused on impacting thermal resiliency of the system can also improve performance and efficiency of system (reducing/eliminating impact of thermal events) in addition to potentially helping system availability at lower cost.

In this context, *efficiency improvement/cost for resiliency improvement* constitutes an effective metric to compare different alternatives or solutions. As an example, a DRAM-only memory system might have an energy-efficiency measure E_{DRAM} and a hybrid Storage Class Memory-DRAM (SCM-DRAM) system with better resiliency might have a measure E_{Hybrid}. If C_{Hybrid} is the incremental cost of supporting the more resilient hybrid system, the new measure would be evaluated as $(E_{Hybrid}E_{DRAM})/C_{Hybrid}$. Different alternatives for hybrid SCM-DRAM designs would then be compared based on their relative values for this measure. On a similar note, different methods to improve thermal resiliency can be compared on their improvement in average system performance or efficiency normalized to the cost of their implementation.

5 HARD-ERROR VULNERABILITIES

This section describes the underlying physical mechanisms which may lead to reliability concerns in advanced integrated circuits (ICs). The considered mechanisms are those which affect the chip itself, including both the transistor level ("front end of line," or FEOL) and wiring levels ("back end of line," or BEOL), but not covering reliability associated with packaging—even though this is a significant area of potential field failures. The following is a nonexhaustive list of common permanent failures in advanced ICs:

(a) **Electromigration (EM)**—A process by which sustained unidirectional current flow experienced by interconnect (wires) results in progressive increase of wire resistance eventually leading to permanent open faults.

(b) **Time-dependent dielectric breakdown (TDDB)**—A process by which sustained gate biases applied to transistor devices or to interconnect dielectrics causes progressive degradation towards oxide breakdown eventually leading to permanent short or stuck-at faults.

(c) **Negative Bias Temperature Instability (NBTI)**—A process by which sustained gate biases applied to transistor devices causes a gradual shift upwards of its threshold voltage and degradation of carrier mobility, causing it to have reduced speed and current-drive capability, eventually leading to permanent circuit failure.

(d) **Hot Carrier Injection (HCI)**—A process by which a transistor device (with sustained switching usage) causes a gradual shift upwards of its threshold voltage and degradation of carrier mobility, causing it to have reduced speed and current-drive capability, eventually leading to permanent circuit failure.

Reliability mechanisms can be broadly divided into two categories. "Random" or "hard" failure mechanisms are by nature statistical. This type of failure is associated with a distinct failure time, however, the failure time is a random variable, different for each similar circuit element even when subject to the same voltage and temperature history. The first two mechanisms in the list above (i.e., EM and TDDB) fall into this category. EM typically follows a log-normal probability distribution of failure times, with $\sigma \approx 0.2$. TDDB typically follows a Weibull distribution with shape parameter ≥ 1 in the case of state-of-the-art gate dielectrics.

The other category of failure is "wearout." This type of mechanism results in a gradual shift of electrical characteristics, which is the same for every similar circuit element subject to the same voltage and temperature history. This may eventually lead to circuit failure if and when the characteristics shift outside of the allowable operating range. This type of gradual, continuous parametric shift can lead, for example, to changes in switching speed which may in turn lead to critical path timing errors, or to reduced noise margins which may lead to register errors especially in circuits operating near their voltage limits. The last two mechanisms above (i.e., NBTI and HCI) are of this type. The reader should note that the term "wearout" is very often used in a different sense, to denote the increasing failure rate near end-of-life (EOL) in the traditional bathtub-shaped reliability curve. This statistical meaning does not consider the underlying physical mechanism.

In both wearout and random failure mechanisms the underlying physics involves a gradual accumulation of damage to the device or circuit element. The fundamental difference is that in the random or hard failure type, this damage is not readily apparent until a certain threshold is reached. For example, in electromigration (EM), metal vacancies appear and grow until finally an open circuit forms, resulting in a sudden increase in metal resistance. There is typically a gradual increase in resistance before the complete open circuit, but this is usually small enough to be ignored in practice. Fig. 3 shows a typical resistance versus time plot for electromigration. Similar characteristics are also seen in TDDB, where the phenomenon known as "progressive" breakdown has been extensively studied [17].

In wearout mechanisms the damage is manifest as a continuous shift in electrical characteristic, lacking any sudden (delayed) onset. Fig. 4 shows typical threshold voltage shift versus time for NBTI. In very small transistors, wearout mechanisms such as threshold voltage shifts are subject to statistical variation because of the discrete nature of charge. Thus, even these "uniform" wearout mechanisms should be treated statistically. This is still a subject of current investigation and the treatment of the statistical distribution of NBTI and HCI has not yet seen widespread implementation in practice [18]. NBTI-induced wearout can be annealed or relaxed when the gate bias is removed, as shown in Fig. 4.

Fig. 5 shows the typical analytical models for each mechanism. This figure highlights the dependence of each mechanism on physical and operational parameters (acceleration factors). Thermal activation energies are listed in the third column. EM and TDDB are the failure modes which are most strongly accelerated by temperature. This makes them likely candidates for reliability limiters associated with hot spots. However, since these are both random statistical processes (described by the failure time distributions given in the figure) the failure of any particular chip cannot be guaranteed, only the probability of failure can be ascertained. In all cases the absolute failure times or parameter shifts depend strongly on structural and material parameters, e.g., insulator thickness, device size, etc.

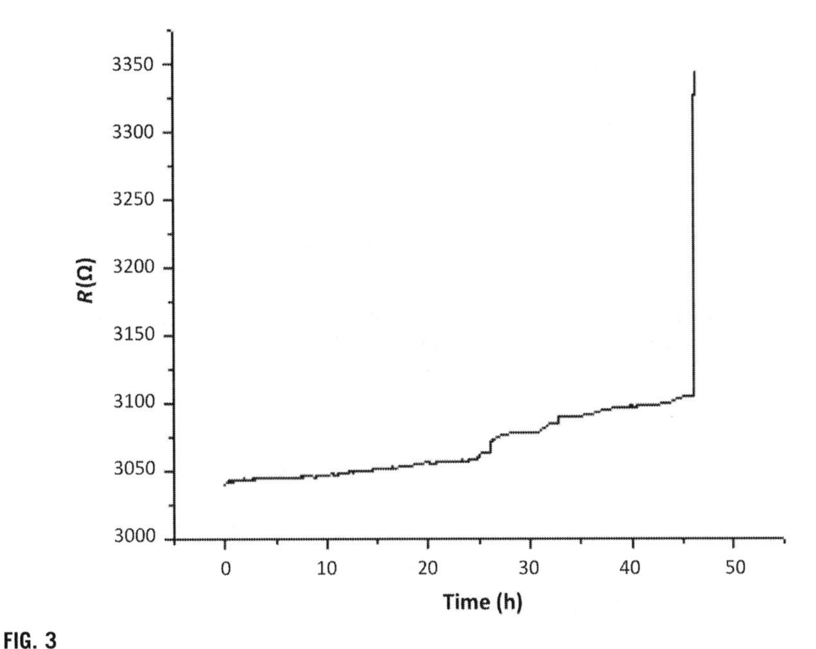

FIG. 3

Interconnect resistivity increase over time, under sustained EM stress.

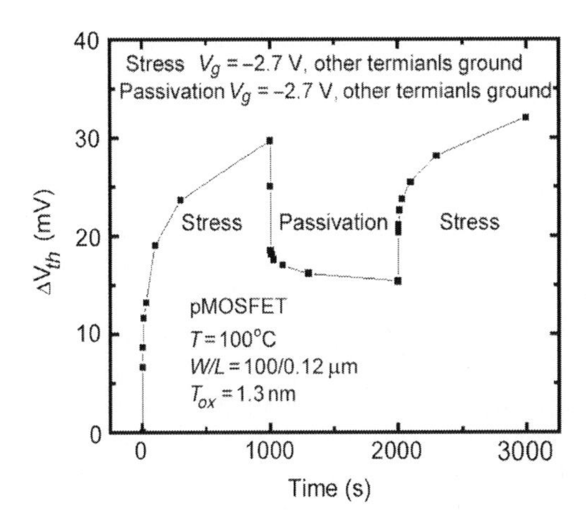

FIG. 4

Typical NBTI-induced threshold voltage stress/recovery characteristics.

EM testing typically entails measurement on various metal lines and via test structures that allow evaluation of electromigration from electron flow either up from a via below the metal line (via depletion), or electron flow down from a via above the metal line (line depletion) under unidirectional current flow. Stresses are typically carried out at elevated temperature (e.g., 250–300°C) using constant

Mechanism	Typical model equations	Parameter definitions	Typical thermal activation energy (eV)	Typical voltage or current dependence
EM	Current corresponding to median failure time t_{use}: $J_{use} = J_a \left(\frac{t_a}{t_{use}}\right)^{1/n} e^{\frac{-E_a}{kT}}$ Distribution of failure times: *lognormal*	J_a stress current t_a failure time at stress J_{use} current at operation t_{use} lifetime E_a activation energy n current density exponent	0.8–0.9	$n = 1.1–1.7$
TDDB	Voltage corresponding to characteristic failure time $V_{use} = V_a \left(\frac{t_a e^{\frac{-Ea}{kT}}}{t_{use}}\right)^{1/n}$ Distribution of failure times: *Weibull*	V_a stress voltage t_a failure time at stress V_{use} voltage at operation t_{use} lifetime E_a activation energy n voltage exponent	0.6	$n = 40–50$
NBTI	Threshold voltage shift: $\Delta V_T \propto (V_{use})^b \times (t_{use})^n \times e^{\frac{-Ea}{kT}}$	V_{use} voltage at operation t_{use} lifetime E_a activation energy b voltage exponent n time exponent	0.1	$b \approx 4$
HCI	Drive current degradation: $\Delta I_d \propto e^{(b \times V_{use})} \times (t_{use})^n \times e^{\frac{-Ea}{kT}}$	V_{use} voltage at operation t_{use} lifetime E_a activation energy b voltage factor n time exponent	0.05	$b = 5–10$

FIG. 5

Analytical models for the considered physical mechanisms associated with reliability vulnerabilities.

current stress J_a of order 10 mA/μm^2. Vulnerable circuits for EM failure are those with highly loaded devices or high duty factors, i.e., which tend to drive DC currents. The EOL criterion is typically a resistance increase $(dR/R) \geq 20\%$ or an excess leakage current (e.g., due to metal extrusion) >1 μA at stress condition. This is a reliability engineering benchmark, but may not necessarily cause circuit failure in every case.

TDDB testing for gate dielectrics is performed by stressing individual transistors under inversion condition (i.e., gate positive for n-fets, and gate negative for p-fets). Stresses are typically carried out at elevated temperature (e.g., 100–180°C) using constant voltage stress V_a of order 2–5 V in order to obtain reasonable failure times. The EOL criterion is typically any small increase in leakage current ("first breakdown") or an excess leakage current of say >10 μA at use condition. This level of leakage has been shown to impact functionality of certain circuits such as SRAM but may not cause all circuit types to fail.

Back-end (intermetal dielectric) TDDB is an area of increasing interest and concern, as the insulator thickness, particularly between the gate conductor and the source or drain contact metals, is now comparable to thicknesses used under the gate only a couple of decades ago. Similar phenomenology and models describe back-end TDDB, but breakdown voltages or fields in these insulators are often limited by extrinsic or integration related issues (etch profiles, metal contamination, etc.). For back-end TDDB, test structures typically comprise comb structures.

The EM and TDDB equations in Fig. 5 give the allowed current density J_{use} (for EM) or voltage V_{use} (for TDDB) corresponding to a specified median fail time t_{use} for a single metal line, via, transistor gate, or other dielectric. The median failure time under accelerated stress conditions is t_a, corresponding to stress current density J_a (for EM) or voltage V_a (for TDDB). The distribution of EM failure times follows log-normal statistics, while TDDB failure times typically follow Weibull statistics, at least in the low-failure rate tail which is of interest.

Since semiconductor technology is typically qualified to support parts per million chip reliability with at least 1E5 lines per chip and $\approx$1E9 transistors per chip, the failure time for a single wire or device exceeds the product lifetime by many orders of magnitude. Therefore, significant current, voltage, or temperature excursions may be required to induce failure on any given device or target circuit.

On the other hand, since NBTI and HCI are uniform wearout modes, with similar threshold voltage shift or current degradation versus time for all transistors which operate at the same voltage and temperature, the potential failure by these mechanisms caused by voltage or temperature excursions should be more readily predictable. As NBTI and HCI degradation cause transistors to become weaker, circuits may be vulnerable to timing errors or to decreased noise immunity. The BTI and HCI equations in Fig. 5 gives the threshold voltage shift (for NBTI) or drain current degradation (for HCI) as a function of usage time t_{use} and voltage V_{use}.

NBTI, like TDDB, is performed by stressing individual p-fet transistors under inversion condition (gate negative). Stresses are typically carried out at elevated temperature (e.g., 80–150°C) using constant voltage stress V_a of order one to two times normal operation voltage in order to obtain measurable shifts without oxide breakdown. A typical EOL criterion is 50 mV threshold voltage shift, although as pointed out earlier this is not a "hard" fail criterion since the shift is continuous and gradual. Vulnerable circuits for NBTI are those low duty cycle (p-fet remains on for long times) since NBTI is exacerbated by continued DC (nonswitching, or transistor "on") conditions.

HCI testing is also performed on individual transistors, but unlike TDDB and NBTI, it requires drain bias in addition to gate bias since the wearout is driven by energetic carriers in the channel. The drain is biased at constant voltage stress V_a of order one to two times normal operation voltage and the gate is typically held either at the same bias or at one-half of the drain voltage. Vulnerable circuits for HCI failure are those with highly loaded devices, and high duty factors, since the damage occurs only during the conducting phase when a gate switches.

The hard error vulnerability models discussed above (see Fig. 5), are physics-based behavioral equations that help us calculate the mean-time-to-failure (MTTF) at the individual component (device or interconnect) level. It needs to be mentioned here that many of those equations evolve over time to capture the unique effects of particular semiconductor technology nodes. In fact, even for the same basic technology era, individual vendors have in-house, customized models that are very specific to their particular foundry. The equations depicted in Fig. 5 are generic examples, representative of technologies of a recent prior era—and as such, they should not be assumed to be accurately representative of particular, current generation foundry technologies.

Also, when it comes to be able to model the effect of such hard errors at the system level, in the context of real application workloads, techniques like RAMP [19–21] at the processor core level and follow-on multicore modeling innovations (e.g., Shin et al. [22–24]) need to be mentioned. The idea is to first collect representative utilization and duty cycle statistics from architecture-level simulators, driven by target application workloads. These are then used in conjunction with device-level physics models as well as device density and circuit-level parameters to deduce the failures in time (FIT) values on a structure-by-structure basis. The unit-specific FITs are then combined appropriately to derive the overall chip FITs. From that, the chip-level MTTF can be derived, under suitable assumptions about the error incidence (distribution) functions.

6 SOFT-ERROR VULNERABILITIES

As CMOS technology scales and more transistors are packed on to the same chip, soft error reliability has become an increasingly important design issue for processors. Soft errors (or single event upsets or transient errors) caused by alpha particles from packaging materials or high energy particle strikes from cosmic rays can flip bits in storage cells or cause logic elements to generate the wrong result. Such errors have become a major challenge to processor design. If a particle strike causes a bit to flip or a piece of logic to generate the wrong result, we refer to it as a *raw* soft error event. Fortunately, not all raw soft errors cause the processor to fail. In a given cycle, only a fraction of the bits in a processor storage structure and some of the logic structures will affect the final program output. A raw error event that does not affect these critical bits or logic structures has no adverse effect on the program outcome and is said to be *masked*. For example, a soft error in the branch prediction unit or in an idle functional unit will not cause the program to fail.

Applications running on a given computing platform exhibit a wide range of tolerance to soft errors affecting the underlying hardware. For example, many image processing algorithms are quite tolerant to noise, since isolated bit-flips in image data frequently go unnoticed by the end-user or decision engine. Mukherjee et al. [25] introduced the term *architectural vulnerability factor* (AVF) to quantify the architectural masking of raw soft errors in a processor structure. The AVF of a structure is effectively the probability that a visible error (failure) will occur, given a raw error event in the structure [25]. The AVF can be calculated as the percentage of time the structure contains *architecturally correct execution* (ACE) bits (i.e., the bits that affect the final program output). Thus, for a storage cell, the AVF is the percentage of cycles that this cell contains ACE bits. For a logic structure, the AVF is the percentage of cycles that it processes ACE bits or instructions.

The AVF of a structure directly determines its MTTF [26]—the smaller the AVF, the larger the MTTF and vice versa. It is therefore important to accurately estimate the AVF in the design stage to meet the reliability goal of the system. Many soft error protection schemes have significant space, performance, and/or energy overheads; e.g., ECC, redundant units, etc. Designing a processor without accurate knowledge of the AVF risks over- or underdesign. An AVF-oblivious design must consider the worst case, and so could incur unnecessary overhead. Conversely, a design that underestimates the AVF would not meet the desired reliability goal.

Soft errors may be caused by various events including neutrons from cosmic particle incidence, alpha particles from trace radioactive content in packaging materials, and inductive noise effects (Ldi/dt) on the chip supply voltage resulting from aggressive forms of dynamic power management. As technology continues scaling down, early estimation of soft errors in SRAM cells and latch and logic elements becomes even more critical. Chip design must begin with a consideration of system-level MTTF targets, and the design methodology must be able to estimate or set bounds for chip-level failure rates with reasonable accuracy in order to avoid in-field system quality problems. A balanced combination of circuit- and logic-level innovations and architecture- and software-level solutions is necessary to achieve the required resiliency to single-event upsets (SEUs). In particular, there is the need for a comprehensive understanding of the vulnerabilities associated with various units on the chip with regard to workload behavior. When such information is available, appropriate approaches—such as selective duplication, SER-tolerant latch design adoption, and error-correcting code (ECC) and parity protection of SER hotspots[1] —may be used for efficient error resiliency.

6.1 APPLICATION CHARACTERIZATION THROUGH FAULT INJECTION

One approach to assess the level of resiliency of a system is by artificially injecting faults and analyzing the system's resulting behavior. When carefully determined, fault injection campaigns can significantly increase the coverage of early fault tolerance and resilience testing. Fault injection can be applied at different levels in the hardware-software stack—in particular, application-level fault injection (AFI) is an effective approach given that it is relatively easy and flexible to manipulate an application's architectural state. Fig. 6 presents a high-level overview of a possible *fault injector*.

[1]*SER hotspot* refers to a region of the chip that is deemed to be highly vulnerable to bit flips in an element (latch, combinational logic, or SRAM). An upset in this region is much more likely to cause a program failure than other adjoining areas. The SER hotspot profile across the chip floorplan may vary with the executing workload; however, it is quite likely that some of the SER hotspots may be largely invariant with regard to the workload. For example, portions of the instruction decode unit (through which all program instructions must pass) may well turn out to be SER hotspots across a wide range of input workloads.

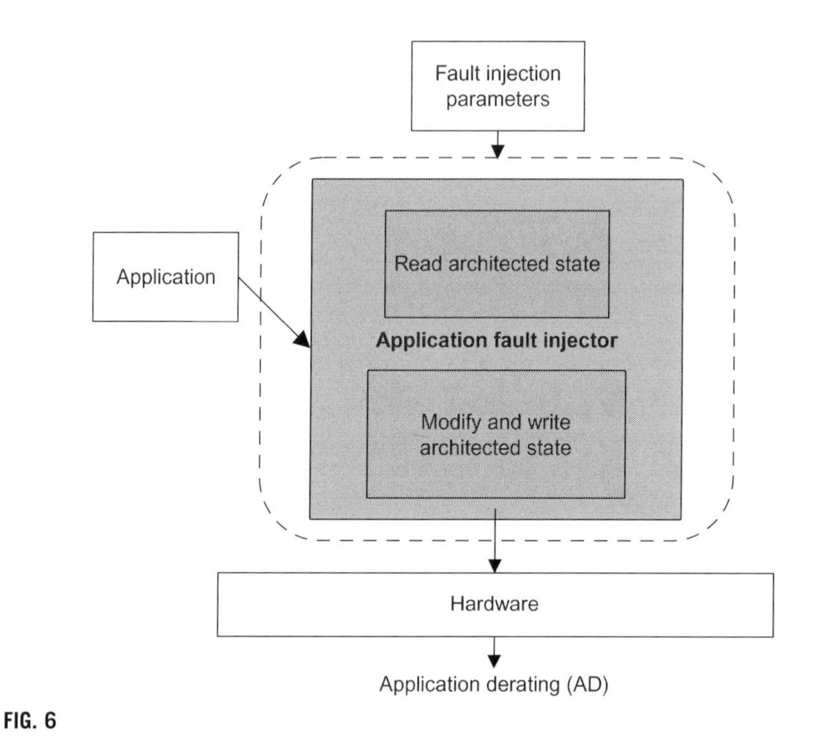

FIG. 6

High-level block diagram of a possible AFI framework.

In Unix/Linux scenarios, it is possible to resort to tools like the "process trace" (*ptrace*) debugging facility to put together a framework in which any target application can be compiled and run under user-controllable fault injection directives. The ptrace-centered facility provides a mechanism by which a parent process can observe and control the execution of another process—namely the application execution process that is targeted for fault injection. The parent process can examine and change the architected register and memory state of the monitored application execution process.

A fault injection facility like the one just described is designed to identify the most vulnerable regions of a targeted processor chip from the perspective of transient and permanent fault injections. As an example, we next describe experiments where pseudorandom fault injections were targeted to corrupt register state, instruction memory state, and data memory state of the application running on a POWER7 machine under the AIX (IBM Unix) operating system. A well-known challenge in fault injection experiments is the determination of the sensitivity to the number of injections necessary and sufficient to capture the true strength of the experiment. Fig. 7 presents the masking saturation rate of three exemplary workloads as the number of faults injected increases. Clearly, for these workloads in this particular example, 5000 injections were shown to be sufficient for a stabilizing point.

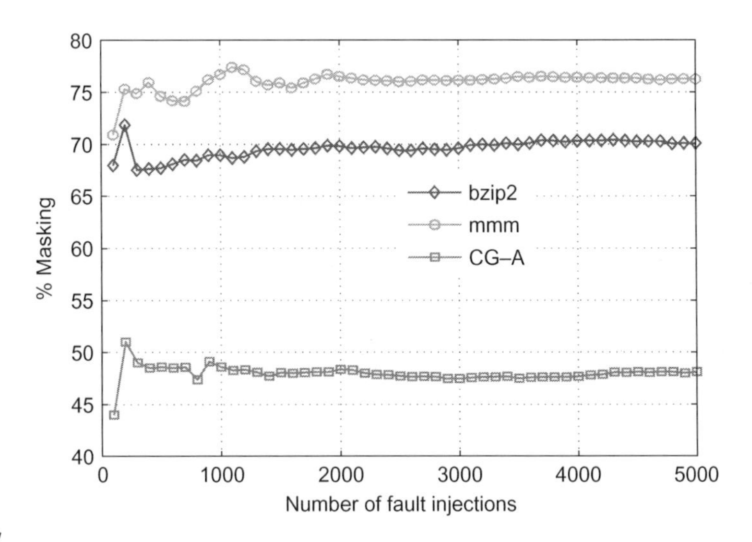

FIG. 7

Masking saturation rate for *CG-A*, *mmm*, and *bzip2*.

As stated before, in this example a total of 5000 single-bit, pseudorandom fault injection experiments were made into the architected register space in each controlled experiment (e.g., using various compiler optimization levels). Each such injection leads to one of the following outcomes:

- The bit that is flipped via the injection has no effect on the program execution profile, including the final program output or final data memory state. These fault injections are categorized as being *fully masked*.
- The injected bit-flip results in a silent data corruption (SDC)—i.e., the program output or the final data memory state shows an error, when compared to the fault-free, "golden" run of the program.
- The injected error results in a program crash, where the operating system terminates the program due to a detected runtime error (e.g., divide by zero exception or segmentation fault, illegal memory reference, etc.).
- The injected bit-flip results in a "hung" state, where there is no forward progress of the program execution; effectively, in practice such a situation would require a user-initiated program termination or even a physical machine reboot.

From the measured percentage of each event's occurrence, the corresponding failure rate can be estimated, provided the total number of random experiments (e.g., 5000) is deemed to be large enough to draw statistically meaningful inferences. Fig. 8 shows snapshot results (analyzing the effect of register state fault injections) obtained using AFI for two example applications: *mmm* in Fig. 8A (which is a kernel within the *dgemm* family) and bzip2 in Fig. 8B. The data shown in Fig. 8 leads one to

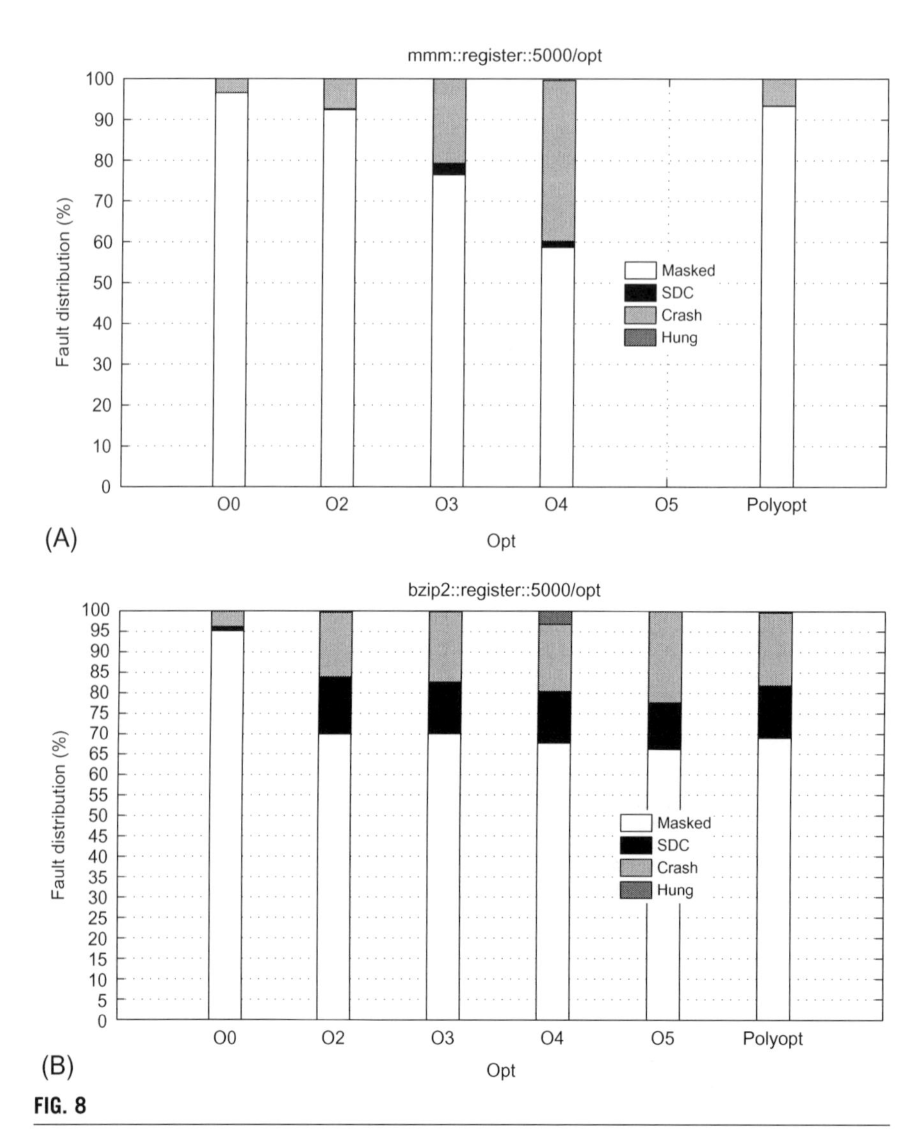

FIG. 8

Fault injection experimental results for different compiler optimization levels. (A) *mmm* benchmark. (B) *bzip2* benchmarks.

conclude that vulnerability to random bitflips worsens with the degree of compiler optimization—at least for these two example applications.

7 MICROBENCHMARK GENERATION

A systematic set of microbenchmarks is needed to serve as specific stressers in order to sensitize a targeted processor chip to targeted failure modes under harsh environments. The idea is to run such specialized microbenchmarks and observe the onset of vulnerabilities to robust operation. Microbenchmarks are targeted at the processor so that deficiencies can be identified or diagnosed in:

1. traditional architecture performance (e.g., instructions per cycle, cache misses);
2. power or energy related metrics;
3. temperature and lifetime reliability metrics;
4. resilience under transient errors induced by high energy particle strikes, voltage noise events, and thermal hot spots; etc.

Microbenchmarks can certainly be developed manually with detailed knowledge of the target architecture. However, from a productivity viewpoint there is a clear value in automating the generation of microbenchmarks through a framework. A microbenchmark generation framework that allows developers to quickly convert an idea of a microbenchmark into a working benchmark that performs the desired action is necessary. The following sections will take a deep dive into a microbenchmark generation process and capabilities of an automated framework. We will also present a simple example to make concepts clear.

7.1 OVERVIEW

In a microbenchmark generation framework, flexibility and generality are main design constraints since the nature of situations on which microbenchmarks can be useful is huge. We want the user to fully control the code being generated at assembly level. In addition, we want the user to specify high level (e.g., loads per instruction) or dynamic properties (e.g., instructions per cycle ratio) that the microbenchmark should have. Moreover, we want the user to quickly search the design space, when looking for a solution with the specifications.

In a microbenchmark generation process, microbenchmarks can have a set of properties which are *static* and *dynamic*. Microbenchmarks that fulfill a set of static properties can be directly generated since the static properties do not depend on the environment in which the microbenchmark is deployed. These types of properties include instruction distribution, code and data footprint, dependency distance, branch patterns, and data access patterns. In contrast, the generation of microbenchmarks with a given set of dynamic properties is a complex task. The dynamic properties are directly affected by the static microbenchmark properties as well as the architecture on which the microbenchmark is run. Examples of dynamic properties include instructions per cycle, memory hit/miss ratios, power, or temperature.

In general, it is hard to statically ensure dynamic properties of a microbenchmark. However, in some situations, using a deep knowledge of the underlying architecture and assuming a constrained execution environment, one can statically ensure the dynamic properties. Otherwise, to check whether dynamic properties are satisfied, simulation on a simulator or measurement on a real setup is needed. In that scenario, since the user can only control the static properties of a microbenchmark, a search for the design space is needed to find a solution.

Fig. 9 shows the high level picture of a microbenchmark generation process. In step number one, the user provides a set of properties. In the second step, if the properties are abstract (e.g., integer unit at 70% utilization), they are translated into architectural properties using the property driver. If the properties are not abstract, they can be directly forwarded to the next step, which is the microbenchmark synthesizer. The synthesizer takes the properties and generates an abstract representation of the microbenchmark with the properties that can be statically defined. Other parameters that are required to generate the microbenchmarks are assigned by using the models implemented in the architecture back-end. In this step, the call flow graph and basic blocks are created. Assignment of instructions, dependencies, memory patterns, and branch patterns are performed. The "architecture back-end" consists of three components: (a) definition of the instruction set architecture via op code syntax table, as well as the high-level parametric definition of the processor microarchitecture; (b) an analytical reference model that is able to calculate (precisely or within specified bounds) the performance and unit-level utilization levels of a candidate loop microbenchmark; (c) a (micro)architecture translator segment that is responsible for final (micro)architecture-specific consolidation and integration of the microbenchmark program. Finally, in the fourth step the property evaluator checks whether the microbenchmark fulfills the required properties. For that purpose, the framework

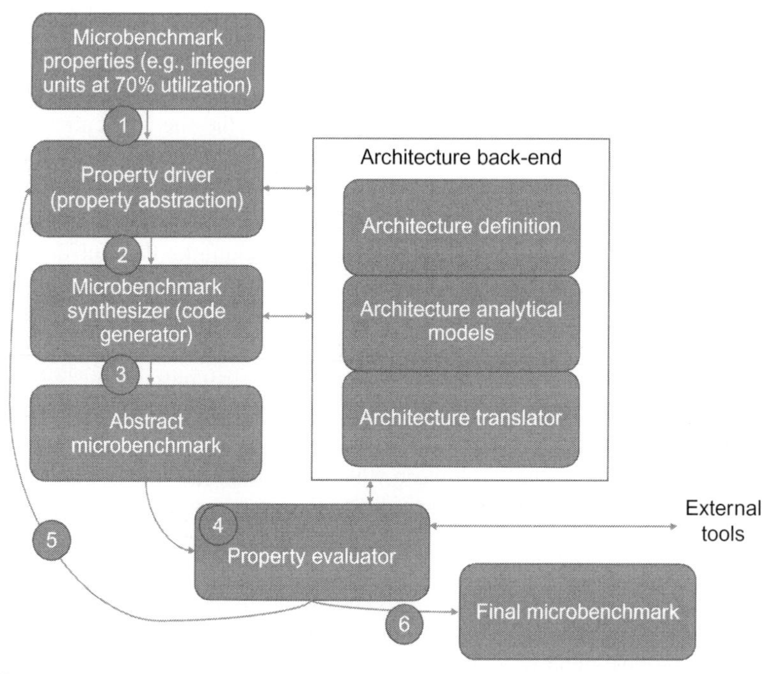

FIG. 9

High-level description of a microbenchmark generation process.

can rely on a simulator, real execution on a machine or analytical models provided by the architecture back-end. If the microbenchmark fulfills the target properties, the final code is generated (step 6). Otherwise, the property evaluator modifies the input parameters of the code generator and an iterative process is followed until the search process finds the desired solution (step 5).

These steps outline the workflow for a general use case. However, different cases require different features and steps. Moreover, for the cases where the user requires full-control of the code being generated, the framework can provide an application programming interface (API). The API allows the user to access all the abstractions and control steps defined in the workflow. Overall, the microbenchmark generation framework provides for an iterative generate and test methodology that is able to quickly zoom in on an acceptable microbenchmark that meets the specified requirements.

7.2 EXAMPLE OF A MICROBENCHMARK GENERATION PROCESS

Figs. 10 and 11 show a sample script that generates 20 random microbenchmarks and a sample of the generated code, respectively. The script in Fig. 10 highlights the modularity and flexibility of a microbenchmark generation framework. The script uses the basic abstractions provided by the framework to specify the process that needs to be done in order to generate the microbenchmarks. The code wrapper object is in charge of implementing the code abstraction layer. This abstraction layer provides a language independent API to dump microbenchmarks, as well as an abstraction level in order to implement different code layouts. Instances of this class control the declaration of variables and functions, as well as the prologues and epilogues before and after the microbenchmark control flow graph. Moreover, it also controls how the instructions are generated. In our example, the wrapper that is gathered in line 9 of Fig. 10 which is passed to the synthesizer in line 15, is in charge of variable and function declaration (lines 1–10), control flow graph prologue, control flow graph epilogue, and function finalization. We can see that this wrapper generates C code featuring a main function with an endless loop using PowerPC instructions. This is the reason why it is named as CInfPpc (Fig. 10, line 9). The microbenchmark synthesizer is designed in a multipass fashion, which means that synthesizer processes the benchmark representation several times. Each pass, represented by an instance of the pass class, takes the result of the previous pass as the input, and creates an intermediate output. In this way, the (intermediate) benchmark is modified by multiple passes until the final pass generates the final code. This can be observed in Fig. 10. In line 15, a synthesizer object is created and in lines 20, 23, 26, 29, 32, and 35 several passes are added. Finally, in line 39, the benchmark is synthesized and then it is saved to a file (line 42). Note that this design permits users to specify the transformations (passes), as well as the order in which they are performed. This ensures the flexibility of the framework.

```
1   ...<some initializations>...
2
3   # get the architecture object
4   arch = MicrobenchmarkGenerationFramework.get.architecture("ProcessorX")
5
6   #get the code wrapper, in this case we take the 'CInfPpc'
7   #which generates a 'main' function with an infinite loop
8   #in C language using PowerPC assembly instructions.
9   wrapper = MicrobenchmarkGenerationFramework.code.get_wrapper("CInfPpc")
10
11  # define a random function
12  rnd = lambda: random.randrange(0,(1<<64)-1)
13
14  # create the microbenchmark synthesizer
15  synth = MicrobenchmarkGenerationFramework.code.Synthesizer(arch,wrapper())
16
17  #add the passes we want to apply in order to synthesize
18  #microbenchmarks
19  #pass 1: Init registers to random values
20  synth.add_pass(MicrobenchmarkGenerationFramework.passes.init.INIT_Registers(rnd))
21
22  #pass 2: Add a single basic block of size 'size'
23  synth.add_pass(MicrobenchmarkGenerationFramework.passes.cfg.CFG_SingleBBL(size))
24
25  #pass 3: Fill the basic blocks using random instructions
26  synth.add_pass(MicrobenchmarkGenerationFramework.passes.ins.INS_RandomInstruction())
27
28  #pass 4: Add random access pattern
29  synth.add_pass(MicrobenchmarkGenerationFramework.passes.mem.MEM_RandomMemModel())
30
31  #pass 5: Set the dependency distance between instructions
32  synth.add_pass(MicrobenchmarkGenerationFramework.passes.dep.DEP_RandomDependency())
33
34  #pass 6: Initialize immediate values to random values
35  synth.add_pass(MicrobenchmarkGenerationFramework.passes.init.INIT_Immediates(rnd))
36
37  for idx in range(0, 20):
38  #generate the microbenchmark (applies the specified passes)
39  ubench = synth.synthesize()
40
41  # save the benchmark
42  synth.save("%s/random=%s"% (outputdir,idx), ubench)
```

FIG. 10

Sample script for random microbenchmark generation.

8 POWER AND PERFORMANCE MEASUREMENT AND MODELING

8.1 IN-BAND VERSUS OUT-OF-BAND DATA COLLECTION

From a power and performance standpoint, a system's status can be described by a variety of sensors and/or performance counters that usually provide power consumption, temperature, and performance information of the different system components (CPUs, memory and storage modules, network, etc.). There are two approaches to

```
1    ...<some includes>...
2
3   /* Memory arrays */
4   char L1D[32768]  __attribute__((aligned (268435456)));
5   char L2[262144]  __attribute__((aligned (268435456)));
6   char L3[4194304] __attribute__((aligned (268435456)));
7   Char MM[536870912] __attribute__((aligned (268435456)));
8
9   int main(int argc, char **argv, char **envp)
10  {
11  /*register initialization for:
12      -base and index registers reserved for accessing memory arrays*/
13  __asm__(" lis 3, L1D@highest ");
14  __asm__(" ori 3,3,L1D@higher ");
15  __asm__(" rldicr 3,3,32,31 ");
16  __asm__(" oris 3,3,L1D@h ");
17  __asm__(" ori 3,3,L1D@l ");
18  < other initializations>
19  /*loop starts here */
20  __asm__(" infloop: ");
21  __asm__(" addi 9,9,8288 ");
22  __asm__(" xvrdpi 62,63 ");
23  __asm__(" lbzu 20, 20420(3) ");
24  <several instructions>
25  __asm__(" xsnmsubmdp  56,55,54 ");
26  __asm__(" vcmpgtuw 11,10,19 ");
27  __asm__(" addi 3,3,-7827 ");
28  __asm__(" xori 26,27,25460 ");
29  /* memory access guard instructions:
30   e.g. check/set address registers (e.g. register 3) */
31  __asm__(" b infloop ");
32  }
```

FIG. 11

Sample generated microbenchmark.

access such information on line: it can be done either *in-band* or *out-of-band*. We say that a system is sensed in-band when the collected information is generated by the same component that the information refers to; like, for example, CPU power consumption or thermal information generated *internally* by the CPU itself in some specific way. This is the case of the on-chip controller (OCC) in POWER-based systems [27]—a co-processor embedded directly on the main processor die—as well as Intel's Power Control Unit (PCU) and associated on-board power meter capabilities [28]. On the other hand, if the collected information is obtained in a way that does not involve the component under measurement, then we say that system information is gathered out-of-band. This may be the case of CPU power consumption estimated by an external process (running on the same system or even on a different one) using some "proxy" CPU information (like performance counters). IBM's Automated Measurement of Systems for Energy and Temperature Reporting (AMESTER) is an example of out-of-band system measurement [29]. AMESTER constitutes an interface to remotely monitor POWER-based systems through the service processor available in those systems. The service processor (which is different from the main system processor) accesses a variety of power, temperature, and other sensors that monitor the entire system status and deliver them to the remote AMESTER client.

In practice, resorting to in-band or out-of-band system measurement will depend on (1) the available data collection capabilities of the system in question, (2) the performance perturbation, and (3) the expected information granularity and correlation with the actual execution. For example, on-chip power metering capabilities are relatively novel and may not be widely available, in which case chip power consumption could be estimated through the use of proxy information—for example, performance counters to gauge processor activity and derive corresponding power figures. Even though in-band measurement may look more appealing over out-of-band approaches, it may also perturb system performance or interfere with the running workloads since the component being measured has to "produce" the requested information in some way. Finally, the correlation (or lack thereof) of the collected data and the actual measured events should be taken into account. Out-of-band measurement techniques usually perform poorer in terms of correlating the measurements with actual events with precision, in particular when the measurement is conducted remotely.

Regardless if a system is measured in-band or out-of-band, *processor performance counters* constitute a key source of information when it comes to characterize the run-time behavior of a CPU from performance, power, and thermal perspectives. Hence they are vastly used to model power consumption and temperature when power and temperature measurement from sensors is not possible. The critical part in this process is, however, the selection of the right set of counters that correlates with real values with enough accuracy—a problem extensively studied by several researchers and groups, like for example [30–33].

8.2 PROCESSOR PERFORMANCE COUNTERS

Processor performance counters are hardware counters built into a processor to "count" certain *events* that take place at the CPU level, like the number of cycles and instructions that a program executed, its associated cache misses, accesses to off-chip memory, among several other things. The events that can be tracked with performance counters are usually simple ones, but combined in the right way can provide extremely useful insights of a programs behavior and, therefore, constitute a valuable tool for debugging purposes. As it is explained by Elkin and Indukuru in [34], the first step in optimizing an application is characterizing how well the application runs on the target system. The fundamental intensive metric used to characterize the performance of any given workload is *cycles per instruction* (CPI)—the average number of clock cycles (or fractions of a cycle) needed to complete an instruction. The CPI is a measure of processor performance: the lower it is, the more effectively the processor hardware is being kept busy. Elkin and Indukuru also provide a detailed description of the *performance monitoring unit* (PMU) built into

IBM's POWER7 systems, which comprises of six thread-level *performance monitor counters* (PMC_i). PMC_1 to PMC_4 are programmable, PMC_5 counts nonidle completed instructions and PMC_6 counts non idle cycles. Other processor manufacturers offer similar counting capabilities, like Intel's Performance Counter Monitor [35] and ARM's Performance Monitoring Unit [36].

Processor performance counters can also be effectively used to drive run-time power and thermal management decisions in the CPU. As an example, the *Power-Aware Management of Processor Actuators* algorithm (PAMPA) proposed by Vega et al. [37] aims to provide robust chip-level power management by coordinating the operation of dynamic voltage and frequency scaling (DVFS), core folding and per-core power gating (PCPG) using CPU utilization information derived from a few performance counters. More specifically, PAMPA collects on-line hardware events information from the PM_RUN_CYC performance counter available in POWER7+ systems [38]. PM_RUN_CYC counts the nonidle processor cycles at physical thread level. In other words, it filters out those processor cycles during which a particular physical thread is idle and, therefore, constitutes a suitable proxy to estimate physical thread-level utilization.

8.3 POWER MODELING

The measurement of real, calibrated power consumption in hardware is difficult—the availability of on-chip power metering is a fairly new feature in today's systems. Power modeling constitutes an alternative to real power measurement at the expense of *accuracy* and *performance*. If the lack of accuracy and the rate at which power is estimated can both be kept at acceptable levels, then power modeling becomes attractive when real power measurement is not feasible. As an example, IBM's POWER7 does not have the necessary internal power measurement circuits and, therefore, the amount of power that each core consumes is estimated [39]. More specifically, POWER7 implements a hardware mechanism in the form of a "power proxy" by using a specially architected, programmably weighted counter-based architecture that monitors activities and forms an aggregate value. Activities are carefully selected with an understanding of the processor's microarchitecture, such that they correlate maximally with active power consumption.

Power modeling approaches like the one just described for POWER7 are intended to provide power consumption estimations when the system is in production. Prior to that, during the concept phase, early-stage *accurate* power estimation is key to assess and predict the feasibility of a system (or a system component) from power, thermal, and efficiency standpoints. In particular, architectural power models are widely used to estimate the power consumption of a microprocessor, where high-level architectural and microarchitectural parameters (e.g., cache sizes, page size, pipeline depth/width) and activity factors (e.g., cache accesses, total instructions) are specified to the power modeling tool, which abstracts away the underlying implementation details. These high-level abstractions, which represent a trade-off between detail and flexibility/ease of use, enable an architect to quickly evaluate design decisions and explore various design spaces. Wattch [40] and McPAT [41] are two well-known examples of these models. Both of these tools are analytical, meaning that they use analytical equations of capacitance to model dynamic power. In contrast, empirical models, like PowerTimer [42] and ALPS [43], use precharacterized power data and equations from existing designs. For structures like control logic that are difficult to model analytically, empirical models and/or fudge factors are often used instead. The differences between analytical and empirical models have been described in past work [44, 45].

9 SUMMARY

This chapter presents an introduction to the fundamentals of reliable and power-aware systems from a general standpoint. In the first part of the chapter, the concepts of *error*, *fault*, and *failure*, the *resolution phases* of resilient systems, and the definition and associated metrics of *hard* and *soft errors* are discussed. The second part of the chapter presents two effective approaches to stress a system from resilience and power-awareness standpoints—namely *fault injection* and *microbenchmarking*. Finally, the last part of the chapter briefly introduces power and performance measurement and modeling basic ideas.

The goal of this chapter is to provide the reader with a basic understanding and overview to board the next chapters where these notions are considered within the specific domain of *rugged* systems. In this case, reliability issues become even more critical due to the inherent *unfriendliness* of harsh environments (like extreme temperature and radiation levels, very low power and energy budgets, strict fault tolerance and security constraints, among others). For example, NASA's Curiosity rover (exploring the "Red Planet" as of the time of writing) uses two RAD750 radiation-hardened microprocessors in its on-board computer [46]. The engineers who designed this microprocessor had to consider the high radiation levels and wide temperature fluctuations in Mars to ensure mission success. Similarly, on-board computers design for unmanned aerial vehicles (i.e., *drones*) pose hurdles due to the tight power budget to perform mission-critical tasks, which must be successfully accomplished under a wide range of environmental conditions. Interplanetary rovers and drones are just two examples of harsh environment-capable embedded systems—many of the day-to-day devices that we use and rely on are subject to similar constraints, in some cases with critical consequences when they are not met. For example, cars, trucks, and even motorcycles are becoming smarter, and in some cases autonomous (*driverless*). Highly reliable, low-power embedded chips for harsh environments are the key enablers for autonomous and semiautonomous vehicles, and it is not difficult to imagine the safety consequences if the on-board control systems fail or are improperly operated. These topics are treated and discussed next in the following chapters.

REFERENCES

[1] A. Avizienis, J.C. Laprie, B. Randell, C. Landwehr, Basic concepts and taxonomy of dependable and secure computing, IEEE Trans. Dependable Secure Comput. 1 (1) (2004) 11–33.

[2] J.M. Emmert, C.E. Stroud, M. Abramovici, Online fault tolerance for FPGA logic blocks, IEEE Trans. Very Large Scale Integr. VLSI Syst. 15 (2) (2007) 216–226.

[3] R. Hyman Jr., K. Bhattacharya, N. Ranganathan, Redundancy mining for soft error detection in multicore processors, IEEE Trans. Comput. 60 (8) (2011) 1114–1125.

[4] K. Paulsson, M. Hubner, J. Becker, Strategies to on-line failure recovery in self-adaptive systems based on dynamic and partial reconfiguration, in: Proceedings of the First NASA/ESA Conference on Adaptive Hardware and Systems (AHS 2006), 2006, pp. 288–291.

[5] W. Rao, C. Yang, R. Karri, A. Orailoglu, Toward future systems with nanoscale devices: overcoming the reliability challenge, Computer 44 (2) (2011) 46–53.

[6] Dependable embedded systems, http://spp1500.itec.kit.edu.

[7] M. Berg, C. Poivey, D. Petrick, D. Espinosa, A. Lesea, K.A. LaBel, M. Friendlich, H. Kim, A. Phan, Effectiveness of internal versus external SEU scrubbing mitigation strategies in a Xilinx FPGA: design, test, and analysis, IEEE Trans. Nucl. Sci. 55 (4) (2008) 2259–2266.

[8] B. Hargreaves, H. Hult, S. Reda, Within-die process variations: how accurately can they be statistically modeled? in: Proceedings of the Asia and South Pacific Design Automation Conference (ASPDAC 2008), 2008, pp. 524–530.

[9] M. Malek, A comparison connection assignment for diagnosis of multiprocessor systems, in: Proceedings of the Seventh Annual Symposium on Computer Architecture (ISCA 1980), 1980, pp. 31–36.

[10] F.P. Preparata, G. Metze, R.T. Chien, On the connection assignment problem of diagnosable systems, IEEE Trans. Electron. Comput. EC-16 (6) (1967) 848–854.

[11] R.F. DeMara, K. Zhang, C.A. Sharma, Autonomic fault-handling and refurbishment using throughput-driven assessment, Appl. Soft Comput. 11 (2) (2011) 1588–1599.

[12] S. Mitra, N.R. Saxena, E.J. McCluskey, Common-mode failures in redundant VLSI systems: a survey, IEEE Trans. Reliab. 49 (3) (2000) 285–295.

[13] C. Carmichael, Triple module redundancy design techniques for Virtex FPGAs, Xilinx Application Note: Virtex Series XAPP197 (v1.0.1), 2006.http://www.xilinx.com/support/documentation/application_notes/xapp197.pdf.

[14] F. Lima Kastensmidt, L. Sterpone, L. Carro, M. Sonza Reorda, On the optimal design of triple modular redundancy logic for SRAM-based FPGAs, in: Proceedings of the Conference on Design, Automation and Test in Europe—Volume 2 (DATE 2005), 2005, pp. 1290–1295.

[15] G.W. Greenwood, On the practicality of using intrinsic reconfiguration for fault recovery, IEEE Trans. Evol. Comput. 9 (4) (2005) 398–405.

[16] J.C. Laprie, Dependable computing and fault tolerance: concepts and terminology, in: Proceedings of the 25th International Symposium on Fault-Tolerant Computing—Highlights From Twenty-Five Years, 1995, pp. 2–11.

[17] S. Lombardo, J. Stathis, B. Linder, K.L. Pey, F. Palumbo, C.H. Tung, Dielectric breakdown mechanisms in gate oxides, J. Appl. Phys. (2005) 98, pp. 121301-1 – 121301-36.

[18] S.E. Rauch, The statistics of NBTI-induced VT and β mismatch shifts in pMOSFETs, IEEE Trans. Device Mater. Reliab. 2 (4) (2002) 89–93.
[19] J. Srinivasan, Lifetime reliability aware microprocessors, Ph.D. thesis, University of Illinois at Urbana-Champaign, 2006.
[20] J. Srinivasan, S. Adve, P. Bose, J. Rivers, C.K. Hu, RAMP: a model for reliability aware microprocessor design, IBM Research Report, 2003. http://domino.watson.ibm.com/library/CyberDig.nsf/papers/DFFC51D1FD991F0D85256E13005B95A3/$File/rc23048.pdf RC 23048.
[21] J. Srinivasan, S.V. Adve, P. Bose, J.A. Rivers, Exploiting structural duplication for lifetime reliability enhancement, in: Proceedings of the 32nd Annual International Symposium on Computer Architecture (ISCA 2005), 2005, pp. 520–531.
[22] J. Shin, Lifetime reliability studies for microprocessor chip architecture, Ph.D. thesis, University of Southern California, 2008.
[23] J. Shin, V. Zyuban, P. Bose, T.M. Pinkston, A proactive wearout recovery approach for exploiting microarchitectural redundancy to extend cache SRAM lifetime, in: Proceedings of the 35th Annual International Symposium on Computer Architecture (ISCA 2008), 2008, pp. 353–362.
[24] J. Shin, V. Zyuban, Z. Hu, J.A. Rivers, P. Bose, A framework for architecture-level lifetime reliability modeling, in: Proceedings of the 37th Annual IEEE/IFIP International Conference on Dependable Systems and Networks (DSN 2007), 2007, pp. 534–543.
[25] S.S. Mukherjee, C. Weaver, J. Emer, S.K. Reinhardt, T. Austin, A systematic methodology to compute the architectural vulnerability factors for a high-performance microprocessor, in: Proceedings of the 36th Annual IEEE/ACM International Symposium on Microarchitecture (MICRO 36), 2003, pp. 29–40.
[26] X. Li, S.V. Adve, P. Bose, J.A. Rivers, Architecture-level soft error analysis: examining the limits of common assumptions, in: Proceedings of the 37th Annual IEEE/IFIP International Conference on Dependable Systems and Networks (DSN '07), 2007, pp. 266–275.
[27] OpenPOWER, On chip controller (OCC), http://openpowerfoundation.org/blogs/on-chip-controller-occ/.
[28] E. Rotem, Power management architecture of the 2nd generation Intel Core™ microarchitecture, formerly codenamed Sandy Bridge, in: Proceedings of the 23rd Hot Chips Symposium, 2011.
[29] T. Rosedahl, C. Lefurgy, POWER8 on chip controller: measuring and managing power consumption, http://hpm.ornl.gov/documents/HPM2015:Rosedahl.pdf.
[30] F. Bellosa, The benefits of event-driven energy accounting in power-sensitive systems, in: Proceedings of the Ninth Workshop on ACM SIGOPS European Workshop (EW 200), 2000, pp. 37–42.
[31] G. Contreras, M. Martonosi, Power prediction for Intel XScale® processors using performance monitoring unit events, in: Proceedings of the 10th ACM/IEEE International Symposium on Low Power Electronics and Design (ISLPED 2005), 2005, pp. 221–226.
[32] C. Isci, M. Martonosi, Runtime power monitoring in high-end processors: methodology and empirical data, in: Proceedings of the 36th Annual IEEE/ACM International Symposium on Microarchitecture (MICRO 2003), 2003, pp. 93–104.
[33] R. Joseph, M. Martonosi, Run-time power estimation in high performance microprocessors, in: Proceedings of the Sixth ACM/IEEE International Symposium on Low Power Electronics and Design (ISLPED 2001), 2001, pp. 135–140.

[34] B. Elkin, V. Indukuru, Commonly used metrics for performance analysis—POWER7, https://www.power.org/documentation/commonly-used-metrics-for-performance-analysis/.

[35] T. Willhalm, R. Dementiev, P. Fay, Intel performance counter monitor—a better way to measure CPU utilization, http://www.intel.com/software/pcm.

[36] ARM Ltd., Using the performance monitoring unit (PMU) and the event counters in DS-5, http://ds.arm.com/developer-resources/tutorials/using-the-performance-monitoring-unit-pmu-event-counters-in-ds-5/.

[37] A. Vega, A. Buyuktosunoglu, H. Hanson, P. Bose, S. Ramani, Crank it up or dial it down: coordinated multiprocessor frequency and folding control, in: Proceedings of the 46th Annual IEEE/ACM International Symposium on Microarchitecture (MICRO 2013), Davis, California, 2013, pp. 210–221.

[38] A. Mericas, B. Elkin, V.R. Indukuru, Comprehensive PMU event reference—POWER7, http://www.power.org/documentation/comprehensive-pmu-event-reference-power7/.

[39] M. Floyd, M. Allen-Ware, K. Rajamani, B. Brock, C. Lefurgy, A.J. Drake, L. Pesantez, T. Gloekler, J.A. Tierno, P. Bose, A. Buyuktosunoglu, Introducing the adaptive energy management features of the POWER7 chip, IEEE Micro 31 (2) (2011) 60–75.

[40] D. Brooks, V. Tiwari, M. Martonosi, Wattch: a framework for architectural-level power analysis and optimizations, in: Proceedings of the 27th Annual International Symposium on Computer Architecture (ISCA 2000), 2000, pp. 83–94.

[41] S. Li, J.H. Ahn, R.D. Strong, J.B. Brockman, D.M. Tullsen, N.P. Jouppi, McPAT: an integrated power, area, and timing modeling framework for multicore and manycore architectures, in: Proceedings of the 42nd Annual IEEE/ACM International Symposium on Microarchitecture (MICRO 2009), 2009, pp. 469–480.

[42] D. Brooks, P. Bose, V. Srinivasan, M.K. Gschwind, P.G. Emma, M.G. Rosenfield, New methodology for early-stage, microarchitecture-level power-performance analysis of microprocessors, IBM J. Res. Dev. 47 (5–6) (2003) 653–670.

[43] S.H. Gunther, F. Binns, D.M. Carmean, J.C. Hall, Managing the impact of increasing microprocessor power consumption, Intel Technol. J. 5 (1) (2001).

[44] D. Brooks, P. Bose, M. Martonosi, Power-performance simulation: design and validation strategies, ACM SIGMETRICS Perform. Eval. Rev. 31 (4) (2004) 13–18.

[45] X. Liang, K. Turgay, D. Brooks, Architectural power models for SRAM and CAM structures based on hybrid analytical/empirical techniques, in: Proceedings of the 2007 IEEE/ACM International Conference on Computer-Aided Design (ICCAD 2007), 2007, pp. 824–830.

[46] W. Harwood, Slow, but rugged, Curiosity's computer was built for Mars, 2012 http://www.cnet.com/news/slow-but-rugged-curiositys-computer-was-built-for-mars/.

CHAPTER

Real-time considerations for rugged embedded systems

3

A. Tumeo*, M. Ceriani†, G. Palermo†, M. Minutoli*, V.G. Castellana*, F. Ferrandi†

*Pacific Northwest National Laboratory, Richland, WA, United States**

Polytechnic University of Milan, Milano, Italy†

Embedded systems are defined as computer systems that perform a dedicated function. They are usually located within a larger mechanical or electrical system. Because of their specificity, embedded systems have substantially different design constraints with respect to conventional desktop computing systems.

Being designed to perform specific tasks, embedded systems do not have a fixed list of characteristics, but they may present at least one or more of the following characterizing aspects at the same time.

1. Embedded systems are application and domain specific: they are designed taking into consideration a precise application or a well defined domain, and only address those requirements. If moved to another domain or application, an embedded system may only provide suboptimal results.
2. Embedded systems may need to be reactive and have real-time requirements. Embedded systems usually interact with the real world, acquiring information through sensors and other input devices. Sometimes, they are required to *react* to any measured variation with some type of action. Variation may be periodic or aperiodic (i.e., not regularly happening). In many cases, embedded systems have to act, or react, to events in predefined timeframes (i.e., before a specific deadline), so that the action performed is effective or useful. They should, in other words, provide real-time operation. Real-time operation requires the ability to perform deterministic scheduling or, at least, to be able to quantify as much as possible uncertainty. Not all embedded systems must provide real-time operations, and not in all cases do real-time embedded systems have only hard deadlines. For example, in automotive, the antilock brake system (ABS) has hard real-time requirements, because it must activate as soon as there is pressure on the brakes, while the audio system may skip some samples (soft deadline) only impacting the quality of the entertainment without creating safety concerns. A more detailed coverage of embedded systems for the automotive domain is discussed in Chapter 8.

Rugged Embedded Systems. http://dx.doi.org/10.1016/B978-0-12-802459-1.00003-8

3. They may have small size and low weight. Depending on the application, embedded systems are required to fit in predetermined dimensions and not add too much weight. This is true for consumer products, such as cellular phones, as well as for control systems that have to fit, for example, on satellites. Size and weight obviously impact on the portability or mobility of the whole system.
4. Embedded systems are often power constrained. They should operate with very limited power budgets. In some cases (e.g., cellular phones, mobile devices), they are battery powered, and as such limiting power consumption provides longer operation times. Even on stationary systems, reducing power consumption may be critical for reducing size, weight, and cooling requirements.
5. Embedded systems may be distributed. A system may be a combination of different, independent embedded systems that provide an overall functionality (e.g., a vending machine), or a set of similar embedded systems communicating together or controlled by a supervisor (e.g., a sensor network).
6. Cost and time-to-market. Embedded systems may have cost constraints, connected to both design and manufacturing. Design costs usually are one time costs, while manufacturing costs are repetitive. Depending on the target application, for high-volume designs manufacturing costs dominate, for low-volume designs engineering costs are more relevant. Thus, design time may be traded off with the cost to manufacture a product. However, also the time-to-market plays a crucial role. In the consumer market, a product has to come out before the competitors to gather the majority of the profits or profits for a longer time. In mission critical applications, systems have to solve a problem at the right time, and may have to rapidly adapt to changes. For example, in high-frequency trading, algorithms have to adapt to changes in a few weeks. Software defined radio military applications may require to support new protocols. In other applications, systems have to adapt to new form factors.
7. Embedded systems may have reliability requirements. An embedded system should be able to identify faults and, in several cases, keep operating correctly even in case of errors. For example, control systems on planes, satellites, and space probes should guarantee specific levels of reliability to avoid as much as possible unrecoverable errors. Reliability is often obtained by implementing redundant hardware and/or software components. In some cases a graceful degradation of the performance may be allowed to provide correct operation in presence of faults. It is possible to obtain measures (typically statistical) of the mean time between failures and mean time to repair it.
8. Embedded systems may have security requirements. Depending on the application, an embedded system may need to maintain information security. An embedded system may have to maintain confidentiality of data and application, i.e., it should not allow unauthorized disclosure. An embedded system should provide integrity of data and applications, not permitting unauthorized modifications. Finally, an embedded system should provide availability of data and applications, or part of them, only to authorized users.

9. An embedded system should be safe. The concept of safety is connected with the fact that, while it is functioning, it should not create any risk for the operator and/or the system that it may be controlling. In many environments, if an embedded system fails, any type of damage to the system or the external environment are not tolerated.
10. They may have to operate in harsh environment. Modern embedded systems may not necessarily operate in a controlled environment. It is actually very common for an embedded system to operate in an external environment with adverse operating conditions. The environment may be dusty (e.g., the lunar soil), or wet (e.g., the oceans), or with high temperatures (e.g., a combustion chamber). The system may need to withstand vibration, shock, lighting, power supply fluctuations, water, corrosion, fire and, in general, physical abuse. Furthermore, there may be limited or no possibilities for check ups and maintenance (e.g., space applications). The ability to operate in harsh environment can heavily influence the other design characteristics, requiring trade offs in some of the characterizing aspects to guarantee continuity of operations.

1 OPERATING IN HARSH ENVIRONMENTS

A computing system is said to be *rugged* when it is specifically designed to operate reliably in such an environment or conditions as described in the previous point 10. An important consideration is the fact that a rugged system is designed from the inception to sustain the conditions of the environment where it is envisioned to operate. Hence, there is a significant focus on its construction. The design philosophy is to provide a controlled environment for the electronics, and the electronics themselves may go through a very specific selection process that guarantees higher reliability in difficult conditions with respect to commodity components.

Manufacturers design rugged versions of commodity general-purpose systems such as laptops, tablets, and smartphones. These generally are characterized by stronger and thicker casing than regular products. They usually are vibration, shock, drop, dust, and water proof. They are also designed such that they can tolerate humidity, solar radiation, altitude, and temperature extremes. They try to limit as much as possible moving parts, such as removing spinning hard disks and replacing them with solid state disks, or eliminating fans. They may also implement some features that make easier accessing of some specific functions of the system. For example, phones may include push-to-talk functions, phones and tablets barcode readers, and laptops quick launch buttons for specific applications.

Currently, there is not a single standard that measures a single "level" of *ruggedization*. There is, however, a large set of *certifications* that cover military use, temperature, waterproofing, ignition explosive atmosphere, electromagnetic compatibility, and more.

Table 1 presents a nonexhaustive list of some common certifications and standards, discussing the aspects that they cover.

Table 1 List of Certification Standards

MIL-STD-810	A military standard, issued in 1962, that establishes a set of tests for determining equipment suitability for military operations
MIL-STD-461	A military standard that establishes how to test equipment for electromagnetic compatibility
MIL-S-901	A military standard for equipment mounted on ships. It features two levels. Grade A: items essential to the safety and continued combat capability of the ship. Grade B: non essential to safety and combat capability, but hazards for personnel.
IEEE 1156.1-1993	IEEE Standard Microcomputer Environmental Specifications for Computer Modules. Provides information on minimum environmental withstand condition for modules that may make up larger systems and are purported to have a rated environmental performance level.
IEEE 1613	Computers in electrical substations used to concentrate data or communicate with SCADA (supervisory control and data acquisition, a type of industrial control system) systems follow IEEE 1613 "Standard Environmental and Testing Requirements for Communications Networking Devices in Electric Power Substations."
IP	The IP Code (International Protection Marking) classifies and rates the degree of protection against intrusion, dust, accidental contact, and water by mechanical casings and electrical enclosures. The code begins with the IP indication, followed by a mandatory single numeral (from 0 to 6) representing solid particle protection, a mandatory single numeral (from 0 to 9) representing the liquid ingress protection, and one or more optional single letters representing other protections. Older specifications of the code also included a numeral to specify mechanical impact resistance, but this is no longer in use and has been replaced by the IK code.
IS	The IS (Intrinsic Safety) is an approach to design electrical equipment for hazardous areas that limits the energy available for ignition (i.e., to make them explosion proof). The techniques usually involve reducing or eliminating internal sparking, controlling component temperatures, and eliminating component spacing that would allow dust to short a circuit.
TEMPEST	A certification referring to spying on information systems through leaking emanations, including unintentional radio or electrical signals, sounds, and vibrations. It covers methods to spy upon others and methods to shield equipment against spying. TEMPEST certified systems implement protection through distance (from walls, other wires, other equipment), shielding (of wires, instruments, and buildings), filtering, and masking.

Continued

Table 1 List of Certification Standards *–cont'd*

ATEX	Two EU directives that describe what equipment and work environment are allowed in an environment with an explosive atmosphere. The ATEX 95 equipment directive 94/9/EC, Equipment and protective systems intended for use in potentially explosive atmospheres, is for the manufacturers of the equipment. The ATEX 137 *workplace* directive 99/92/EC, Minimum requirements for improving the safety and health protection of workers potentially at risk from explosive atmospheres, is for the users of the equipment.
NEMA	The National Electrical Manufacturers Association (NEMA) defines enclosure types in its standard number 250. These types are similar to the IP Code ratings, but also require additional product features and tests, including functionality in icing conditions, enclosures for hazardous areas, knock-outs for cable connections, etc.
IK	The IK Code is an international numeric classification for the degrees of protection provided by enclosures for electrical equipment against external mechanical impacts. It was originally defined in the European Standard BS EN50102, and after its adoption as an international standard was renumbered to EN62262. It uses a two numeral code that directly depends on the impact energy. The standard specifies how enclosure should be mounted during tests, the atmospheric conditions, the number of impacts and their distribution, and the characteristics of the hammer that produce the energy levels.
EN50155	An international standard that covers electronic equipment used on rolling stock for railway applications. It covers operating temperature range (−25 to 70°C), resistance to humidity, shocks, vibrations, radiation, and other parameters.
ISO 26262	An international standard for functional safety of electrical and/or electronic systems in production of automobiles. It is a risk-based safety standard, where risk of hazardous operational situations is qualitatively assessed. The safety measures are defined to avoid or control systematic failures, and to detect or control random hardware failures, or mitigate their effects. ISO 26262 provides an automotive safety lifecycle, covers functional safety aspects of the entire development process, provides a risk-based approach for determining risk classes specific to automotive systems, uses the risk classes to specific safety requirements of an item and achieving an acceptable residual risks, and provides requirements for validation and confirmation to ensure that an acceptable level of safety is being achieved. ISO 26262 will be discussed in more detail in Chapter 8.

Rugged computers, however, today represent only a very limited number of systems designed for harsh environments. The majority of these systems are, in fact, embedded, as a direct consequence of delegating more and more control functions for a variety of electrical, mechanical, and physical systems to computing systems. As discussed in the previous section in characteristic 10, embedded systems often have to operate in harsh environments. However, designing an embedded system for a harsh environment has a direct impact on all the previously enumerated characteristics. Take for example the case of automotive applications. A system controlling the ABS has to sustain vibrations. It may also need to sustain impact as much as possible. Depending on the position of the system on the chassis, it may require also being able to sustain dirt and water. As such, it requires an appropriate enclosure. Although a car designer has some flexibility, weight and size cannot be neglected, as these may need to fit in standard, mass-produced, frames. It must be verifiable and serviceable. Inside there is more structure. If there are components and/or plug-in cards, these need to be individually supported and secured, so that they do not pop out in case of shock events. As the system needs to be encased, cooling to guarantee correct operations requires additional care. Any heat sinks require to be secured, and there is the need to guarantee airflow if necessary. These mainly are physical characteristics that require making an embedded system rugged, and thus able to sustain operation in a harsh environment.

There are, however, an increasing number of cases where also the firmware and software components need a design that is able to operate in harsh environments. Typically, real-time embedded systems have, for some tasks, hard deadlines that have to be respected. Depending on the specific application and, thus, the conditions in which the system is working, a failure could be critical. At the same time, some tasks may have soft deadlines or, simply, aperiodic behaviors, only requiring best effort completion but without interfering with the execution of the tasks with hard deadlines. Providing these types of behaviors involve approaches that span both the hardware design and the software stack.

In the rest of this chapter we present one of such case studies [1]. We consider a possible architecture template for an automotive control system, and design a prototype of such a system. We then discuss a real-time scheduling algorithm, describing its ability to guarantee execution of periodic tasks with hard deadlines but, at the same time, manage aperiodic tasks in a best effort policy. This allows completing aperiodic tasks without disrupting real-time responsiveness of the system. We discuss the hardware and software features implemented in the prototype to enable these functionalities, and then present a brief study that shows how dynamic effects of the system can impact the efficiency of the system. This indicates that a proper system design must take into account these issues by, for example, overprovisioning resources and/or computational capabilities.

2 CASE STUDY: A FIELD PROGRAMMABLE GATE ARRAY PROTOTYPE FOR THE VALIDATION OF REAL-TIME ALGORITHMS

Multiprocessor systems-on-chip, composed of several processing elements, including multiple general-purpose cores, and on-chip memories, have become the standard for implementing embedded systems. Thanks to the large number of processing elements, these systems potentially allow a better management of periodic workloads and can react faster to external, aperiodic events.

However, the adoption of these powerful architectures in real-time systems opens several problems concerning scheduling strategies [2]. It is well known that optimal scheduling for multiprocessor systems is a NP-Hard problem [3]. On real systems, it is not trivial to coordinate and correctly distribute tasks among the different processors. This is even more complex when external peripherals trigger interrupts that require an efficient handling. An important example is the case of automotive applications, where several periodic tasks that check the status of sensors and other mechanisms run in parallel with tasks triggered by external events (e.g., security warnings, system setup, etc.). Efficient hardware solutions must support a low overhead real-time scheduler to allow the design of a usable system. Many commercial real-time operating systems still rely on single processor architectures for manageability purposes or adopt simple priority-based preemptive scheduling in multiprocessor solutions [4–6].

Real-time scheduling algorithms for multiprocessors are usually divided into two classes: local and global schedulers. Local schedulers rely on a prepartitioning of the tasks on the different processors, and then try to schedule them in the best way possible on the preassigned elements using well-known uniprocessor scheduling algorithms [7–9]. These are usually employed to schedule periodic tasks, while aperiodic tasks are allowed to migrate to any processor [10].

Global schedulers, instead, try to find the best allocation on all the processors available in the system at the same time. However, these types of multiprocessor global schedulers do not usually deal with aperiodic tasks [11–13].

Among all the real-time scheduling algorithms proposed in the recent years for multiprocessor systems an interesting solution is the *multiprocessor dual priority* (MPDP) *scheduling algorithm* [14]. Thanks to its hybrid local-global nature, it guarantees periodic task deadlines (local), and it allows serving aperiodic requests with good average response time (global), which is crucial for reactive systems. It has low memory usage and low computational overhead because it is based on the offline computation of the worst case response time for periodic tasks. This makes it suitable for efficient implementation on small embedded systems.

Field programmable gate arrays (FPGAs) are an interesting design alternative for system prototyping. In some cases, they can also be the target implementation technology for certain mission critical applications when the production volume is medium to low.

We choose to implement an existing algorithm for real-time task scheduling and then study its performance on a realistic architecture prototyped with FPGA technology, rather than by simulation. We developed several hardware devices to support reactive behavior and allow inter processor communication and implemented on top of it the MPDP algorithm [14].

3 ARCHITECTURE

This section presents the architecture of our real-time multiprocessor system on FPGA. We will first describe the basic architecture and then give some details on the multiprocessor interrupt controller design that is used to distribute activation signal for the scheduler and to trigger the execution of sporadic tasks.

3.1 PROTOTYPE

We designed our prototype with the Xilinx Embedded Developer Kit (EDK) 8.2 and synthesized it with Xilinx ISE version 8.2 on a Virtex-II PRO XC2VP30 Speed Grade-7 FPGA targeting a frequency of 50 MHz.

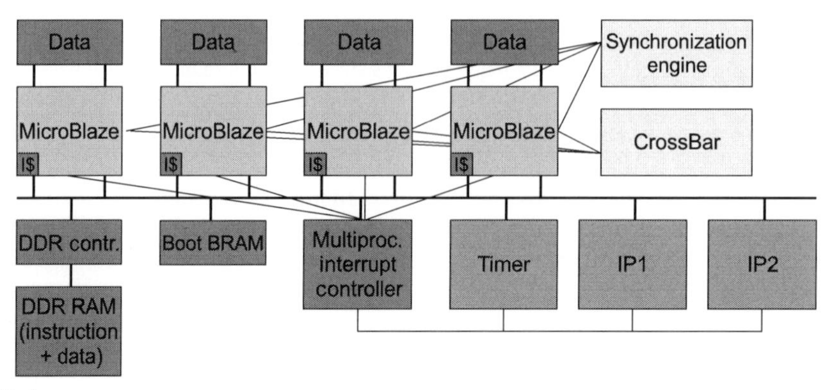

FIG. 1

Architectural overview of our real-time system prototype.

This architecture, shown in Fig. 1, is composed of several Xilinx MicroBlaze processors connected to a shared on-chip peripheral bus (OPB). Each processor has access to a local memory for private data (stack and heap of the currently executing thread), implemented through Block RAMs (BRAMs), and to an external shared double data rate RAM memory for shared instructions and data. Instruction cache is implemented for each processor, bringing down access latency from 12 to 1 clock cycle in case of hit. Local memories, which have the same latency of caches, are instead used for data.

A shared BRAMs is connected to the OPB for boot purposes. The system adopts an ad-hoc coprocessor (Synchronization Engine) [15] that provides hardware support for lock and barrier synchronization primitives and a CrossBar module that allows interprocessor communication for small data passing without using the shared bus. The system features a multiprocessor interrupt controller [16] specifically designed for this architecture. This controller distributes to all the processors in the system interrupt signals of various nature. It forwards the signal triggered by the system timer, that determines the scheduling period and starts the scheduling cycle, to an available processor. It connects to the peripherals so they can launch interrupts signal to any processor and trigger the start of aperiodic tasks for subsequent elaboration.

Peripherals can be interfaces to sensors and data acquisition systems, like, for example, controller area network (CAN) interfaces, widely used in automotive applications. They are connected to the shared OPB and seen by all the processors, while their interrupt lines are connected to the multiprocessor interrupt controller, which in turn forwards them to the MicroBlaze soft-cores.

3.2 MULTIPROCESSOR INTERRUPT CONTROLLER

The multiprocessor interrupt controller is an innovative design [16] that allows supporting several features in a multiprocessor architecture designed with the Xilinx toolchain.

The MicroBlaze soft-core has a very simple interrupt management scheme. It only has a single interrupt input, so if multiple peripherals that generate interrupt are implemented in the system, an interrupt controller is required. However, when multiple processors are used, the standard interrupt controller integrated in the Embedded Developer Kit is ineffective, since it only permits to propagate multiple interrupts to a single processor. Our multiprocessor interrupt controller design, instead, allows connecting them to multiple processors introducing several useful features:

1. it *distributes* the interrupts coming from the peripherals to free processors in the system, allowing the parallel execution of several interrupt service routines;
2. it allows a peripheral to be *booked* by a specific processor, meaning that only a specific processor will receive and handle the interrupt coming from a specific peripheral;
3. it supports *interrupts multicasting and broadcasting* to propagate a single interrupt signal to more than one processor;
4. it supports *interprocessor interrupts* to allow any processor to stop the execution of another processor.

Interrupt distribution allows exploiting the parallelism of a multiprocessor system to manage interrupt services routines, enabling the launch of concurrent interrupt handlers when many interrupts are generated at the same time.

This effectively permits a more reactive system to contemporary external events. Our design adopts a fixed priority scheme with timeout. When an interrupt signal is propagated to a processor, it has a predefined deadline to acknowledge its management. If the processor is already handling an interrupt, it cannot answer because interrupt reception is disabled. So, when the timeout fires, the interrupt line to that processor is disabled, and the interrupt is propagated to the subsequent processor in the priority list. Booking can very useful when dynamic thread allocation is used. In fact, if a processor offloads a function to an intellectual property (IP) core, we may want that the same processor that started the computation manage the read-back of the results. The booking functionality allows propagating interrupt signals coming from IP cores only to specific processors. Interprocessor interrupts allow processors to communicate. They can be useful, for example, for synchronization or for starting a context switch with thread migration from a processor to another. Broadcast and multicast are important to propagate the same interrupt signal to more than one processing element, like, for example, a global timer that triggers the scheduling on all the processors.

These features are exposed to the operating system with very simple primitives that are seamlessly integrated in the Xilinx toolchain. The interrupt controller is connected to the OPB and, when the processors access its registers for configuration and acknowledgments, mutual exclusion is used. Thus, controller management is sequential, but the execution of the interrupt handlers is parallel.

4 REAL-TIME SUPPORT

We now briefly describe how the MPDP works, describe the implementation details of the real-time microkernel, and detail how it interacts with the target architecture.

4.1 MPDP ALGORITHM

The scheduling approach adopted in our design is based on the MPDP algorithm [14]. This solution adapts the dual priority model for uniprocessor systems to multiprocessor architectures with shared memory.

The dual priority model [17] for single processor systems provides an efficient method to responsively schedule sporadic tasks, while retaining the offline guarantees for crucial periodic tasks obtained with fixed priority. This model splits priorities into three bands, Upper, Middle, and Lower. Hard tasks are assigned two priorities, one from the Upper and one from the Lower band. Soft tasks have a priority in the middle band.

A hard task is released with a priority in the Lower Band, and it may be preempted by other hard tasks with higher priorities in the same band, by soft tasks in the middle band or by any hard task executing at an upper band priority. However, at a fixed time, called promotion time, its priority is promoted to the upper band. From that moment, it could be preempted only by other hard tasks that have been promoted in the upper band and have higher priority.

A priori guarantees for hard periodic tasks are obtained through offline computation of worst case response, using the analysis derived from fixed priority scheduling [18]. The key is to determine the correct promotion time U_i for each task i. Promotion time is $0 \leq U_i \leq D_i$, where D_i is the deadline for task i, and it is computed with the recurrence relation:

$$W_i^{m+1} = C_i + \sum_{j \in hp(i)} \left\lceil \frac{W_i^m}{T_j} \right\rceil C_j$$

This relation is also used as a schedulability test. W_i is the length of a priority level busy period for task i, starting at time 0 and culminating in the completion of the task. T represents the period and C the worst case execution time. In the worst case, task i is unable to execute any computations at its lower band priority, but when promoted to the upper band, it will be subjected to interference only from tasks with higher priority in this band (i.e., all the tasks with priority $hp(i)$).

Let us assume that the task i has been released at $-U_i$ and time 0 corresponds to its promotion time. Thus, promotion time is $U_i = D_i - W_i$. Iteration starts with $w_i^0 = 0$ and ends when $w_i^{m+1} = w_i^m$ or $w_i^{m+1} > D_i - U_i$, in which case task i is unschedulable.

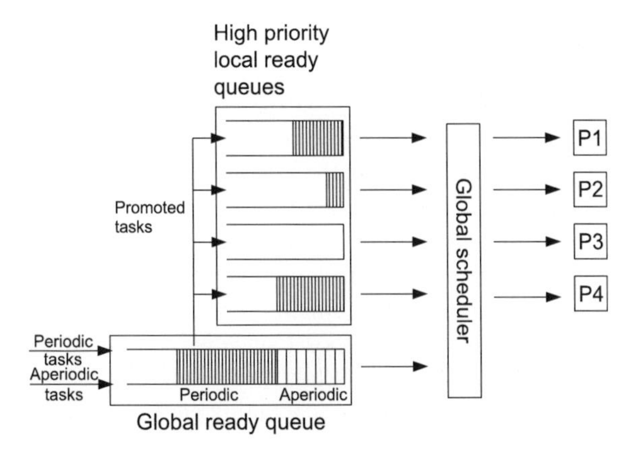

FIG. 2

Overview of the MPDP queues [14].

This model can be easily adapted to multiprocessor systems. The approach, proposed in Ref. [14] and summarized in Fig. 2, is a hybrid between local and global scheduling. Periodic tasks before promotion can be executed on any processors (global scheduling), while after promotion they are executed on a predefined processing element (local scheduling). Initially, periodic tasks are statically distributed among the processors.

The uniprocessor formula is used to compute worst case response time of periodic tasks on a single processor. During runtime, when a task arrives, being it periodic or aperiodic, it is queued in a global ready queue (GRQ). Aperiodic tasks have higher priority than periodic tasks and are queued in FIFO order. Periodic tasks are sorted according to their fixed low priority. The global scheduler selects the first N tasks from this queue to execute on the processors. There are N high-priority local ready queues (HPLRQ), used to queue promoted periodic tasks for each processor. When a periodic task is promoted, it is moved from the GRQ to the respective HPLRQ.

If there are any tasks in this queue, a processor is not allowed to execute tasks from the GRQ. A promotion implies a change in priority and can cause preemption, while the task migrates to its original processor. This scheme permits to advance the periodic work, guaranteeing at the same time that the system can satisfy aperiodic demands without missing the hard periodic deadlines. Thus it seems suitable for implementation on a real-time system, which must perform periodic tasks while at the same time react with good response times to external events.

In Ref. [14] this algorithm is presented only from a theoretical point of view, with an efficiency analysis of the algorithm that assumes negligible system overheads. However, task migration and context switching have a cost. Moreover, on a real shared memory multiprocessor system there are physical overheads determined by contention on the shared resources. Our objective is to propose a realistic multiprocessor system implemented on FPGA that adopts this algorithm, testing it with a real workload for automotive applications and showing the impact of all these aspects.

4.2 IMPLEMENTATION DETAILS

Our MPDP kernel implementation exploits the underlying memory model of the target architecture. In fact, the private data of the task are allocated in the fast, local memory of the processors, and are moved to shared memory and again in local memory when context changes and task switching occurs. Scheduling phases are triggered by a system timer interrupt. The multiprocessor interrupt controller assigns this interrupt to a processor that is currently free from handling other interrupts.

A single processor performs the scheduling phase. The others can continue their work while the task allocation is performed. Our implementation slightly differs from the original MPDP algorithm proposal since we use two different queues for periodic tasks in low priority (Periodic Ready Queue) and aperiodic tasks (Aperiodic Ready Queue), which makes the global scheduling easier and faster. Furthermore, effectively being on a closed system, we need to park periodic tasks while they have completed their execution and are waiting for the next release. We thus use a Waiting Periodic Queue, in which periodic tasks are inserted ordered by proximity to release time. Each scheduling cycle, periodic tasks that have reached release time leave the Waiting Periodic Queue and enter the Periodic Ready Queue, ready for actual scheduling on the multiprocessor system.

Then, periodic tasks in the Periodic Ready Queue are checked for promotion. If a promotion happens, the periodic task is moved to the high-priority local queue of its target processor, in a position determined by its high-priority value. At this point, task assignment can be performed according to the MPDP model. Processors with tasks in their high-priority local queue get the highest in priority from their specific queue. If any, sporadic tasks (which have middle band priorities) are assigned to other processors according to their arrival order (oldest tasks are scheduled first). Finally, if there are still free processors, nonpromoted periodic tasks (which have lower band priorities) are allocated. After the complete task allocation is performed, the scheduling processor triggers interprocessor interrupts to all the processors that have received new tasks, starting a context change. If a task is allocated on the same processor it was currently running on, the processor is not interrupted and can continue its work. If the task is allocated to the scheduling processor itself before the scheduling cycle has been changed, the scheduling routine exits, launching the context change. Note that, before promotion, periodic tasks can be assigned to any processor, while after promotion they can execute only on the predefined (at design time) processor. Thus, it could be possible that two processors switch their tasks. If a processor completes execution of its current task, it will not wait until the next scheduling cycle, but it will automatically check if there is an available task to run, following the priority rules.

Tasks contexts consist of the register file of the MicroBlaze processor and the stack. During context switching, the contexts are saved in shared memory, stored in a vector that contains a location for each task runnable in the system. The context switch primitive, when executed, loads the register file into the processor and the stack into the local memory. The implementation of the context switching mechanisms takes care of the exiting from the interrupt handling state and the stack relocation in local memories.

Sporadic tasks are triggered by interrupts from peripherals or external devices. When these interrupts occur, the interrupt controller distributes them to processors that are not handling other interrupts. Thus, if a processor is executing the scheduling cycle, or it is executing a context switch, it will not be burdened by the sporadic task release. Furthermore, multiple sporadic task releases can be managed at the same time.

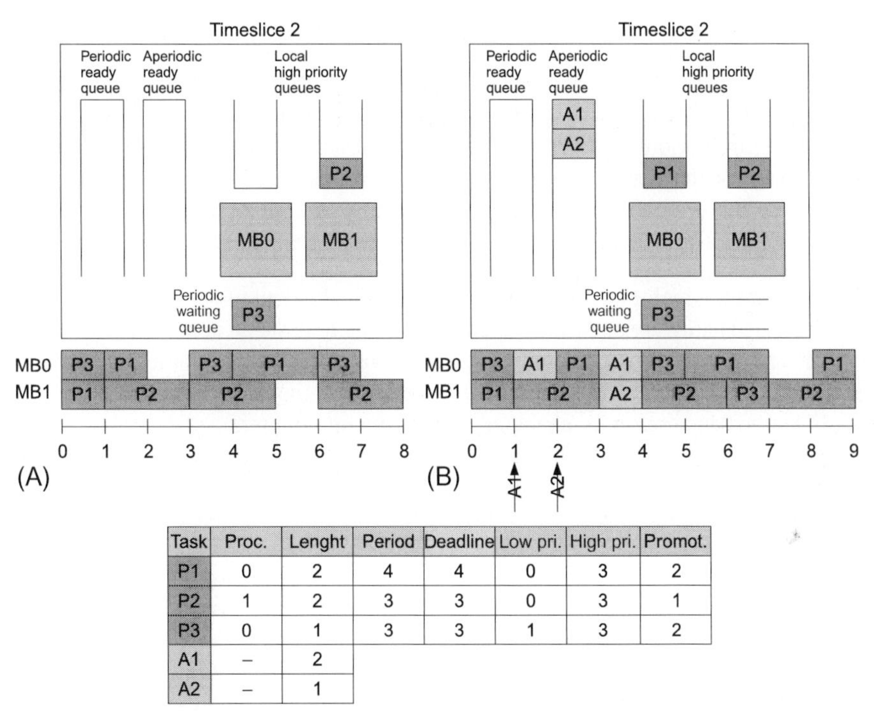

Task	Proc.	Lenght	Period	Deadline	Low pri.	High pri.	Promot.
P1	0	2	4	4	0	3	2
P2	1	2	3	3	0	3	1
P3	0	1	3	3	1	3	2
A1	–	2					
A2	–	1					

FIG. 3

A sample schedule with three periodic and two aperiodic tasks on a dual MicroBlaze architecture. The status of the queues without and with aperiodic workload is shown respectively in (A) and (B).

Fig. 3 shows an example of scheduling on a dual processor architecture with three periodic and two aperiodic tasks. The table reports the basic information of the tasks required for the scheduling. Priorities can be 0 and 1 for periodic tasks in low priority mode and 3 in high priority. Aperiodic tasks are thus positioned with priority 2. Schedule A shows that without aperiodic tasks, we have an available slot in timeslice 2 on MicroBlaze 1. However, we can see that to guarantee completion before time-slice 3, task P2 has been promoted to high priority.

Schedule B adds the two aperiodic tasks, which arrive at the beginning of timeslices 1 and 2. Part of task A1 is executed as soon as it arrives, since P1 in time-slice 1 is in low priority. However, at timeslice 2, P1 gets promoted to its high priority, A1 is interrupted and P1 completed. A2 arrives at timeslice 2 but it is inserted in the queue after A2 and waits for the completion of the higher priority promoted periodic tasks.

5 EVALUATION

To test our dual priority real-time multiprocessor system we adapted the automotive set of the MiBench benchmark suite [19]. MiBench is a collection of commercially representative embedded benchmarks. Among them, we decided to port the automotive set as it is more representative for a real-time and reactive system. In such a system, there are not only sets of critical periodic tasks that must meet their deadline (e.g., to check that some sensors are in functional order), but also many sporadic tasks (like the ones triggered by security systems) that must be served in the best way possible. The benchmark set basically presents four groups of applications:

- *basicmath*, which performs simple mathematical calculations not supported by dedicated hardware and can be used to calculate road speed or other vector values (three programs: square roots, first derivative, angle conversion);
- *bitcount*, which tests bit manipulation abilities of the processors and is
- linked to sensor activity checking (five different counters);
- *qsort*, which executes sorting of vectors, useful to organize data and priorities;
- *susan*, which is an image recognition package that can recognize corners or edges and can smooth an image, useful for quality assurance video systems or car navigation systems.

We used a mix of datasets (small and large) for the different benchmarks launched on the system. We selected the workload in order to obtain a variable periodic utilization, and measured the response time of an aperiodic task triggered by an interrupt. The benchmarks have been executed varying the number of processors of the target platform.

We run a total of 19 tasks on the system, 18 periodic and 1 aperiodic. The aperiodic task is the susan benchmark with the large dataset. This choice is justified by the fact that some of the services performed by susan can be connected to car navigation systems and are triggered by aperiodic interrupts that, for example, can signal the arrival of the image to analyze, from the cameras. All the other applications are executed as periodic benchmarks running in parallel on the system with different datasets (small and large). Periodic utilization is determined by varying the periods of the applications in accordance to their critical deadline.

The worst case response times of the tasks have been determined by taking into account an overhead for the context switching and considering the most complex datasets. Promotion time and schedulability have been calculated using the recurrent formula through an in house tool that takes in input worst case execution times, period and deadlines of the tasks, and produce the task tables with processor assignments and all the required information for both our target architecture and the simulator.

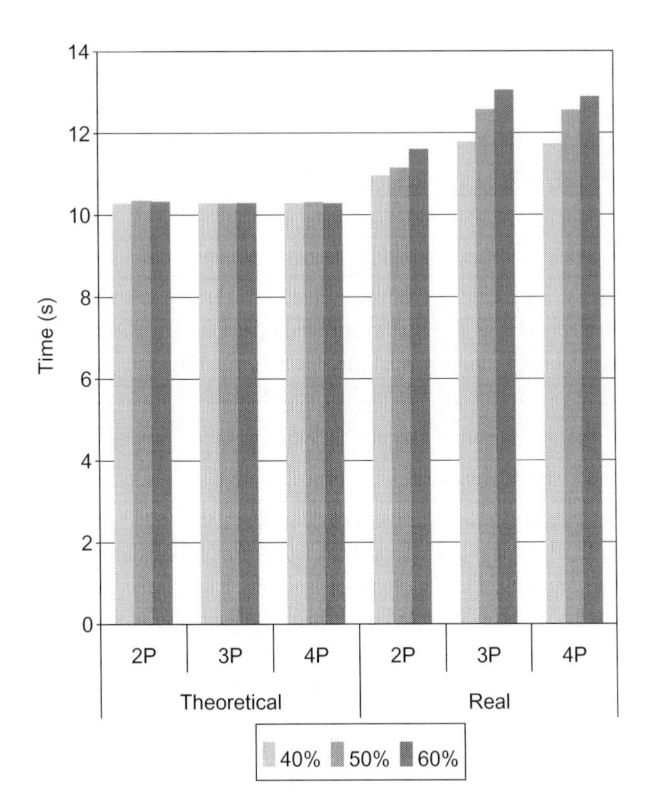

FIG. 4

Response time seconds of an aperiodic task on our system with different periodic utilization.

Fig. 4 shows the average response time of the selected aperiodic task on architectures from two to four processors, with a periodic utilization of the systems from 40% to 60%.

As MPDP adopts a best effort approach to serve aperiodic tasks, utilization near half the capacity of the system is an appropriate balanced choice between exploitation of the system and the responsiveness to external events. The theoretical data for architectures with two, three, and four processors are calculated with a simulator that adopts the same approach of the scheduling kernel of the target architecture, considering a small overhead (2%) for context switching and contention. The scheduling phase is triggered each 0.1 second by the system timer. The aperiodic task, on a single processor architecture, should execute in 10.26 seconds with the given dataset at 50 MHz.

The data shows that, with these utilizations, the system is not fully loaded and the algorithm should execute the aperiodic task with very limited response times, almost near the execution time, with all the architectures, with the only overheads of context switching when moving the task on free processors (10.32 seconds in the worst case).

However, the results on the prototype show that this assumption is not true. It is easy to see that the real 2-processor architecture is respectively 7%, 8%, and 12% slower in response times than the simulated architecture with 40%, 50%, and

60% periodic utilization respectively. Aperiodic response times should grow with periodic utilization when the system is heavily loaded: it is clear that context switching and contention, in particular on the shared memory and bus, are an important constraint for our architecture. The trend is confirmed when analyzing the 3-processor solution: here the prototype is 15%, 22%, and 27% slower than the expected response times. With more processors, contention is higher, and the system loses responsiveness. In particular, task switching, with movements of contexts and stacks for many applications from and to shared memory, generates consistent traffic, even with a clever implementation of the algorithm that limits switching only when necessary.

Nevertheless, with 4 processors, our prototype gives almost the same results obtained with 3 MicroBlazes, even slightly better. This is a clear indication that the bus and memory access patterns have settled, and with these utilizations the behavior is almost the same. Note that an utilization of 50% on a system with 4 processors corresponds to a workload that is double with respect to a 2 processor system with an utilization of 50%. In fact, 4 processors used half of the time can allocate twice the tasks of two processors used half of the time.

This means that, even if the aperiodic response time is worse, the system is anyway doing much more periodic work. On 4 processors, with a 60%—utilization workload, our architecture can reach a response time of 12.88 seconds, 25% worse than the optimal response time obtained in simulation, definitely a good result considering the characteristics of the MicroBlaze soft-core, the number of tasks running simultaneously on the system, and their periods.

Overall, these results show that designing a real-time system able to provide effective reactive behavior requires taking into consideration the dynamic architectural effects. Using a realistic prototype is an empirical way for studying and validating the effectiveness of real-time scheduling algorithms, providing a way to quantify if and how much a system should be overprovisioned to guarantee certain properties. This may be critical when the system targets harsh environments, and may be difficult to upgrade or modify on the field. Fast prototyping may represent an appealing opportunity, because as the complexity of the systems increases, the complexity in developing and proving formal models that enable establishing a priori guarantees for specific behaviors increases too, and remains open research.

6 CONCLUSIONS

In this chapter we summarized the characteristics of embedded systems for harsh environments. Because embedded systems are, by definition, designed to perform specific tasks, characteristics may vary depending on the task. We have, however, listed some of the features that may appear. We have also discussed what it means for such embedded systems to operate in a harsh environment, and what design points have to be considered by developers to make the system "rugged." We have, in particular, discussed both the physical and the software aspects. If physical construction is an obvious aspect, because a rugged system may be subject to worse environmental conditions and as such requires higher "physical robustness," the software aspect is less obvious, but equally important. We have provided the list of the most important and famous international standards for rugged systems that cover both the hardware and the software aspects.

To provide an example of the importance of the software aspects, we have presented a case study based on a realistic FPGA prototype to validate a real-time scheduling algorithm, mimicking an automotive control system with a set of benchmarks from the area. The case study shows the importance of designing operating system software for rugged systems considering the dynamic effects of the architecture, explaining why a system may be designed with higher tolerances and validated in realistic conditions to verify that certain properties are met.

REFERENCES

[1] A. Tumeo, M. Branca, L. Camerini, M. Ceriani, M. Monchiero, G. Palermo, et al., A dual-priority real-time multiprocessor system on FPGA for automotive applications, in: DATE '08: Conference on Design, Automation and Test in Europe, 2008, pp. 1039–1044.

[2] J.A. Stankovic, M. Spuri, M. Di Natale, G.C. Buttazzo, Implications of classical scheduling results for real-time systems, Computer 28 (6) (1995) 6–25.

[3] M.R. Garey, D.S. Johnson, Complexity results for multiprocessor scheduling under resource constraints, in: J.A. Stankovic, K. Ramamritham (Eds.), Tutorial: Hard Real-Time Systems, IEEE Computer Society Press, Los Alamitos, CA, 1989, pp. 205–219.

[4] Mentor, Nucleos RTOS, Retrieved from https://www.mentor.com/embedded-software/nucleus/.

[5] RTEMS, Real-Time Executive for Multiprocessor Systems (RTEMS). Retrieved from https://www.rtems.org.

[6] WindRiver, VxWORKS. Retrieved from: http://windriver.com/products/vxworks/.

[7] G. Fohler, Joint scheduling of distributed complex periodic and hard aperiodic tasks in statically scheduled systems, in: Real-Time Systems Symposium, IEEE, 1995, pp. 152–161.

[8] K. Ramamritham, J.A. Stankovic, W. Zhao, Distributed scheduling of tasks with deadlines and resource requirements, IEEE Trans. Comput. 38 (8) (1989) 1110–1123.

[9] S. Saez, J. Vila, A. Crespo, Soft aperiodic task scheduling on hard real-time multiprocessor systems, in: RTCSA '99: Sixth International Conference on Real-Time Computing Systems and Applications, 1999, pp. 424–427.

[10] J.M. Banús, A. Arenas, An efficient scheme to allocate soft-aperiodic tasks in multiprocessor hard real-time systems, in: PDPTA '02: International Conference on Parallel and Distributed Processing Techniques and Applications, 2002, pp. 809–815.

[11] B. Andersson, S. Baruah, J. Jonson, Static-priority scheduling on multiprocessors, in: RTSS 2001: 22nd IEEE Real-Time Systems Symposium, IEEE, London, 2001, pp. 193–202.

[12] J. Goossens, S. Funk, S. Baruah, Edf scheduling on multiprocessor platforms: some (perhaps) counterintuitive observations, in: RealTime Computing Systems and Applications Symposium, 2002.

[13] B.A. Jonsson, Fixed-priority preemptive multiprocessor scheduling: to partition or not to partition, in: Seventh International Conference on Real-Time Computing Systems and Applications, IEEE, 2000, pp. 337–346.

[14] J. Banus, A. Arenas, J. Labarta, Dual priority algorithm to schedule real-time tasks in a shared memory multiprocessor, in: IPDPS: International Parallel and Distributed Processing Symposium, IEEE, 2003.

[15] A. Tumeo, C. Pilato, G. Palermo, F. Ferrandi, D. Sciuto, HW/SW methodologies for synchronization in FPGA multiprocessors, in: FPGA, 2009, pp. 265–268.

[16] A. Tumeo, M. Branca, L. Camerini, M. Monchiero, G. Palermo, F. Ferrandi, et al., An interrupt controller for FPGA-based multiprocessors, in: ICSAMOS, IEEE, 2007, pp. 82–87.

[17] R. Davis, A. Wellings, Dual priority scheduling, in: RTSS'95: 16th Real-Time Systems Symposium, IEEE, 1995, pp. 100–109.

[18] A.N. Audsley, A. Burns, M. Richardson, K. Tindell, Applying new scheduling theory to static priority pre-emptive scheduling, Softw. Eng. J. 8 (5) (1993) 284–292.

[19] M.R. Guthaus, J.S. Ringenberg, D. Ernst, T.M. Austin, T. Mudge, R.B. Brown, Mibench: a free, commercially representative embedded benchmark suite, in: WWC-4: 2001 IEEE International Workshop on Workload Characterization, 2001, pp. 3–14.

CHAPTER

Emerging resilience techniques for embedded devices

4

R.F. DeMara, N. Imran, R.A. Ashraf

University of Central Florida, Orlando, FL, United States

Rugged Embedded Systems. http://dx.doi.org/10.1016/B978-0-12-802459-1.00004-X

1 ADVANCING BEYOND STATIC REDUNDANCY AND TRADITIONAL FAULT-TOLERANCE TECHNIQUES

1.1 COMPARISON OF TECHNIQUES

Reliability of field programmable gate array (FPGA)-based designs [1] can be achieved in various ways. Table 1 provides a comparison of previous approaches toward fault handling (FH) in FPGA-based systems. *Passive recovery* techniques, such as triple modular redundancy (TMR), are commonly used although they incur significant area and power overheads. The TMR technique involves a triplication of system components, all of which are active simultaneously. The recovery capability is limited to occurrences of faults that occur within a single module. This limitation of TMR can be overcome using *self-repair* [2, 3] approaches to increase sustainability, such as refurbishing the failed module using *jiggling* [4]. Other *active recovery* techniques incorporate control schemes which realize intelligent actions to cope with a failure. Evolutionary techniques [5] avoid the area overhead of predesigned spares and can repair the circuit at the granularity of individual logic blocks, yet lack a guarantee that a recovery would be obtained within a bounded time period. They may require hundreds of genetic algorithm (GA) iterations (generations) before finding an optimal solution, thus undesirably extending the recovery time. On the other hand, operation is bounded by the maximum number of evaluations. Evolvable hardware techniques presented in the literature that rely on modifications in current FPGA device structure, such as custom hybrid analog/digital fabrics [6] and tunable circuits [7], offer other dimensions of refinement. In addition, a fitness evaluation function must be defined a priori to select the best individuals in a population, which may in turn necessitate knowledge of the input-output logic functionality. Yet, it is possible to avoid both of these complications. Altogether, they allow evaluation of the actual inputs, instead of exhaustive or pseudoexhaustive test vectors, on any commercial off-the-shelf FPGA with partial reconfiguration (PR) capability.

As depicted by the taxonomy in Fig. 1, one approach to reduce overheads associated with TMR is to employ the comparison diagnosis model with a pair of units in an adaptable CED arrangement subjected to the same inputs. For example, the competitive runtime reconfiguration (CRR) [8] scheme uses an initial population of functionally identical (same input output behavior), yet physically distinct (alternative design or place-and-route realization) FPGA configurations which are produced at design time. At runtime, these individuals compete for selection to a CED arrangement based on a fitness function favoring fault-free behavior. Hence, any physical resource exhibiting an operationally significant fault decreases the fitness of those configurations which use it. Through runtime competition, the presence of the fault becomes occluded from the visibility of subsequent operations.

CRR utilizes a temporal voting approach whereby the outputs of two competing instances are compared at any instant and alternative pairings are considered over time. The presence or absence of a discrepancy is used to adjust the fitness of both individuals without rendering any judgment at that instant on which individual is actually faulty. The faulty, or later exonerated, configuration is determined over time through other

Table 1 Comparison of Fault-Tolerance Techniques for SRAM-Based FPGAs

Approach	Area Requirement	Basis for Recovery	Detection Latency	Number of Reconfigs	Additional Components Required	Granularity	Guarantee of Improvement
TMR	Threefold	Requires two datapaths are operational	Negligible	Not applicable	Two of three majority voter	Function or resource level	100% for single fault, 0% thereafter
Evolutionary hardware	Minimal: design-time allocated PLBs	Redundancy and competitive selection	Not applicable	Only when fault is present; nondeterministic	GA engine	Logic blocks	No
CRR	Duplex	Recovery complexity	Negligible	Only when fault is present; varies	CRR controller	Function level	No
Online recovery (roving STARs, online BIST)	Roving area	Available spares	Significant: linear in number of PLBs	Continuous reconfiguration	Test vector generator, Output response analyzer	Logic blocks	Yes
Emerging adaptive architectures	*Uniplex*	*Priority of functionality*	*Negligible*	*Only when fault is present; linear in number of functions*	*Reconfig. controller*	*Computational functions*	*Yes*

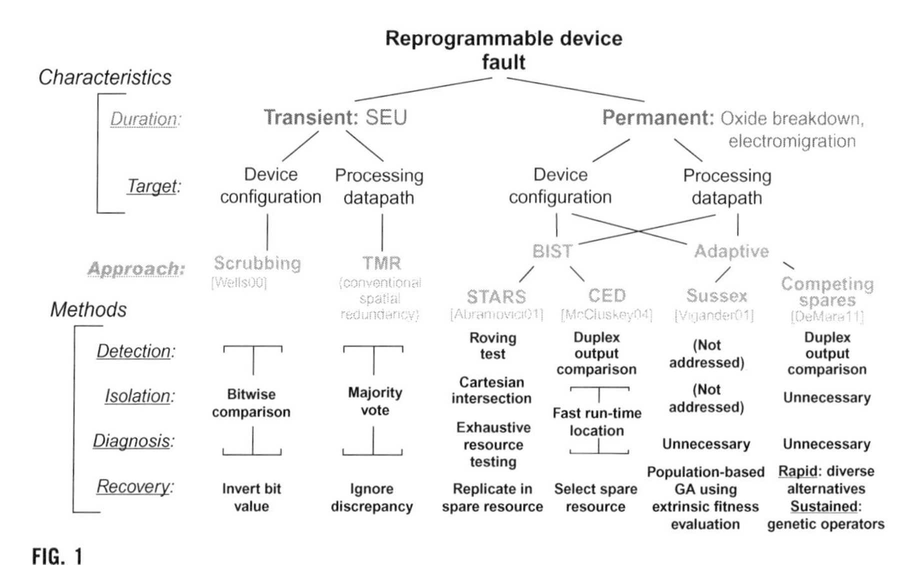

FIG. 1

Fault-handling techniques for SRAM-based FPGAs.

pairings of competing configurations. The competitive process is applied repeatedly to form a strong consensus across a diverse pool of alternatives. Under CRR, the FPGA's outputs are compared before they leave the reconfigurable device so anomaly detection occurs on the first erroneous output and detection latency is negligible.

To illustrate the process, consider the FPGA physical arrangement shown in Fig. 2. It depicts two competing half-configurations labeled functional logic left (L) and functional logic right (R) loaded in tandem. Additionally, CRR can have a third instance activated as warm or cold spare to avoid reconfiguration overhead on throughput. Whereas, in case of TMR and Jiggling approaches, the overhead of third module is always there. The competing configurations in CRR can be stored in the same EEPROM Flash device required to boot the SRAM FPGA to keep device count low or can be maintained in an external RAM. Both configurations are evolved at runtime to refurbish the functionality of the circuit under repair.

Other runtime testing methods, such as online Built-in Self Test (BIST) techniques [9] offer the advantages of a roving test, which checks a subset of the chip's resources while keeping the remaining nontested resources in operation. Resource testing typically involves pseudoexhaustive input-space testing of the FPGA resources to identify faults, while functional testing methods check the fitness of the datapath functions [10]. For example, the Self-Testing AReas (STARs) approach [11], employs a pair of blocks configured with identical operating modes which are subjected to resource-oriented test patterns. This keeps a relatively small area of the device off-line which is being tested, while the rest of the device is online and continues its operation. STARs compares the output of each programmable logic block (PLB) to that of an identically configured PLB. This utilizes the property that a

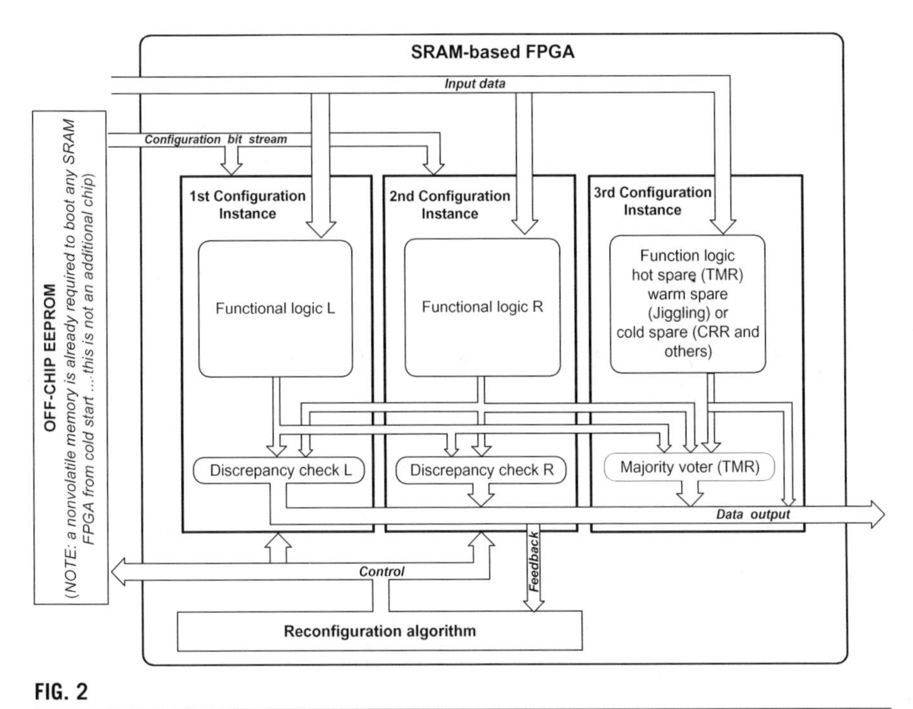

FIG. 2

Tandem CRR arrangement compared with TMR and jiggling approaches in SRAM-based reconfigurable FPGA.

discrepancy between the output flags the PLB as being *suspect* as outlined by Dutt et al.'s *roving tester (ROTE)* technique [10]. In the STARs approach, each block-under-test is successively evaluated in multiple reconfiguration modes, and when a block is completely tested then the testing area is advanced to the next block in the device. To facilitate reconfigurability in order to relocate the system logic, there is a provision to temporarily stop the system operation by controlling the system clock. The recovery in STARs is achieved by remapping lost functionality to logic and interconnect resources which were diagnosed as healthy.

In contrast, adaptive architectures like Fault Demotion using Reconfigurable Slack (FaDReS) [13] and Priority Using Resource Escalation (PURE) [14] perform functional testing of the resources at higher granularity by comparing outputs of processing elements (PEs) which execute functions that comprise a signal processing algorithm. This allows resources to be tested implicitly within the context of their use, without requiring an explicit model of each PE's function. In the proposed dynamic resource allocation schemes, the testing components remain part of the functional datapath until otherwise demanded by the FH procedure. Upon anomaly detection, these resources are designated for fault diagnosis (FD) purposes. Later, upon the completion of FD and failure recovery, the reconfigurable slacks (RSs) may then be recommissioned to perform priority functions in the throughput datapath.

Fig. 3 [15] gives an overview of the adaptive architectural approach toward realizing fault-resilient embedded systems. It has the following advantages:

- Oblivious fault detection: Intrinsic measurement of applications health metric using feedback loop. Simplex operation during majority of mission time.
- Desirable fault isolation (FI): System is kept online while concurrent error detection is performed using actual runtime inputs. No need for test vectors.
- Degraded quality versus energy consumption: Resources computing least priority functions can be reconfigured. Throughput is application regulated.
- Energy-aware FH: A simplex configuration is shown to be sufficient for applications such as discrete cosine transform (DCT) when a health-metric such as PSNR is available.
- Graceful degradation during diagnosis: Degradation spanned a few frames, during which time a partial throughput is available, as an intrinsic provision of degraded mode.
- Priority-aware fault recovery (FR): Healthy resources are utilized for most significant computations.

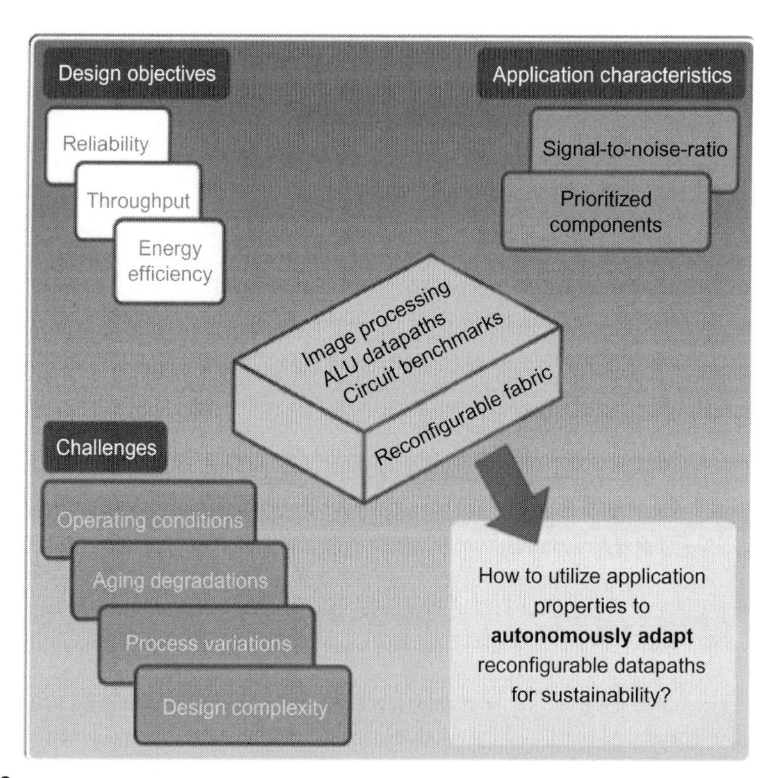

FIG. 3

Toward realizing rugged embedded systems in the continuum of quality and resiliency.

1.1.1 Desirable characteristics

Fig. 4 illustrates these characteristics of an ideal autonomous recovery technique and the techniques presented in this chapter within the context of existing approaches toward achieving these goals. A favorable fault-resilience technique minimally impacts the throughput datapath. In addition, the area-overhead of a FH controller, δ should be very low as compared to the baseline area where δ is fixed, independent of the datapath complexity, and a fraction of the size of the datapath.

1.1.2 Sustainability metrics

Sustainability metrics provide a qualitative comparison of the benefits provided by various methods. In conjunction with the overheads, these metrics provide insight into the tradeoffs involved in choosing a specific method.

Fault exploitation

By nature, all FH methods are able to bypass faulty resources. Methods that are capable of fault exploitation can reuse residual functionality in fault-affected elements. Such methods increase the effective size of the resource pool by virtue of recycling faulty resources. For example, using online BIST methods, if a LUT has a cell with stuck-at-0 fault, then the LUT can still be used to implement a function that requires a 0 to be in that cell.

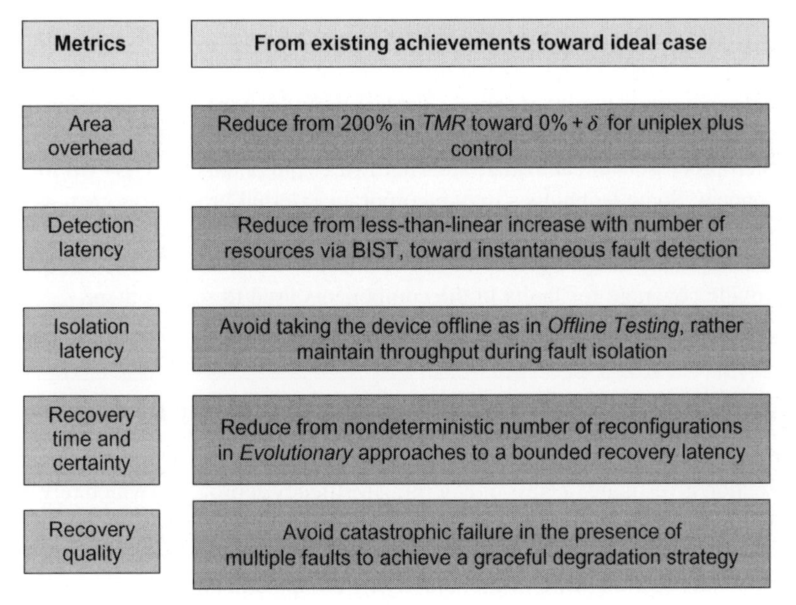

Metrics
From existing achievements toward ideal case
Area overhead
Reduce from 200% in TMR toward 0% + δ for uniplex plus control
Detection latency
Reduce from less-than-linear increase with number of resources via BIST, toward instantaneous fault detection
Isolation latency
Avoid taking the device offline as in Offline Testing, rather maintain throughput during fault isolation
Recovery time and certainty
Reduce from nondeterministic number of reconfigurations in Evolutionary approaches to a bounded recovery latency
Recovery quality
Avoid catastrophic failure in the presence of multiple faults to achieve a graceful degradation strategy

FIG. 4

Characteristics of an ideal autonomous recovery technique.

Recovery granularity

Recovery granularity defines the smallest set of FPGA resources in which faults can be handled. The set of resources identified as being fault affected are excluded from further utilization by the methods that are not capable of fault exploitation.

Fault capacity

Each method is capable of handling a varying number and types of faults. Fault capacity defines the number of fault-free resource units required to ensure that the system continues to function if there is a single additional fault, irrespective of the location of the faults. The fault capacity for the surveyed methods is listed in Table 1 and discussed subsequently.

Fault coverage

Fault coverage defines the ability of a method to handle faults in various components of the FPGA. Some of the surveyed methods can distinguish between transient and permanent faults. All methods surveyed handle faults occurring within logic resources. None of the surveyed methods handles faults occurring within I/O blocks. Only some of the methods surveyed, however, address faults occurring within interconnect resources. Since the interconnect resources can compose up to 80% of the FPGA area [71], the probability of a fault occurring within interconnect is higher than that of an occurrence within PLB logic resources.

Critical components

Critical components are those FPGA components that are essential for failure-free operation of the FH method. If the FH techniques do not provide coverage for these components, a single fault in one of these can cause the failure of the FH technique, thus defining a single point of failure. Table 1 provides a list of components that are critical to each technique. FPGAs such as the Virtex-4 XC4VFX140 include embedded microprocessors or can realize such hardware equivalents with its PLB logic and interconnect. Some techniques may incorporate external procedures such as a custom place and route mechanism within the device. While such instantiations are not external to the actual device, they are critical components if the technique does not provide coverage for faults in the components used to realize them.

2 AUTONOMOUS HARDWARE-ORIENTED MITIGATION TECHNIQUES FOR SURVIVABLE SYSTEMS

This section investigates a class of emerging resiliency approaches which rely on two interacting and self-calibrating FH processes. The first process is the continuous use of *intrinsic evaluation* of the computing system's throughput. The second process is the *autonomic restructuring* of the underlying computational resources to facilitate the desired level of application performance, despite the presence of faults. These two processes can cooperate to maintain a continuum of output quality via online exploration of the search space of resource configurations which maintain application throughput, and thus adaptively realize resilience.

The most intriguing of these survivability techniques do not require fault models. Instead, they rely on measurement of an application-layer *quality of service (QoS)* metric. A QoS metric quantifies the suitability of the instantaneous system behavior on a continuum spanning from "completely meeting correct computational criteria" to "providing at least some useful output" to "failing insufficient for operational use." Judicious selection of a proper *goodput*-oriented [16] QoS metric constitutes a vital design step to realizing autonomous survivable systems.

We illustrate how goodput-oriented QoS metrics allow in-the-loop control of hardware-oriented mitigation techniques. Suitable metrics, such as peak signal-to-noise ratio (PSNR), are described in the context of hardware resiliency as an application throughput *health metric*. The FR mechanism can be realized and guided by maintaining a threshold-level value for the health metric by adapting the use of the available processing resources. Moreover, the FH procedures themselves remain oblivious to the fault specifics. For instance, we describe two FH approaches whereby the need for FI and FD processes become subsumed within an appropriate reaction to changes in the health metric's value over a recent evaluation window. Thus, we illustrate how new resource prioritization approaches using QoS-oriented health metrics can realize survivable computing systems without the need for detailed fault models.

2.1 FUNCTIONAL DIAGNOSIS OF RECONFIGURABLE FABRICS

The primary focus of this section is to define and analyze emerging approaches which rely on intrinsic evaluation and autonomous restructuring of computing fabrics to increase resilience. Fig. 5 depicts a continuum of granularity and adaptation approaches for sustainable reconfigurable hardware. At the left end of the colored spectrum is *static redundancy* while the other extreme is *unguided evolvable hardware*. Both of these extreme approaches exhibit challenges which make it difficult to obtain a survivable datapath. The perspective in this section is that the middle zone between these two extremes is more advantageous for obtaining most of static redundancy's predictability and low runtime overhead, while also obtaining most of evolvable hardware's maximal sustainability and graceful degradation capability. Namely, using novel hybrid approaches at the proper level of granularity can minimize challenges of cessation of operation, brittleness of repair, and control complexity, while providing better bounds on recovery time.

Starting at the top right-hand side of Fig. 5, it is seen that the coarseness of granularity is least at the logic level and increases to modules and processing cores as one approaches a static redundancy method. Meanwhile, a range of periodic logic and routing adaptations are feasible. In this section, three techniques in the middle of the continuum are discussed. The two techniques which utilize module-level reconfiguration are the FaDReS and PURE approaches of mitigating hardware faults in reconfigurable fabric. The schemes offer various advantages in terms of continuous operation, power

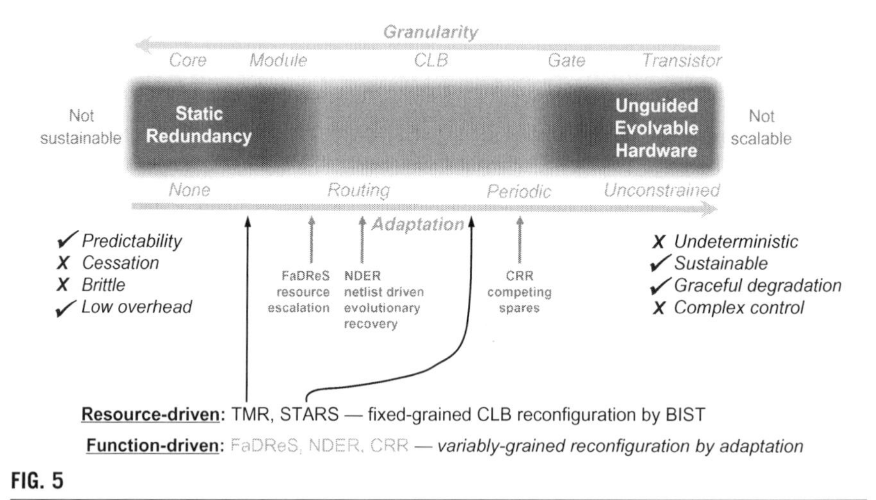

FIG. 5

Continuum of datapath reconfiguration for fault handling.

consumption, and area-overhead while improving reliability. For netlist-driven evolutionary recovery (NDER), netlist information is used to reduce the evolvable hardware search space. This illustrates an example of *pruned refurbishment*.

A system-level block diagram is shown in Fig. 6 which identifies the roles of reconfigurable logic fabric and on-chip processor core of a typical FPGA device. Within the reconfigurable logic fabric, the desired processing function such as a DCT or Advanced Encryption Standard (AES) core, is realized by the PEs which comprise a processing array. These PEs are reconfigurable at runtime in two ways. First, they can be assigned alternative functions. Functional assignment is performed to leverage priority inherent in the computation to mitigate performance-impacting phenomena such as extrinsic fault sources, aging-induced degradations, or manufacturing-induced process variations. Second, the input data can be rerouted among PEs as necessary by the reconfiguration controller. These reconfigurations are only initiated periodically, for example when adverse events such as aging-induced failures occur based on perturbations to the health metric. The functional remapping is performed by fetching alternative partial configuration bitstreams for the PEs which are stored in an external memory device such as compact flash. A configuration port, such as the Internal Configuration Access Port (ICAP) on Xilinx FPGAs, provides an interface for the reconfiguration controller to instantiate the PEs with the bitstreams used to perform computational functions in the processing array. The input data used by the PEs, such as input video frames, resides in a DRAM Data Memory that is also accessible by the on-chip processor core. Together these components support the data and reconfiguration flows needed to realize a runtime adaptive approach for resilient architectures.

Relaxing the requirement of test vectors for fault detection can realize a significant reduction in the testing overhead as compared to BIST-based approaches. To realize FD and recovery, PURE utilizes runtime reconfigurability by considering priorities

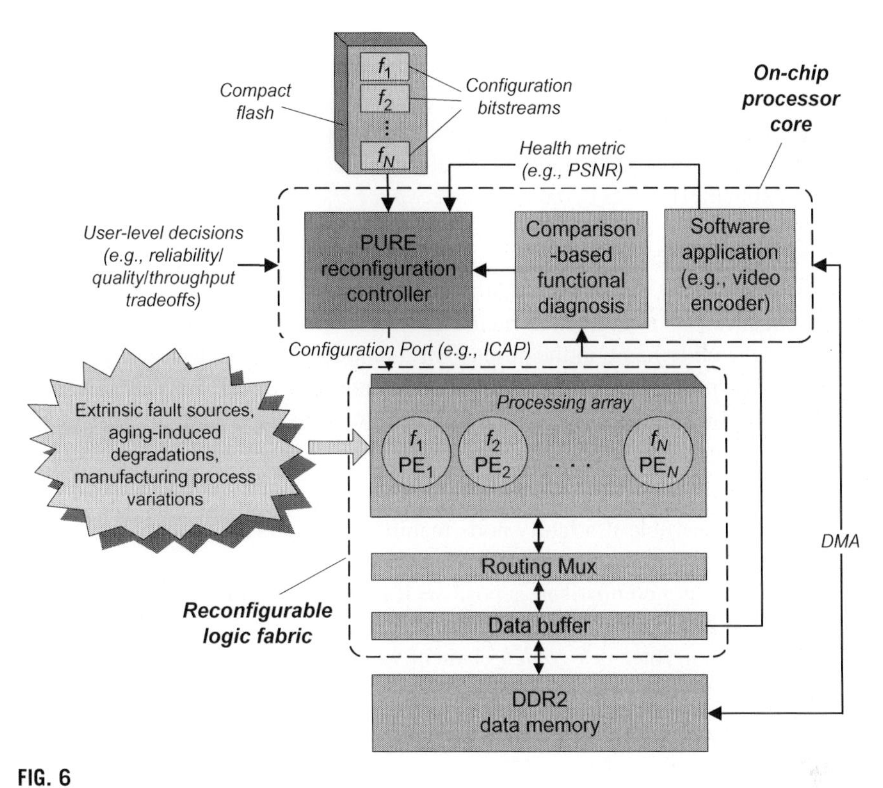

FIG. 6

Self-adapting resource escalation of the FPGA device.

in the underlying computation. To reassign the function executed by an identified faulty PE, either a design time configured spare is engaged into the active path, or the least-priority PE is utilized by multiplexing the input-output data. Functional reassignment is realized by the PE-to-physical-fabric mapping. This mapping is encoded in a configuration bit file which maps the PE function on to a subset of the available FPGA resources. Resource instantiation is realized at configuration time when a bit file is deployed by loading it from external memory into the FPGA configuration logic memory. In addition, the faulty PE is configured with a blank bitstream to cease switching activity which otherwise would incur additional power consumption.

In a broad sense, the provision of resilience in reconfigurable architectures for signal processing can take advantage of a shift from a conventional precisely-valued computing model toward a significance-driven approximate computing model [17–19]. This significance-driven model provides inherent support for a continuum of operational performance which is compatible with the concepts of signal quality and noise. In this way, PURE recasts the reliability issues of contemporary nanoscale logic devices in terms of the significance associated with these computations.

It may be noted that the methods discussed here concentrate on local permanent damage rather than soft errors which can be effectively mitigated by *Scrubbing*

[20–22]. Scrubbing can be effective for SEUs/soft errors in the configuration memory, however it cannot accommodate local permanent damage due to stuck-at faults, electromigration (EM), and bias temperature instability (BTI) aging effects which require an alternate configuration to avoid the faulty resources.

Similar to previous approaches, the techniques discussed in this chapter progress through explicit FH stages of *anomaly detection*, *FD*, and *FR*. Fault detection can be either performed by continuously observing a system health metric like signal-to-noise ratio (SNR), or checking the processing nodes in an iterative fashion, as will be discussed shortly. For example, in the case of a video encoder, the PSNR of a video sequence provides a health metric for a uniplex arrangement without redundancy. On the other hand, in the absence of a uniplex health metric such as PSNR, the designer can tradeoff the use of periodic temporal CED [23, 24] or spatial CED [25] redundant computations based on throughput, cost, and reliability constraints.

The following assumptions are made in the functional-level FD scheme:

1. faults are of permanent nature,
2. a fault is observable if a faulty node manifests a discrepant output at least once in a given *evaluation window* period,
3. the outcome of a comparison is positive if at least one of the nodes in a CED pair has an observable fault, and
4. the comparator/voter is a *golden* element which can be relied upon for fault-free operation.

In the event of failure to identify a healthy PE after multiple reconfigurations, the slack is updated to a different PE as described by the specific reconfiguration sequencing algorithms discussed in the forthcoming section below. When a healthy RS is found in a given slack update iteration *s*, it indicates that the previously selected slacks were faulty.

In the recovery phase, the priority of functions can be taken into account while recovering from fault scenarios. For the DCT case, the PE that is computing the DC coefficient is the most important as it indicates the average over a set of samples. Next most important is the so-called AC_0 coefficient, followed next by the AC_1 coefficient, and so on. For an application with equally important processing nodes, the priority vector is initialized with all ones.

The priorities are assigned at design time considering the application properties for the presented use-case. An interesting future work could be to compute the priority values at runtime. This would be useful for applications which cannot be characterized by priorities at design time, or to better utilize the input signal characteristics at runtime, such an approach can be very promising to realize runtime adaptable architectures.

Given a circuit under test (CUT), the objective is to identify faulty nodes as soon as possible while maintaining throughput during the FD phase. For this purpose, the proposed diagnosis schedule demotes the predicted fitness of a node under test (NUT) based upon their discrepancy history. In the following, we describe variations of the FH phase starting with a *divide-and-conquer* approach.

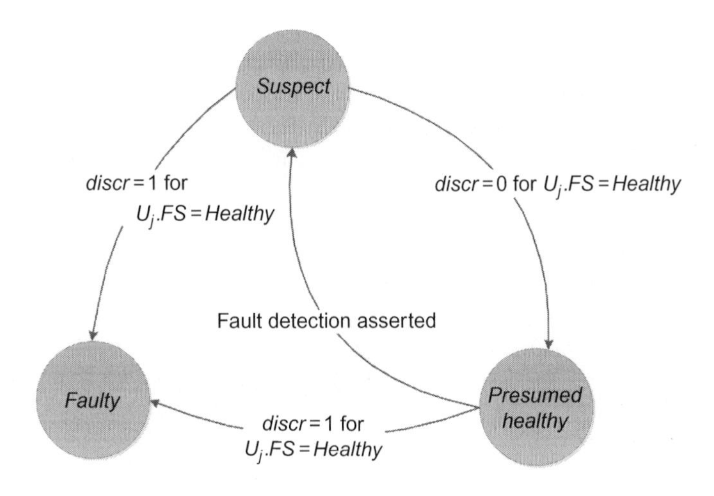

FIG. 7

Fitness states of a design configuration during a circuit's life time.

To construct a functional diagnosis approach to record discrepancies, consider a state transition diagram method based upon the *Suspect, Faulty, and Healthy (SFH)* fitness transitions of the configuration bitstreams. An individual configuration undergoes different transitions in its fitness state throughout the life of a circuit. After fault detection, the fitness state of every configuration is *Suspect*. If two configurations show complete agreement in a given *Evaluation Period, E*, both are declared as *Presumed Healthy*. However, if a *Suspect* configuration exhibits discrepancy with a *Healthy* one, it is marked as *Faulty*. The state transition diagram is illustrated in Fig. 7. The objective of the state transitions' flow is to identify *Healthy* configurations in a pool of *Suspect* configurations which in turn helps to identify *Faulty* items. The problem is similar to the counterfeit coin identification problem [26] with a restriction that only two coins can be tested at a time.

2.1.1 Reconfiguration Algorithm 1: Divide-and-conquer method

Group testing schemes [26–28] have been successfully employed to solve many FI problems in which the number of defective items is much smaller than the size of the overall suspect pool. The problem at hand has an analogy to the group testing paradigm, yet with some important distinctions. Although, the task here is to identify defective elements in a pool of computational resources, we do not pose an assumption about the presence of a known-to-be-healthy functional output element for testing individual nodes. This assertion makes it infeasible to apply a hierarchical testing approach in which testing up to the last single item is performed by a known healthy item. Therefore, the PURE approach also relies upon the comparison diagnosis model or an NMR voting model to isolate faulty elements.

Two different scenarios can be identified in which this hierarchical divide-and-conquer strategy may be more appealing to employ than the two algorithms discussed in further sections:

- If there are no restrictions on throughput or availability during the FH phase, then halving of the suspect pool [28] offers logarithmic time diagnosis latency, or
- if *fault confinement* is desirable, that is, limiting the influence of the fault as soon as possible, then it becomes advantageous to cut off the suspect nodes from the active throughput path. Then, those nodes can be used to check the health of the active nodes. This scenario is pessimistic, and applies to the case when fault rate is high and a large number of nodes become defective before the FH scheme is initiated. A more optimistic approach is to keep the active nodes in the processing datapath while performing a diagnosis process as we discuss in the next sections.

Algorithm 1 defines the diagnosis process. The choice of algorithm in an application depends upon the designer's preferences about diagnosis latency, throughput availability requirement, and area/power trade-offs. The following algorithms present such trade-offs.

2.1.2 Reconfiguration Algorithm2: FaDReS

FaDReS achieves dynamic prioritization of available resources by demoting faulty slacks to the least priority functions [13]. Compared to divide-and-conquer, it attempts to avoid excessive reconfiguration of the processing datapath. Namely, whenever a redundant PE is not available then a lower priority functional module can be utilized. The output from the vacated RS is compared against functional modules in the datapath providing normal throughput. The discrepancy in output of

ALGORITHM 1 DIVIDE-AND-CONQUER FAULT DIAGNOSIS ALGORITHM (WITHOUT RECOVERY).

Require: N
Ensure: $\hat{\Phi}$

1: Partition V into two equal-sized disjoint sets V_a and V_s
2: Designate the set V_a as FE and V_s as RS
3: Perform concurrent comparison to the same inputs for various edges of the bipartite graph represented by connectivity matrix **C**
4: Update the Syndrome Matrix $\mathbf{\Psi}$ based upon comparisons outcome
5: Iterate step-1 to step-4 $log(N)$ times
6: Given $\mathbf{\Psi}$, isolate the faulty nodes:
$\hat{\phi}_i \leftarrow 0$ and $\hat{\phi}_j \leftarrow 0$, if $c_{ij} = 1$, and $\psi_{ij} = 0$
$\hat{\phi}_i \leftarrow 1$ if $\hat{\phi}_j = 0$, $c_{ij} = 1$, and $\psi_{ij} = 1$

identical functional modules isolates the permanent fault. Thus, the FaDRes algorithm iteratively evaluates the functional modules while keeping them in the datapath, as well as slack resources used for checking.

In the FaDReS technique, once the system detects a fault, the FI phase is initiated by reconfiguring multiple PEs with the same functionality for discrepancy checking. The dynamic reconfiguration property of FPGA is employed to create the redundancy needed for FI during runtime. If there is no redundant PE available, then the lower priority functional module can be vacated. The output from the vacated RS is concurrently checked for discrepancy with that of functional modules in the datapath providing normal throughput. The discrepancy in the output of the same functional modules reveals the permanent fault. The proposed algorithm iteratively evaluates the functional modules while keeping them in the datapath, as well as the slack resources used for checking. Dynamically configurable slacks can isolate up to *N-2* faulty PEs in a system with *N* PEs. DMR and TMR techniques have been successfully adapted to utilize dynamic reconfiguration capability of FPGAs in previous work [24]. However, the presented algorithm employs flexible dynamic redundancy and the isolation process can be accelerated by increasing the number of RS.

Hardware organization in FaDReS technique

A baseline reallocation strategy of fault demotion for mission-critical applications is discussed that permits recovery emphasizing small throughput degradations via an RS with stepwise PR demotion. Nonetheless, more efficient reallocation strategies such as Divide and Conquer, and others, can be employed when their impact on throughput is acceptable [29]. The hardware architecture used to evaluate FaDReS is implemented on a Xilinx ML410 development board which contains a Virtex-4 FX60 FPGA. As a case study, the H.263 video encoder application is run on the on-chip processor. All the subblocks except the DCT block [30] are implemented in software, the latter being implemented in hardware. Each PE in the DCT block is responsible for computing one coefficient of the 8-point 1-D DCT.

Anomaly detection, isolation, and recovery

The anomaly detection methodology used observes the PSNR of the recovered frames. The PSNR is computed based upon the difference between input frame and the image in the frame buffer [31]. Once this measure is below a threshold δ, anomaly detection is asserted. In this way, continuous exhaustive testing of the resources is avoided. In addition, avoiding redundancy during the normal operation is beneficial in terms of power and area requirements. Once a fault is detected, the proposed FI and recovery processes are initiated. The FaDReS algorithm for *FI* and *FR* schemes are given below and illustrated by the flowchart in Fig. 8.

The FI scheme is illustrated by an example in Fig. 9. Here, the normal operation is an 8×8 DCT and therefore PE_1 to PE_8 are providing the normal throughput for a 1-D 8-point DCT. Two 1-D DCT operations are performed to compute a 2-D DCT. For this arrangement, PE_9 is required by the system and reconfigured as blank. For this specific situation, assume that two out of nine PEs are faulty. The faulty PEs labeled 1 and 9 are shown by tagged boxes. We choose the blank PE as the RS for discrepancy check with PE_1. Unfortunately the RS itself is faulty, yet we have no a priori knowledge of its fitness state. Therefore, in the first iteration in which the RS is reconfigured eight times for the CED purpose, no information is obtained. Every module's fitness is unknown and marked as *Suspect* (S). We proceed to the second iteration. The RS is moved from PE_9 to PE_8 and CED is performed with the functional modules, sequentially. No discrepancy between PE_8 and PE_2 implies their

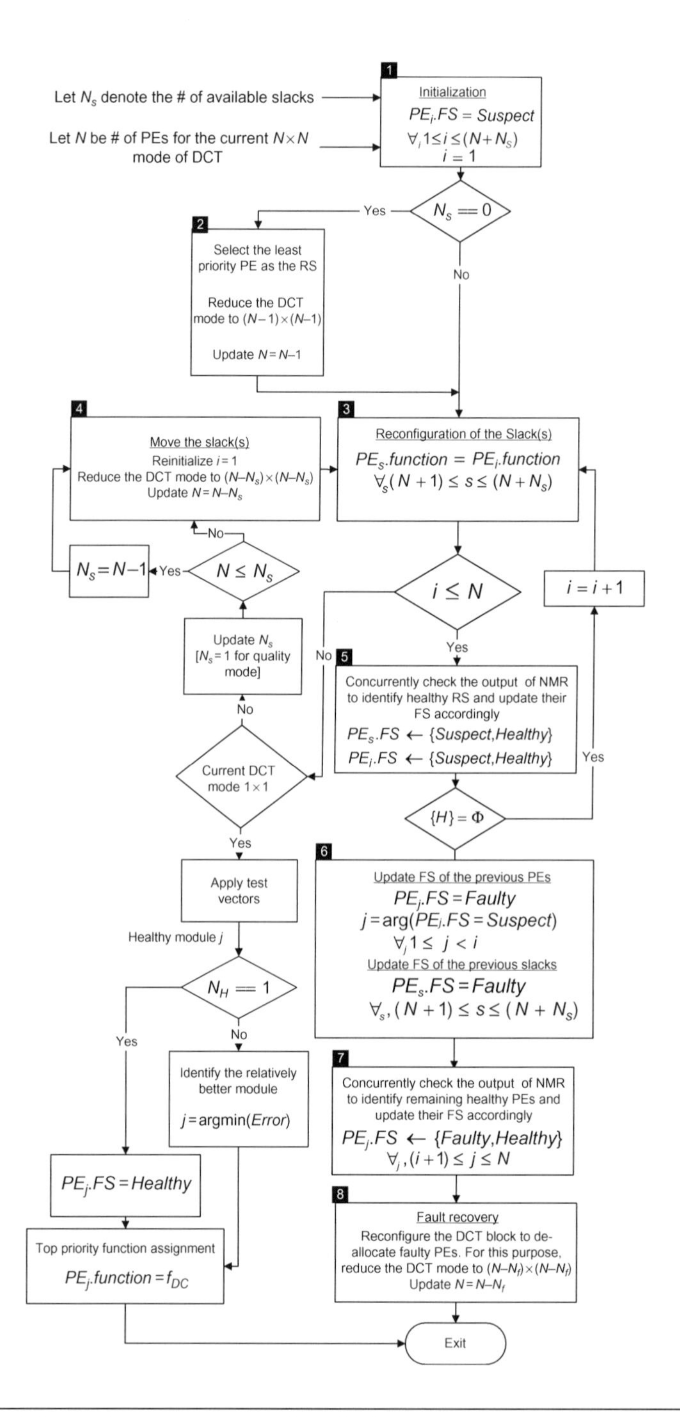

FIG. 8

The fault isolation and recovery process flow chart.

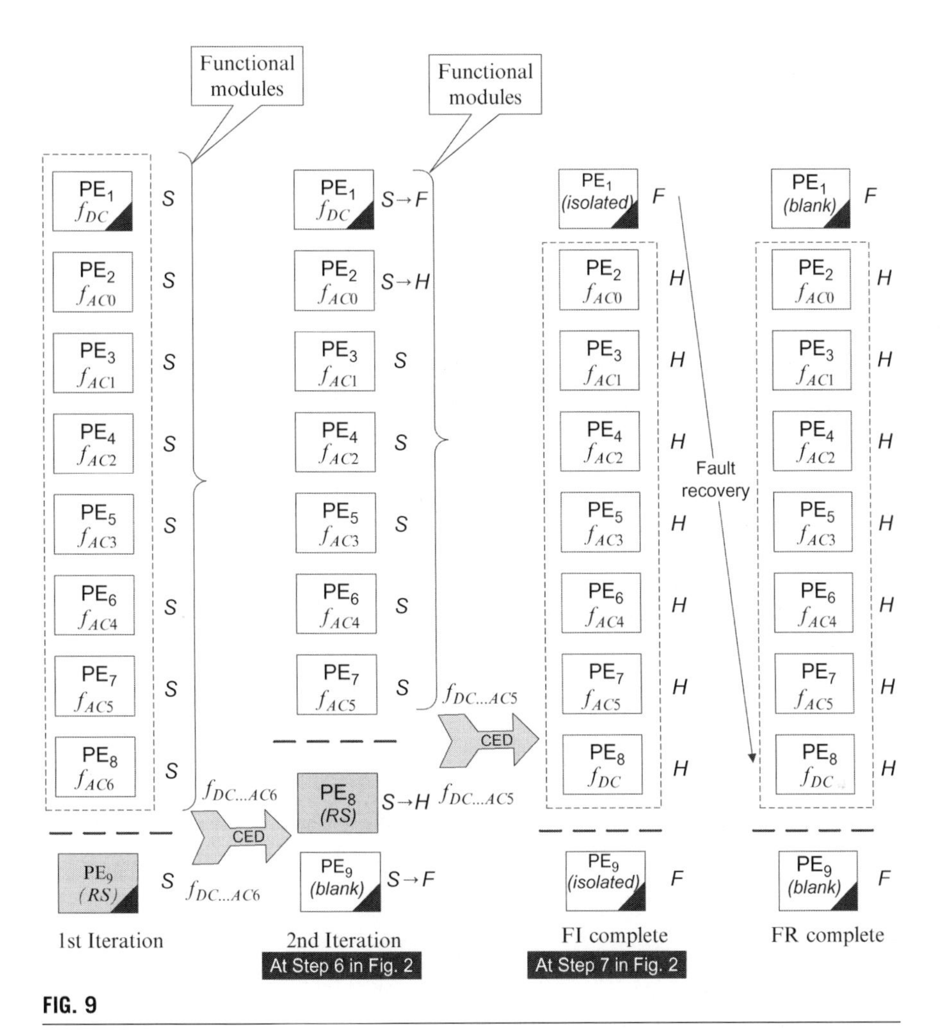

FIG. 9

An example of the fault isolation and recovery scheme.

Healthy (H) nature and therefore, PE_1 and PE_9 are marked as *Faulty* (F) as they did exhibit a discrepancy with the *Healthy* module.

Once the FI process is complete and a healthy RS is identified throughout the evaluation window, that RS is used for error checking of all the functional modules. If there are N healthy PEs available, FR is made without quality degradation. However, if there are fewer healthy modules than the current mode of the DCT, then a graceful degradation strategy is employed where the most important functions are prioritized to be retained in the proven functional resources. In the DCT case, the most influential function to retain is the DC coefficient's computation.

Anomaly detection latency of the proposed approach based upon the PSNR is negligible, whereas latency of isolating faulty PEs is bounded as our algorithm follows a deterministic flow. The idea is to time-multiplex the RS for different functions and compare the output of the RS with those from the active modules in the datapath. A discrepancy between the outputs of two modules results in them remaining in the *Suspect* pool, whereas the agreement marks them as *Healthy*. The algorithm for diagnosis employing dynamic NMR voting on module level is given in Algorithm 2.

2.1.3 Reconfiguration Algorithm3: PURE

PURE achieves dynamic prioritization of available resources by assigning healthy slacks to the highest priority functions. The distinction between the PURE algorithm and FaDReS arises from the fact that after a healthy RS is identified, PURE reconfigures it for priority function computation immediately. An identified healthy RS is used for checking purposes to isolate all other PEs. Thus, Algorithm 3 can be used to prioritize throughput while the FaDReS algorithm depicted in Algorithm 2 can be used to prioritize FD completion.

Fig. 10 shows an illustrative example of the FD in the PURE approach. The fitness state of PEs which are suspect is depicted by rounded-corner blocks. In this example, PE1 and PE7 are afflicted with faults. Upon initialization of the FH algorithm, all PEs are suspect. Then, PE_8 is reconfigured as RS by implementing function f_1 and its output is compared with that of PE_1 to check for any discrepancy. A RS is shown by dashed block. In this example, PE_1 is also faulty; therefore, this first comparison does not provide any useful information about the health of PEs and they remain suspect. Next, PE_8 is reconfigured to second priority function f_2 and its

ALGORITHM 2 FADRES (GREEDY FAULT DIAGNOSIS WITH SUBSEQUENT RECOVERY).

Require: N, N_s, **P**
Ensure: $\hat{\Phi}$
1: Initialize $\hat{\Phi} = [x\ x\ x\ \dots\ x]^T$, $i = 1$, $N_a = N - Ns$
2: Arrange elements of V in ascending order of **P**
3: **while** $(\{k | k \in \hat{\Phi}, k = 0\} = \phi)$ **do**
4: Designate v_s as checker(s) $(N_a + 1) \leq s \leq (N_a + N_s)$; thus $V_s = \{v_s\}$
5: **while** $i \leq N_a$ **do**
6: Reconfigure RS(s) with the same functionality as v_i, $N_{sup} = N_{sup} + 1$
7: Perform NMR majority voting among NUTs when $N_s > 1$, or CED between NUTs when $N_s = 1$, then update Connectivity matrix accordingly, Update the Syndrome matrix $\mathbf{\Psi}$ based upon discrepancy information, $\hat{\phi}_i \leftarrow 0$ for v_i which shows no discrepancy then go to step-12, $\hat{\phi}_i \leftarrow x$ otherwise
8: $i \leftarrow i + 1$
9: **end while**
10: Move the RS by updating $N_a = N_a - N_s$, $N_r = N_r + 1$, Reinitialize $i = 1$
11: **end while**
12: Update the fitness state of the previous RS(s): $\hat{\phi}_j \leftarrow 1$; for $(s + 1) \leq j \leq N$ and $\psi_{j.} = 1$
13: Use a healthy RS to check all other nodes in V_a, $\hat{\phi}_i \leftarrow 0$; if $\hat{\phi}_j = 0$, $c_{ij} = 1$, and $\psi_{ij} = 0$

ALGORITHM 3 PURE (FAULT DIAGNOSIS WITH INTEGRATED PRIORITY-DRIVEN RECOVERY).

Require: N, N_s, $\mathbf{P}$
Ensure:$\hat{\Phi}$, $F*$
1: Initialize $\hat{\Phi} = [x\ x\ x\ \ldots\ x]^T$, $i = 1$, $N_a = N - N_s$
2: Arrange elements of V in ascending order of $\mathbf{P}$
3: **while** $(\{k|k \in \hat{\Phi}, k = x\} \neq \phi)$ //Until all suspect nodes are proven to be healthy **do**
4: **while** $(\{k|k \in \hat{\Phi}_s, k = 0\} = \phi)$ //Identify at least one healthy node in V_s **do**
5: Designate v_s as checker(s) $(N_a + 1) \leq s \leq (N_a + N_s)$; thus $V_s = \{v_s\}$
6: **while** $i \leq N_a$ **do**
7: Reconfigure RS(s) with the same functionality as v_i, $N_{sup} = N_{sup} + 1$
8: Perform NMR majority voting among NUTs when $N_s > 1$, or CED between NUTs when $N_s = 1$, then update Connectivity matrix accordingly, Update the Syndrome matrix $\mathbf{\Psi}$ based upon discrepancy information, $\hat{\phi}_i \leftarrow 0$ for v_i which shows no discrepancy then go to step-13, $\hat{\phi}_i \leftarrow x$ otherwise
9: $i \leftarrow i + 1$
10: **end while**
11: Move the RS by updating $N_a = N_a - N_s$, $N_r = N_r + 1$, Reinitialize $i = 1$
12: **end while**
13: Identify the most prioritized function computing node which is faulty, v_{pf}
14: Use the identified healthy RS to compute a priority function, $F_s^* \leftarrow F_{pf}$ thus RS is removed from V_s and added to V_a
15: **end while**

discrepancy check is performed with PE_2 which implements f_2. An agreement reveals their healthy nature. In addition, it shows that PE_1 was faulty as it had exhibited discrepancy with a healthy PE (i.e., PE_8) in the previous step. As soon as a healthy RS is identified, a faulty PE implementing a priority function is moved to the RS. Thus, PE_1 is configured as blank by downloading a blank bitstream while PE_8 is configured with function f_1 to maintain throughput. Next, PE_7 is chosen as RS whose discrepancy with a healthy PE_2 shows its faulty nature. Lastly, a healthy PE_6 serving as RS accomplishes the diagnosis procedure to update the fitness state of PEs 2 through 5 to healthy. Overall, the FR is achieved by configuring faulty PEs by blank and healthy PEs by functions 1 through 6.

Another scenario can be considered for the above example in which two checker PEs are utilized in the diagnostic stage. As the intermediate results have to be written into data buffer as in Fig. 11, so that the CPU can evaluate for discrepancy check, the data buffer writing timing would be different than the previous scenario. In general, for a given faulty scenario, an increase in N_s can help reducing the latency of diagnosis completion. On the other hand, to improve the FD latency, such a choice of larger N_s can incur more throughput degradation during the diagnosis phase. Thus, the choice of N_s should be made according to maximum tolerable throughput degradation during the diagnosis phase and the desired latency of FH.

For a total of N nodes, there are $N^2 - N$ possible pairings. As evident by Table 2, our FD schemes require significantly fewer comparisons compared to the exhaustive

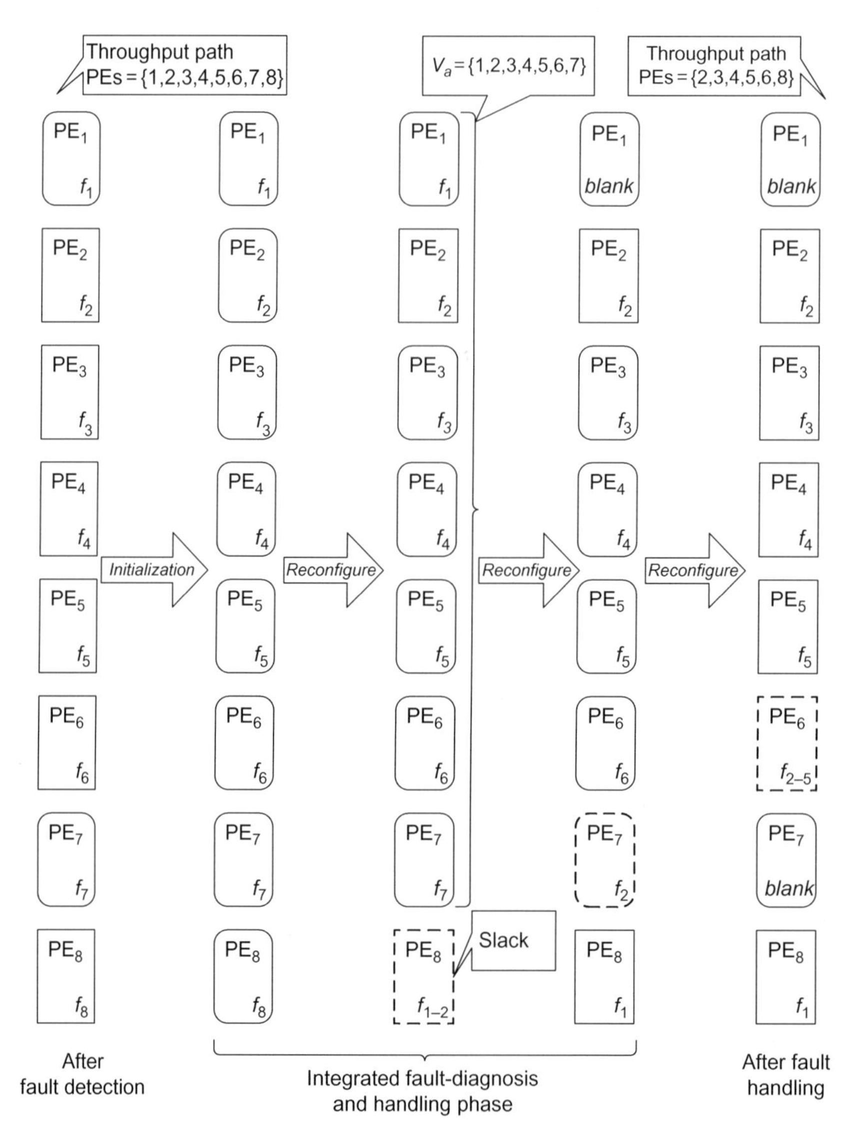

FIG. 10

An example of fault diagnosis in PURE approach.

pair evaluations where the values are scaled to % of total resources available during diagnosis.

PSNR as a Health Metric: PURE adapts the configuration of the processing datapath based on the correctness and performance of recent throughput by incorporating a health metric. PSNR is well-established metric to assess the relative quality of video encoding [32]. The PSNR of a $M \times M$ frame of n pixel-depth is computed

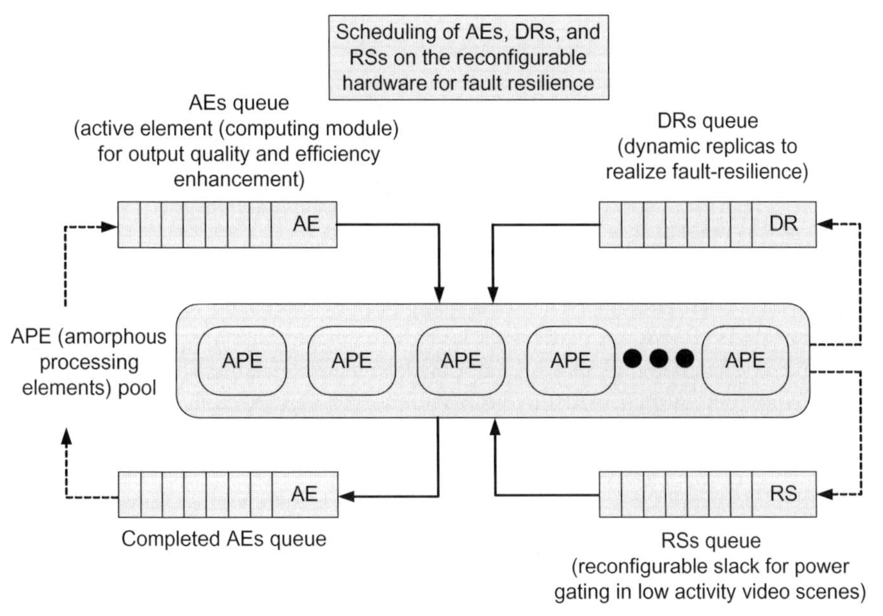

FIG. 11

Flexible configuration of amorphous processing elements (APEs).

Table 2 Latency Versus Throughput Comparisons for the Fault-Handling Schemes in Terms of % Resources Available During Diagnosis

Metric	Approach	Testing Arrangement Instance, *r*		
		1 (%)	2 (%)	3 (%)
N_a during diagnosis	Algo. 1 (divide & conquer)	50	25	12.5
	Algo. 2 (FaDReS)	87.5	75	62.5
	Algo. 3 (PURE)	87.5	75	62.5
No. of bitstream downloads, N_{sup}	Algo. 1 (divide & conquer)	50	50	50
	Algo. 2 (FaDReS)	87.5	75	62.5
	Algo. 3 (PURE)	87.5	75	62.5

based upon the algebraic difference of the input frame and the image in the frame buffer in $O(M^2)$ steps. In the PURE technique, the PSNR of each frame is computed in the background using the On Chip Processor embedded in fabric without decreasing the throughput of the PE array. In the experiments herein, the computation of PSNR was measured to take 4.23 ms for an input image with resolution of 176×144, and thus incurs only 2.79% time utilization of the embedded PowerPC. Likewise, the power consumption overhead during PE reconfiguration is 70 mW considering ICAP [13]. Thus, this approach can be advantageous in terms of power and area requirements by detecting anomalies without incurring redundancy within the PE datapath. Meanwhile, PSNR computations on the processor proceed concurrently with DCT computations in the PE array. PSNR computation is performed as a health metric and is not on the PE array's critical path of the DCT core. Thus, the interval of time between successive calculations of PSNR can be selected independently to be sufficient for health assessment without impacting the DCT core's throughput.

While these previous approaches utilize PSNR for assessing resilient architecture performance, the novelty of the PURE technique is to escalate resources based on their impact on PSNR. In particular, the PURE scheme maintains quality above a certain user-specified tolerance by adapting the datapath. Taking a broad view, a system boundary is defined so that external factors such as environment, occlusions, or signal transmission errors reside outside of the signal processing task. For example within the system boundary of a video encoding task, PSNR reflects the compression quality if the input noise is considered to be part of the input signal. Thus, the PSNR reflects the effectiveness of the signal processing system in terms of its underlying hardware resources. However, even in the absence of faults, PSNR varies depending on the algorithms ability to perform lossy compression and reconstruction in accordance with the nature of the scene's content. For example, in the PURE results shown in Fig. 21 of the following section, PSNR is seen to decline from 33 dB down to 32 dB during frames 1 through 50. When PSNR drops abruptly at frame 51, due to a hardware fault, it triggers the Fault Detection phase of the PURE algorithm.

The PURE algorithm differentiates failure-induced changes in PSNR from ambient changes in PSNR using a user-selected maximum tolerable quality degradation during fault detection (FD), denoted as Δ_{FD}. The quantity Δ_{FD} represents an allowable runtime percentage change in PSNR which would invoke the PURE diagnostic flow. A sliding window of recent PSNR values is used to accommodate differences in the changing nature of the scene's content. Δ_{PSNR} is defined as:

$$\Delta_{PSNR} = 100 \times \frac{(PSNR_{avg} - PSNR_{current})}{PSNR_{avg}} \tag{1}$$

For example, Tables 3 and 4 indicate the feasibility of selecting $\Delta_{FD} = 3\%$ for the `city` input video sequence with a sliding window of six frames. Although the nominal PSNR value may vary, Tables 3 and 4 together show how a desirable Δ_{FD} value could tradeoff both false positive and false negative detections. Finally, selection of the sliding window size can take into account the product of reconfiguration time and frame rate yielding $\lceil T_{recon} \times Framerate \rceil$, e.g., $\lceil 180\ ms \times 30\ fps \rceil = 6$ frames. Table 5 lists the effectiveness of using these detection parameters with a variety of input benchmarks.

Table 6 summarizes the combinations of conditions under which PSNR is a reliable indicator of faults. The first row indicates that when no fault is present and tolerance is not exceeded then FD is not invoked. The last row corresponds to the scenario whereby FD is initiated in response to a fault detected by exceeding detection tolerance. Both of these scenarios invoke the expected response to maintain the quality objective by seeking a repair only when needed. Conditions

Table 3 Effect of $\Delta_{FD} = 3\%$ Tolerance Using Failure-Free Resources for city.qcif, $QP = 5$

Frame	Δ_{PSNR} (%)	Action	Interpretation
7	−0.32	No change	Correct
23	0.26	No change	Correct
...	...	No change	Correct
47	7.53	Reconfiguration triggered	False positive[a]
...	...	No change	Correct
70	2.01	No change	Correct

[a]*Reconfiguration is triggered.*

Table 4 Effect of $\Delta_{FD} = 3\%$ Tolerance Using PEs With 5% Degraded Output for city.qcif, $QP = 5$

Faulty PE	Δ_{PSNR} (%)	Action	Interpretation
1	6.63	Reconfig. triggered at 51	Correct
2	4.07	Reconfig. triggered at 51	Correct
3	4.76	Reconfig. triggered at 52	Correct
4	4.63	Reconfig. triggered at 52	Correct
5	3.99	Reconfig. triggered at 53	Correct
6	3.01	Reconfig. triggered at 55	Correct
7	< 3	No change	False negative[a]
8	< 3	No change	False negative[a]

[a]*Innocuous fault below threshold, reconfiguration is not triggered.*

corresponding to the middle two rows of Table 6 also maintain the desired quality objective, due to the nonintrusive nature of the PURE reconfiguration process. For instance in the second row, PURE still minimizes the impact of inadvertent triggering of reconfiguration by temporarily deallocating the least priority function or reconfiguring the RS. In the third row, the failure is an *innocuous fault* in the sense that it does not manifest a degradation in signal quality sufficient to necessitate repair. In summary, PURE allows the designer to specify the tolerable range of signal degradation by selecting Δ_{FD} to allow fluctuations up to that value without triggering the diagnostic flow. Finally, even though PSNR calculation and the Reconfiguration Controller are not part of the throughput datapath and thus do not impact signal quality, handling of possible faults in these PURE components can be addressed using techniques identified in [33].

2.1.4 Reconfiguration Algorithm 4: FHME

A *fault-handling motion estimation* (FHME) core developed in [34] leverages the priority of functions to guide both power reduction and fault mitigation. A resource allocation scheme is developed to reconfigure the parallelized architecture of the motion estimation (ME) engine utilized at runtime within the video encoder. The

Table 5 Fault Detection Performance ($\Delta_{FD} = 3\%$) Using PEs With 5% Degraded Output at Frame 51, $QP = 5$

	Faulty PE_1			Faulty PE_2		
Sequence	**Trigger Frame**	**Δ_{PSNR} (%)**	**Interpretation**	**Trigger Frame**	**Δ_{PSNR} (%)**	**Interpretation**
Akiyo	51	6.54	Correct, latency = 0 frames	None	–	False negative
Carphone	52	4.81	Correct, latency = 1 frames	52	3.33	Correct, latency = 1 frames
City	51	6.63	Correct, latency = 0 frames	51	4.07	Correct, latency = 0 frames
Claire	53	3.04	Correct, latency = 2 frames	55	3.01	Correct, latency = 4 frames
Football	51	41.55	Correct, latency = 0 frames	51	16.77	Correct, latency = 0 frames

Table 6 Quality-Oriented Fault Diagnosis

Hardware Faults	$\Delta_{PSNR} > \Delta_{FD}$	FD Asserted	Quality Objective Met?
No	No	No	Yes
No	Yes	Yes → False positive	Yes[a]
Yes	No	No → False negative	Yes[b]
Yes	Yes	Yes	Yes

[a]*Small power overhead involved.*
[b]*Innocuous fault.*

concept of directed management of computational resources to achieve *fault resiliency* and *energy efficiency* is illustrated in Fig. 11. The demand and response flow of the component bitstreams is depicted in Fig. 11. An amorphous processing element (APE) can be configured as either an active element (AE) to serve as a processing unit of the ME core, dynamic replica (DR) to concurrently check an AE, or RS to reduce energy consumption when processing low-activity video scenes.

Block-based ME algorithms involve searching for a block in a reference frame that most closely matches with the MB of the current frame. Sum of absolute difference (SAD) is widely used as a matching metric to compare the MB of current frame with that of the reference frame. Typically, instead of evaluating all block positions in a reference frame, a search range S is defined and the search is performed within that search window $[-S, S]$. To pipeline the computation, the data corresponding to various subregions within a search window can be assigned to multiple APEs. As shown in Fig. 12, AE_1 computes the matching metric between current frame's MB and an MB located at the same location in the reference frame. AE_2 computes the matching metric for the reference frame's data located at displacements -1 and $+1$ pixels with respect to the current MB's location, and so on. Overall, an N_a number of AEs operate on the search window data to compute a motion vector (MV). Thus, each APE is comprised of n processing elements (PEs) where $n \in \{1, 2, 4, 8, \ldots\}$. Since the number of APEs in the active datapath directly corresponds to the search window defined, N_a can be reduced for low-motion-activity video input in which a large search range is not required. Fig. 12 illustrates computation of MV spatially along the j-axis in a reference frame's search window with $n = 2$.

This exploits the time varying nature of the input video frames to the ME to adapt the underlying hardware resources. We consider a data parallel architecture of ME in which various APEs concurrently operate on sub-regions of input video frames. Specifically, each APE can be reconfigured to operate on the dataset of any other APE. In addition, an APE can be power-gated for reducing the power consumption of the ME core.

Runtime analysis of time varying characteristics of the input data is beneficial in predicting the computational complexity and hence the required hardware resources. Hardware parallelism combined with software flexibility provides architectural support to deal with these time-varying computing workloads.

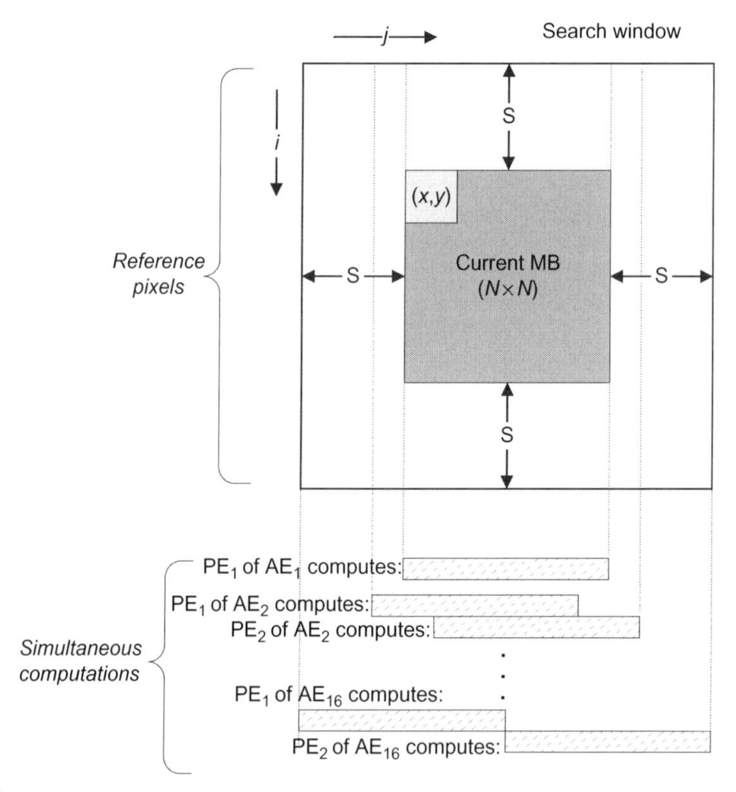

FIG. 12

Computation of a motion vector.

Fault mitigation strategy

As discussed before, multiple APEs are allocated on the computational fabric, where each APE is used for *n* SAD computations per row corresponding to a subregion in the reference frame's search area. In addition, some empty APEs are reserved at design time to provide a reserve for FR. In the absence of such spares due to area or power constraints, some low priority AEs can be vacated to perform more prioritized computations as discussed here.

In video encoders, PSNR can be maintained by holding the QP constant while allowing the bitrate of the encoded bitstream to vary. Our choice of the ME architecture is based upon the intuition that the defectiveness of some of the APEs can be compensated by a reduction in the search range. Such an approach is promising in terms of graceful degradation when confronting faults and mitigating their effects. A decrease in the search range results in reduced coding efficiency as demonstrated by bitrate increase in Fig. 13. On the other hand, its impact on the PSNR measure and hence visual quality of the images is imperceptible as shown by the PSNR curve in Fig. 13. Hence, the bitrate can be used as a health metric to detect hardware faults in

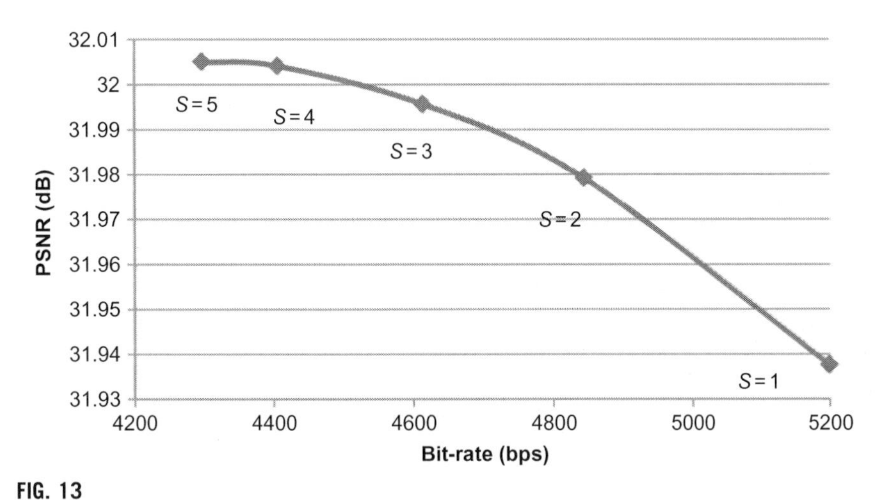

FIG. 13

RD curve showing the effect of increasing search range.

the ME engine. After fault-detection, the scheme proceeds to identify/isolate faulty APEs, and then the FR phase involves avoidance of those isolated APEs. A remapping of tasks from faulty AEs to healthy APEs completes the recovery process.

In the experiments that evaluated the detection, isolation, and recovery techniques, variable bitrate mode is selected for the video encoder. In addition to the scene's high-motion-activity, a failure in ME processing due to hardware faults can also be causal in increasing the bitrate of encoded video stream, thereby degrading overall compression efficiency. The PSNR and bitrate of the encoded bitstream for the `container` input video sequence for various fault-scenarios are shown in Fig. 14A and B, respectively, in which 'd' corresponds to number of faulty APEs.

Detection of hardware faults

As discussed earlier, the bitrate should correspond to the MV values for fault-free operation. For example, a small MV with a large enough search range producing a large bitrate compressed video stream may imply potential hardware errors. Faults in a ME core result in incorrect computation of MVs and hence the SAD increases. To adaptively allocate the resources for power efficiency and fault-mitigation, we monitor the MVs as well as the bitrate for a given search size and QP. When the bitrate of the compressed bitstream from the encoder increases, we evaluate the following two possible scenarios:

1. For the variable bitrate mode (i.e., fixed QP), prediction for P frame is not working quite well due to the high motion activity in the scene, or
2. hardware faults on ME occurred.

Since we do not have a priori knowledge about the cause, we first assume that scenario 1 occurred. If the change in minimum SAD error is significant as a result

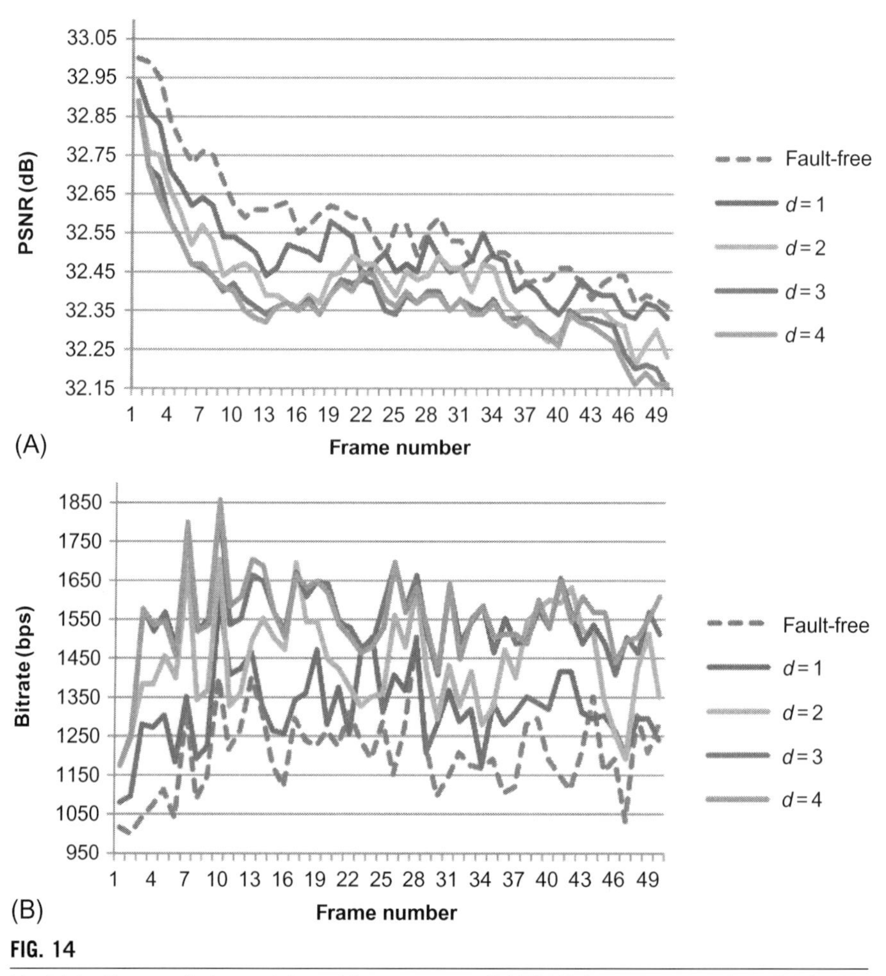

FIG. 14

Fault injection results for container video sequence. (A) PSNR for various number of defective APEs. (B) Bitrate for various number of defective APEs.

of search range increase, then the FH mechanism is not triggered. However, if a widening of the search range still does not correlate with the changes in SAD error, then the FH scheme is triggered to detect and isolate any faulty modules. It is worth mentioning here that the FH scheme runs at a higher layer than the computational layer and does not impact the throughput datapath. The reconfiguration scheme does not contribute to the critical path which would otherwise degrade the performance of the ME core. An additional benefit of keeping it outside the datapath is that the FHME scheme does not impact the functionality of the ME for those video sequences which are outliers to the threshold selected.

Fault diagnosis using dynamic redundancy

After fault detection, the next step is to perform CED at the APE level so that faulty APEs are identified by a pairwise comparison diagnosis procedure. Such a functional testing approach does not require the CUT to be brought offline for input test vector evaluation. Thus, the CUT sustains partially useful throughput even during the diagnosis phase.

Phase 1—Identifying a healthy APE

We assign different priorities to each APE according to the relative importance of the input data which they process. For example, the APE used to search the locations near the predicted MV point in interframe mode has higher priority, while the APE to search the unlikely locations is assigned a lower priority. Initially, some low priority APEs are configured as DRs to serve as checkers for those APEs which are in the throughput datapath. Such a temporary replacement operation of least priority APEs minimally impacts the output's bitrate as shown in Fig. 15. Alternatively, the availability of some healthy RSs obviates the need to temporarily vacate the least priority APEs for diagnostic purposes. Nonetheless, irrespective of availability or unavailability of healthy RSs, the diagnosis algorithm FHME proceeds to identify a healthy APE in the resource pool. It illustrates how dynamic redundancy is employed in the FD process to check the APEs in the throughput datapath.

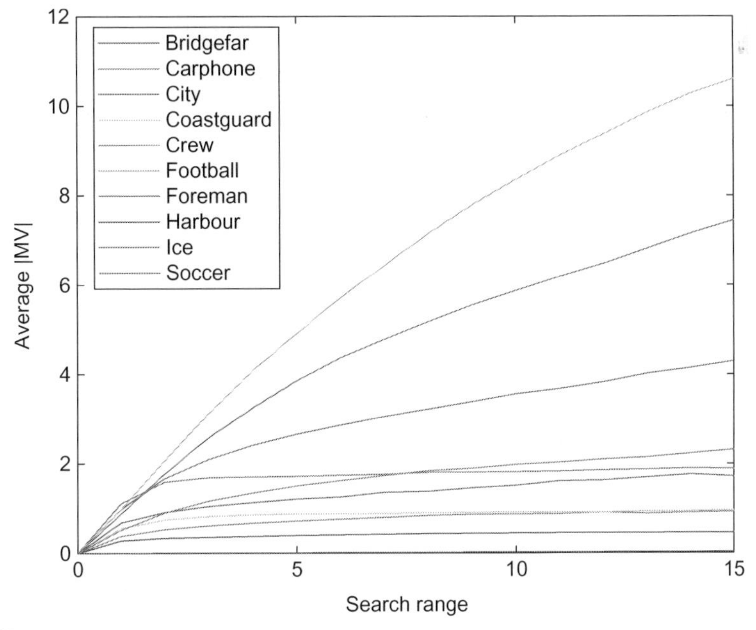

FIG. 15

Effect of search range on motion vector's values for various video sequences.

Phase 2—Isolation of faulty APEs

After at least a single healthy APE is identified, it is used to check the health state of all other APEs in the overall resource pool. In case of multiple healthy identified APEs, the fault-isolation process can be accelerated by concurrently checking the output discrepancy of multiple AEs. For an identified healthy APE_j, the fitness state ϕ_i of a suspect APE_i is updated as follows:

$$\begin{aligned} &\phi_i \leftarrow 0; \\ while_{i \in V_T} \ \{ \ &if \ \ (APE_i.output \neq APE_j.output) \\ &then \quad \phi_i \leftarrow 1; \quad \} \end{aligned}$$

Next, faulty APEs are reconfigured with blank configuration bitstreams thereby eliminating any switching/processing to save power. Lastly, the faulty APEs are removed from the resource pool, i.e., $N \leftarrow N - d$.

Fault recovery

In video encoders, the predicted MV which is calculated by the surrounding MBs' motion vectors becomes the search center since differential coding is used to encode the current MB's MV information to save bits. If the coded MB indication (COD) bit in the output bitstream is set to 1, no further information is transmitted for the given

---

ALGORITHM 4 FHME: COMPARISON-BASED DIAGNOSIS TO IDENTIFY AT LEAST A SINGLE HEALTHY APE.

Require: Current Search Range (S), Number of slacks available for ME (N_{RS}), Number of dynamic replicas to employ in the diagnosis procedure (N_{DR})
Ensure: Identified set of healthy APEs, V_h

1: Initialization: $V_h = \emptyset$, $TCAE = 1$. Current testing candidate APE
2: **if** $N_{DR} < N_{RS}$ **then**
3: $\quad N_{RS} \leftarrow N_{RS} - N_{DR}$. Remove from slack
4: **else**
5: $\quad S \leftarrow S - \dfrac{n(N_{DR} - N_{RS})}{2}$
6: $\quad N_{AE} \leftarrow \dfrac{2S}{n} + 1$. Reduce the search range
7: **end if**
8: **while** $((V_h = \emptyset) \wedge (V_T \neq \emptyset))$ **do**
9: $\quad (APE_j.function \leftarrow APE_{TCAE}.function)\ \forall_j\ where\ N_{AE} < j \leq (N_{AE} + N_{DR})$. Configure N_{DR} APEs as DRs to check the APE_{TCAE}, .e.g., check the zero vector APE first
10: $\quad V_T \leftarrow \{APE_{TCAE}, APE_j\} \forall_j N_{AE} < j \leq N_{AE} + N_{DR}$. Form a pool under test
11: $\quad v = majority(V_T.output)$. Perform majority voting of the pool under test
12: $\quad$ **if** $APE_i.output = v$ **then**
13: $\quad\quad \phi_i \leftarrow 0, V_h \leftarrow APE_i$
14: $\quad$ **else**
15: $\quad\quad \phi_i \leftarrow x; \forall_{i \in V_T}$. Estimate the health status of the pool under test
16: $\quad$ **end if**
17: $\quad$ **if** $TCAE < N_{AE}$ **then**

```
18:       TCAE ← TCAE + 1 . Update the test candidate APE
19:    else
20:       if (N_AE − N_DR) ≥ 1 then
21:          N_AE ← N_AE − N_DR . Configure the AEs as DRs
22:       else
23:          if N_AE > 1 then
24:             N_DR ← N_AE − 1
25:             Na ← 1
26:          else
27:             N_AE ← 0, V_T ← ∅
28:          end if
29:       end if
30:       TCAE = 1 . Recheck the first APE with a different DR
31:    end if
32: end while
```

MB, meaning that it is an inter-MB with MV for the whole block equal to zero and with no coefficient data. This saves considerable bits in static scenes. In the proposed hardware architecture, SAD(0,0) computation, which corresponds to predicted MV location, is assigned to AE_1. Thus, AE_1 performs the most prioritized computational function, i.e., computing the SAD corresponding to zero MV data. AE_{Na} computes the SAD values corresponding to candidate block positions which are farthest from the current block's predicted position.

In the recovery phase, the datapath is reconfigured by selecting healthy APEs using a reconfiguration controller. In addition, AEs priority is also engaged when recovering from a faulty situation. For example, to mitigate the defectiveness of faulty AE_1, its faulty resource is demoted by releasing its preassigned task. A healthy resource from the least priority AE is promoted to perform this important task. Finally, the APEs are relabeled after the reconfigurations so that AEs sorted in ascending order perform computations in descending order of priorities.

2.2 FPGA REFURBISHMENT USING EVOLUTIONARY ALGORITHMS

In this section, FPGA reconfigurability is exploited to realize autonomous FR in mission-critical applications at runtime. NDER technique utilizes design-time information from the circuit netlist to constrain the search space of the evolutionary algorithm by up to 98.1% in terms of the chromosome length representing reconfigurable logic elements. This facilitates refurbishment of relatively large-sized FPGA circuits as compared to previous works which utilize conventional evolutionary approaches. Hence, the scalability issue associated with Evolvable Hardware-based refurbishment is addressed and improved. Moreover, the use of design-time information about the circuit undergoing refurbishment is validated as means to increase the tractability of dynamic evolvable hardware techniques.

The NDER technique isolates the functioning outputs of the FPGA configuration from unwanted evolutionary modification. This allows the EH technique to focus

Table 7 Comparison of Back tracing Approach With Evolutionary Algorithm-Based Fault-Handling Schemes for FPGAs

Approach/ Previous Work	Size of Circuit Used for Experiment	Pruning of GA Search Space	Modular Redundancy	Assumption of Fault-Free Fitness Evaluation
Vigander [35]	4 × 4 Multiplier	No	Triple	Yes
Garvie and Thompson [36]	MCNC benchmark circuit cm42a having 10 LUTs	No	Triple	Yes
CRR [37]	3 × 3 Multiplier; MCNC benchmark circuits having $<$ 20 LUTs [8]	No	Dual	No
RARS [38]	Edge-detector circuit	No	Triple/dual (dynamic)	Yes Yes
Salvador et al. [39]	Image filtering circuit	No	Not addressed	Not addressed
NDER	*MCNC benchmark circuits having up to 1252 LUTs*	*Yes*	*Triple/dual (dynamic)*	*Yes*

exclusively on the discrepant outputs. Thus, the genetic operators which modify the design to avoid faulty resources are only applied to selected elements of the FPGA configuration which are suspected of corrupting the current output. Suspect resource assessment is performed at runtime using design netlist information to form the pool of suspect resources. This significantly decreases the search space for the GA leading to more tractable refurbishment times.

The existing tools and platforms demonstrate that self-healing EH systems are practically viable. Yet, there is a need to propose modifications to the evolutionary algorithm utilized for self-recovery to become suitable for larger circuits and systems. NDER addresses these issues by using properties of the failure syndrome to considerably prune the search space while attaining complete quality of recovery. Table 7 lists a comparison of NDER with other GA-based FH schemes for reconfigurable logic devices.

2.2.1 Fault isolation via back tracing

A set of heuristics are presented to prune the search space for evolutionary algorithms utilized for self-healing in EH. Generally, FD in the form of knowing the location and nature of faults is not a requirement for such algorithms, which is often considered as one of the strengths of this technique [40], but to address the issue of scalable refurbishment of EH systems, it can be leveraged to increase scalability as demonstrated through the NDER technique. The pruning of the search space is achieved by selecting a subset of resources from the entire pool of resources based on both runtime and design-time information. This selection is done based on implication by only the discrepant primary output lines which are implied to be generated by corrupt logic and/ or interconnect resources.

The candidate heuristics to identify the set of suspect resources in a given corrupt FPGA configuration are best explained through a motivating example. Let's consider the FPGA configuration with six primary inputs (PIs) and four primary outputs (POs) as shown in Fig. 16. The PIs are represented through letters a, b, c, d, e, and f whereas the POs are represented with out_1, out_2, out_3, and out_4. The LUTs are represented with capital letters $\{A, B, \ldots, O\}$ which is the complete set of resources. Each LUT is tagged with the index(es) of the PO(s), on which datapath(s) it is used. This information can be obtained by back-tracing from each PO of the circuit and tagging the LUT(s) which are in its datapath. The process is recursively continued from each LUT input until PI is reached. This process is analogous to traversing a tree structure starting from the leaf node, which is the PO in this case, until the root is encountered. For example, LUT D contributes to the generation of both POs out_1 and out_4, whereas LUT K is only responsible for out_4. To illustrate different scenarios in which the suspect resources are selected, two distinct failure cases are presented as shown in Fig. 16. This selection procedure will also be referred to as the process of *"Marking*

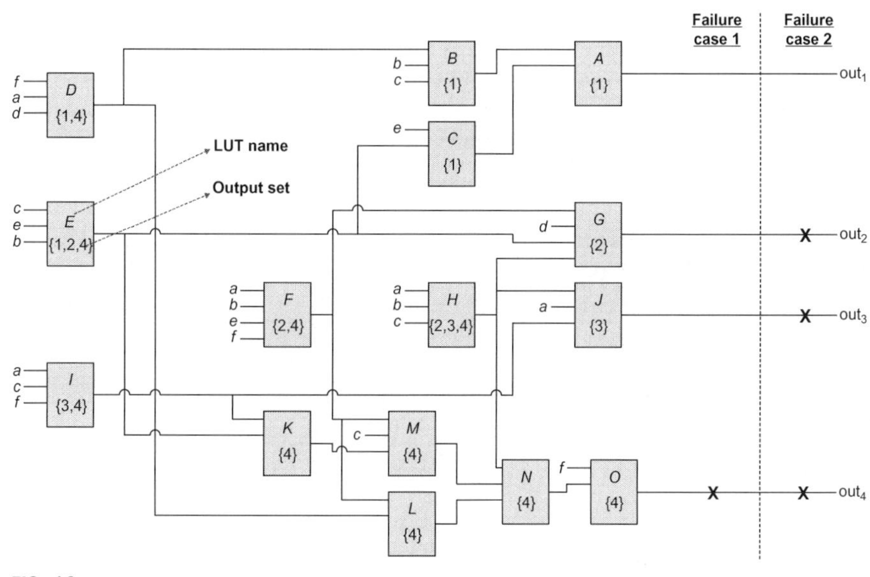

FIG. 16

An example of marking of LUTs using netlist information.

of LUTs." The resources identified using one of the candidate heuristics as presented below are to be utilized during the execution of the evolutionary refurbishment algorithm:

Aggressive Pruning Heuristic (H_A): An aggressive pruning heuristic is adopted in this case where all those LUTs with output set containing any of the noncorrupt outputs are *not marked*. Let's consider the Failure Case 1 as shown in Fig. 16, when only a single output line of the circuit $\{out_4\}$ is identified as corrupt at runtime. Then LUTs $\{K, L, M, N, O\}$ are marked in the FD phase by considering the aggressive pruning heuristic of not marking LUTs which contribute to noncorrupt outputs. For example, LUT I is not selected because it is responsible for POs $\{out_3, out_4\}$, and out_3 is not identified as corrupt. In Failure Case 2 as shown in Fig. 16, multiple circuit output lines are identified as corrupt, which considers an interesting scenario, where initially, only the LUT(s) with output set "*equal*" to the set of corrupt POs is (are) selected. For example, if $\{out_2, out_3, out_4\}$ forms the set of corrupt POs, then only LUT $\{H\}$ is marked as it is the only LUT with maximal output set cardinality of 3 and has matching elements. Based on the demand of the refurbishment algorithm, if more LUTs need to be marked to potentially increase the size of the search space, the next choice is to find all LUTs with output set equal to any of the possible combinations of subsets of cardinality 2 from the set of corrupt POs, i.e., $\{out_2, out_3\}$ or $\{out_2, out_4\}$ or $\{out_3, out_4\}$. Thus, the set of marked LUTs at this iteration is $\{H, F, I\}$. Yet, another iteration of marking is achieved by finding all LUTs with output set equal to $\{out_2\}$ or $\{out_3\}$ or $\{out_4\}$. Thus, the final set of marked LUTs is $\{H, F, I, G, J, K, L, M, N, O\}$. The size of the set of marked LUTs is not increased further based on the selected pruning heuristic.

Exhaustive Pruning Heuristic (H_E): This heuristic is based on marking all those LUTs with an output set having elements of any of the corrupt outputs. For Failure Case 1, all LUTs with output set containing out_4 are marked as shown in Table 8. Similarly, for Failure Case 2, all LUTs with output set containing any of the outputs out_2, out_3, or out_4 are marked. The drawback of this intuitive heuristic is that more number of LUTs tend to be selected as compared to other presented heuristics. On the positive side, the algorithmic realization based on this heuristic is less complex.

Hybrid Pruning Heuristic (H_{AE}): The selection based on this heuristic is an extension of H_A. An additional iteration over H_A is used to mark all the remaining LUTs with output set having elements of any of the corrupt outputs which have *not* been marked in the previous iterations. The LUTs marked in Failure Cases 1 and 2 for H_{AE} are shown in Table 8.

Dynamic Pruning Heuristic (H_D): The selection based on this heuristic is an extension of H_{AE}. The additional iterations of marking are based on observing the status of corrupt POs during the FR process. New potential markings only take place if the cardinality of set of corrupt PO(s) increases from its initial value. Thus, iterations 3, 4, and 5 shown for Failure Case 1 in Table 8 can take place in any order. Note, only LUTs with fan-out of one are marked in these additional iterations. For instance, if $\{out_2\}$ is added to the set of corrupt POs, then LUT $\{G\}$ will be added to the set of marked LUTs. For this particular example, all the LUTs are selected by the

Table 8 Listing of LUTs Marked Through Presented Heuristic Approaches for the Two Failure Cases in the Example presented in Fig. 16

Heuristic	Failure Case 1			Failure Case 2		
	Iteration	Marked LUTs	Pruning Reduction % Nonmarked LUTs	Iteration	Marked LUTs	Pruning Reduction % Nonmarked LUTs
H_A	1	K,L,M,N,O	66.7	1	H	93.3
				2	F,I	80.0
				3	G,J,K,L,M,N,O	33.3
H_E	1	D,E,F,H,I,K,L,M,N,O	33.3	1	D,E,F,G,H,I,J,K,L,M,N,O	20.0
H_{AE}	1	K,L,M,N,O	66.7	1	H	93.3
	2	D,E,F,H,I	33.3	2	F,I	80.0
				3	G,J,K,L,M,N,O	33.3
				4	D,E	20.0
H_D	1	K,L,M,N,O	66.7	1	H	93.3
	2	D,E,F,H,I	33.3	2	F,I	80.0
	3	A,B,C	13.3	3	G,J,K,L,M,N,O	33.3
	4	G	6.7	4	D,E	20.0
	5	J	0.0	5	A,B,C	0.0

end of all iterations for both the failure cases. However, this will not be the case with much bigger circuits with large number of outputs. The execution complexity of H_D is higher than H_{AE} and others. In summary, the heuristics in decreasing order of complexities are listed as: H_D, H_{AE}, H_A, H_E.

Evaluating the Diagnostic Performance The efficacy of the presented heuristics for FD is evaluated through experimentation on circuits selected from the MCNC benchmark suite [41]. The experiments are setup with a single stuck-at fault which is randomly injected at the input of the LUTs in the FPGA circuit. It is to be determined whether the LUTs marked by the heuristics as presented above contain the LUT with the fault. Table 9 lists the observed coverage of faults for multiple experimental runs with benchmark circuits using the first iteration of heuristic H_A. The number of experimental runs as indicated by "*# of Runs*" column, which actually articulates the fault are variable for each circuit as the results are sampled from a total of 700 runs with cases where faults lie on a spare LUT. The percentage of runs with corrupt set having cardinality greater than one ranged from 5.8% to 66.9% according to characteristics of benchmark circuits. The breakdown of marked LUTs in terms of the percentage having fan-out equal to one and fan-out greater than one is also given with more LUTs being marked with fan-out equal to one. The former markings takes place when a single PO is identified as corrupt. In contrast, LUTs having fan-out greater than one are marked as a result of multiple POs being corrupt. The size of the set of marked LUTs can often be reduced in the latter case, although dynamically changed as required. Majority of experimental runs demonstrate successful inclusion of faulty LUT into the set of marked LUTs by heuristic H_A in as soon as the first iteration as indicated by "*Coverage Faulty LUT*" column. The subsequent iterations of heuristic H_A or subsequent application of heuristic H_E provides 100% coverage for cases where the actual faulty LUT is not marked in the first iteration. Thus, full coverage can be attained by using heuristic H_{AE} with a good tradeoff of search space size as compared to H_E as concluded from earlier discussion.

2.2.2 NDER technique

The overall algorithmic flow of the NDER technique is presented in Algorithm 5 with the nomenclature listed in Table 10.

ALGORITHM 5 THE NETLIST-DRIVEN EVOLUTIONARY REFURBISHMENT ALGORITHM.

$t := 0$;
Initialize $P(0) = \{\mathbf{I}_1(0), \ldots, \mathbf{I}_{|P|}(0)\} \in \mathbf{I}^{|P|}$;
Evaluate $P(0) : \{\{\Phi(\mathbf{I}_1(0)), \ldots, \Phi(\mathbf{I}_{|P|}(0))\}, \{\vec{\Omega}_1, \ldots, \vec{\Omega}_{|P|}\}\}$, where $\Phi(\mathbf{I}_k(0))$ is the fitness of an individual $\mathbf{I}_k$ with a value $\in \mathbb{Z}_{m*2^n}$;
$level_k := CountOnes(\vec{\Omega}_k), \ \forall k \in \{1, \ldots, |P|\}$;
Mark$\{\vec{M}_k(0), level'_k\} := Mark\_LUTs(\vec{\Omega}_k, \vec{X}_k, level_k), \ \forall k \in \{1, \ldots, |P|\}$;
$\vec{\lambda}_k := Select\_PIs(\vec{M}_k(0)), \ \forall k \in \{1, \ldots, |P|\}$;

while Exit_Criteria $(P(t)) \neq$ **true do**
for $k = 1$ *to* $|P|$ **do**
Crossover$\{\vec{a'}_k(t), \vec{c'}_k(t), \vec{out'}_k(t)\} := r'_{\{p_c\}}(P(t));$
Mutate$\{\vec{a''}_k(t), \vec{c''}_k(t), \vec{out''}_k(t)\} := m'_{\{p_m\}}(\vec{a'}_k(t), \vec{c'}_k(t), \vec{out'}_k(t));$
end for
Evaluate $P''(t) : \{\Phi(\mathbf{I}''_1(t)), \ldots, \Phi(\mathbf{I}''_{|P|}(t))\};$
for $k = 1$ *to* $|P|$ **do**
if $\Phi(\mathbf{I}''_k(t)) < \Phi(\mathbf{I}_k(t)) * level\_threshold \,\&\&\, level'_k > 0$ **then**
$level'_k = level'_k - 1;$
Mark$\{\vec{M''}_k(t), level''_k\} := Mark\_LUTs(\vec{\Omega}_k, \vec{X}_k, level'_k);$
$\vec{\lambda''}_k := Select_P Is(\vec{M''}_k(t));$
else
$\vec{M''}_k(t) := \vec{M}_k(t);$
$level''_k = level'_k;$
end if
end for
Select $P(t+1) := s_{\{\tau\}}(P''(t))$, where τ is the tournament size;
$t := t + 1;$
end while
return: Refurbished Individual $\mathbf{I}_{final} = (\vec{a}, \vec{c}, \vec{out})$ having $\Phi(\mathbf{I}_{final}) == m * 2^n$ OR $max\{\Phi(\mathbf{I}_1(t)), \ldots, \Phi(\mathbf{I}_{|P|}(t))\};$

Initially, multiple FPGA configurations are utilized to form a population for the GA. This population could be formed with the single under-refurbishment configuration replicated to form a uniform pool, or with functionally equivalent though physically distinct configurations generated at design time to form a diverse pool [8]. However, diverse designs do not necessarily avoid the faulty resources, and in that situation a technique to deal with this is desirable. Herein, a uniform population seed is utilized as it requires less storage and reduced design time effort compared to generating fault-insensitive designs with sufficient coverage. Once synthesized, these configurations are suitable for storage in a fault-resilient nonvolatile memory.

The FPGA configurations are mapped into individuals, represented by $\mathbf{I}_k(t)$, where k uniquely identifies an individual in the population and t represents the time instance, e.g., 0 identifies initial formation time of the population. The fitness of the individuals in the population is evaluated by using the fitness function $\phi(\mathbf{I}_k(t))$. During the initial fitness evaluation, the corrupt output line(s) is(are) also identified and $\vec{\Omega}_k$ is updated with the relevant information for an individual $\mathbf{I}_k$. This is done via bit-by-bit comparison of the POs of the under-refurbishment module with the known healthy module (*golden*) by application of adequate test vectors. These modules are taken offline during the FI phase without affecting the system throughput.

Table 9 Percentage of Marked LUTs and the Probability of Covering Faulty LUT in the First Iteration of Heuristic H_A

Benchmark	Total LUTs	# of Runs	% Runs Corrupt Outputs > 1	Marked LUTs on Avg.		Avg. Pruning Reduction Nonmarked LUTs (%)	Coverage Faulty LUT (%)	Avg. Correctness/ Max. Fitness Exhaustive Evaluation
				Fan-Out = 1 (%)	Fan-Out > 1 (%)			
alu2	185	483	66.9	2.54	18.98	78.5	100	6086/6144
5xp1	49	462	6.5	16.20	0.27	83.6	100	1265/1280
apex4	1252	553	7.4	4.84	0.01	95.2	99.1	9713/9728
bw	68	501	5.8	2.94	0.09	97.0	100	891/896
ex5	352	481	17.3	1.82	0.07	98.1	96.5	16,102/16,128

Table 10 Nomenclature for NDER Algorithm

Symbol	Description
P	Representation of the population
l	Number of LUTs in the FPGA configuration
n	Number of primary inputs of the circuit
m	Number of primary outputs of the circuit
p_c	Crossover rate
p_m	Mutation rate
$\mathbf{I}_k$	Representation of the kth individual in the population
$a_{k,x}$	Logical content of a $LUT_x \in$ individual k having a 16-bit unsigned binary value, $\forall x \in \{1, \ldots, l\}$
$c_{k,y}$	Connection of a $LUT_x \in$ individual k having an integer value $\in \{1, \ldots, n, n+1, \ldots, n+l\}$, $\forall x \in \{1, \ldots, l\}$ & $y \in \{1, \ldots, l * 4\}$
$out_{k,i}$	Primary output $i \in$ individual k having an integer value $\in \{1, \ldots, n, n+1, \ldots, n+l\}$, $\forall i \in \{1, \ldots, m\}$
$\Omega_{k,i}$	Corrupt output flag corresponding to primary output $i \in$ individual k having a binary value $\in \{1, 0\}$, $\forall i \in \{1, \ldots, m\}$
$\lambda_{k,i}$	Primary inputs $i \in$ individual k having a binary value $\in \{1, 0\}$, $\forall i \in \{1, \ldots, n\}$
$M_{k,x}$	Marking of a $LUT_x \in$ individual k having a binary value $\in \{1, 0\}$, $\forall x \in \{1, \ldots, l\}$
$level_k$	Iteration of marking of LUTs for an individual k
$X_{k,x}$	Datapath usage corresponding to m primary outputs for a $LUT_x \in$ Individual k having a m-bit vector, $\forall x \in \{1, \ldots, l\}$

The under-refurbishment module is configured with the individuals from the population as dictated by the algorithm.

The index(es) of the corrupt PO(s) and design netlist information contained in $\vec{X_k}$ is used to *mark* LUTs for the individuals of the population. Here, the netlist information corresponds to the contribution of each LUT to the datapath of PO(s) as described earlier in Section 2.2.1. Then, the information about all the marked LUTs for an individual $\mathbf{I}_k$ is updated in $\underset{k}{\vec{M}}$. The set of marked LUTs is then utilized to form the subset of PIs whose corresponding test patterns are generated for fitness evaluation. The evolutionary loop begins with the application of GA operations of *crossover* and *mutation* to the individuals in the population based on user-defined rates p_c and p_m respectively, and evaluating the fitness of the resulting off-spring population. The mutation and crossover operators are designed within the NDER framework such that only marked LUTs are modified. After the application of these operators an off-spring $\mathbf{I}''_k$ will undergo the process of marking of LUTs if and only if its fitness is less than a user-defined percentage (*level_threshold*) of the fitness of its corresponding genetically unaltered individual $\mathbf{I}_k$. New LUTs may be marked for individuals according to adopted heuristic during each invocation as dictated by its current iteration or *level*. Finally, *selection* phase is used to form the population which is to be utilized in the next generation. This work employs *Tournament*-based selection. The loop is terminated when user-defined criterion is met.

2.2.3 Evaluating the efficacy of NDER approach

Experiments were devised to gauge the efficacy of heuristic-based FD in reducing the search space for evolutionary refurbishment of FPGA-based EH systems. Further, scalability in terms of the size of circuits refurbished by NDER is assessed. Experiments are also conducted with conventional GA refurbishment technique for a comparison in terms of the fitness and recovery times achieved. In conventional evolutionary refurbishment (CER), the GA implementation involves no marking of LUTs to limit the search space and standard GA operators are employed. The goal of conducted experiments was to attain 100% FR as it represents the most demanding case.

Benchmark circuits with various number of output lines and sizes in terms of the number of LUTs are selected from the MCNC benchmark suite [41]. The largest sized circuit is `apex4`, which has 1252 LUTs. Whereas, `ex5` has the highest number of output lines, i.e., 63. The I/O characteristics of the circuits are listed within the parenthesis having inputs first as follows: `alu2(10, 6)`, `5xp1(7, 10)`, `apex4(9, 19)`, `bw(5, 28)`, and `ex5(8, 63)`. The circuits are synthesized and mapped using the open source ABC tool [42]. Then, Xilinx ISE 9.2i is utilized for technology mapping, placement and routing on a Xilinx Virtex-4 FPGA device. The placement process is constrained through the User Constraints File with the post-place and route model generated to randomly allocate spare LUTs (having a fan-out of zero) which are approximately 10% of the total LUTs as shown in Fig. 17. In addition, redundancy is also exploited at unutilized inputs of LUTs by allocating valid nets as illustrated in Fig. 18. This redundancy can also be exploited by the GA to recycle a faulty LUT by appropriately altering its logical content.

Single stuck-at fault model is adopted to model local permanent damage [36]. Therefore, the fault is randomly injected by asserting a 1 or 0 at one of randomly

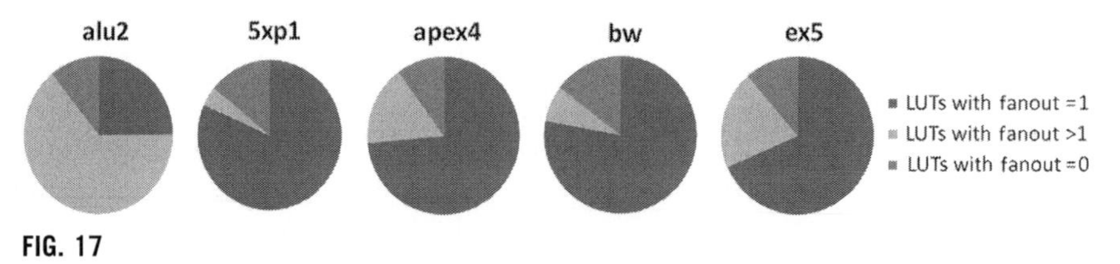

FIG. 17

Distribution of LUTs having fanout equal to one and greater than one, and spare LUTs.

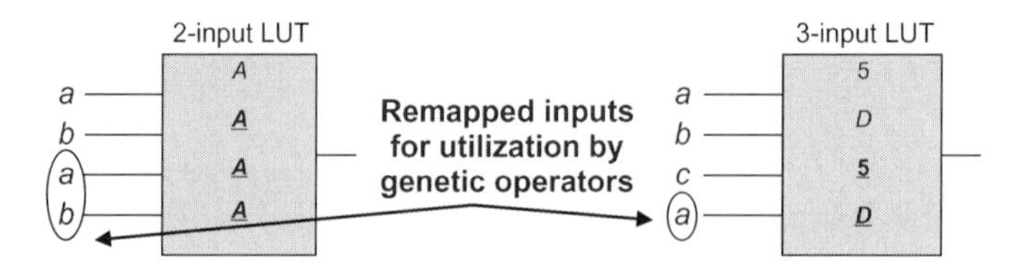

FIG. 18

Design time allocation of redundancy in NDER.

chosen inputs of a LUT in the post-place and route model of the circuit. Afterwards, the GA individuals are encoded for both NDER and CER techniques. The GA refurbishment experiments are conducted using these designs on a simulator able to evaluate the functionality of LUT-based FPGA circuits. Further, both the conventional and proposed GAs are implemented in software.

A finite population standard GA is utilized for both NDER and CER. A population size of 10 is maintained for all the experiments. The crossover rate and tournament size are fixed at 0.6 and 4, respectively. The NDER parameter *level_threshold* varies from 0.96 to 0.99. Mutation rates vary from 0.07 to 0.09 for the experiments. These GA settings are based on similar previous works [38, 43] and validated through experiments. This parameter set produced the best tradeoff in GA performance and memory requirements to store the individuals of the population.

The recovery of `ex5` benchmark having a total of 352 LUTs and 63 output lines is achieved in as few as 784 generations on average over 38 experimental runs. Further, the recovery of largest sized benchmark circuit, i.e., `apex4` with 19 POs is achieved in as few as 633 generations on average over 20 experimental runs. Thus, the scalability of the NDER approach is established as successful refurbishment is achieved for circuits with various sizes and with varying number of output lines.

The speedups achieved in refurbishment times with NDER as compared to CER for the selected benchmarks are reported in Table 11. It is obtained by the ratio of time taken by CER (t_{CER}) and time taken by NDER (t_{NDER}). The results are significant as the recovery times for CER did not achieve useful recovery and indicate it likely requires significantly more generations corresponding to increased speedups. Thus, these estimates are conservative. Finally, high fan-out of suspect resources has been shown to be useful to reduce recovery times achieved via NDER. This effect can be observed due to the intersection of the failed components effects on the overall circuit correctness. In particular, failures resulting from multiple disagreements were shown to decrease recovery times up to 23.4-fold.

Difficult-to-Recover Circuit Characteristics: The NDER experiments conducted indicate some circuit characteristics listed below which make them difficult to recover:

Table 11 Speedup in Recovery Time Achieved Via NDER

Benchmark	Speedup t_{CER}/t_{NDER}
alu2	$\geq$ 17.5
5xp1	$\geq$ 140.1
apex4	$\geq$ 39.5
bw	$\geq$ 404.1
ex5	$\geq$ 25.5

(a) *Circuits with small number of primary outputs:* This results in large percentage of overall LUTs to be selected for evolution in most cases. For example, alu2 and 5xp1 benchmarks take relatively large average number of generations to refurbish.

(b) *Circuits synthesized with large resource sharing:* This can result in observation of common failure signatures at the primary outputs. Hence, it is difficult to limit the pool of marked LUTs. This may be catered by adequate circuit synthesis at design time. For example, this scenario is observed with the alu2 benchmark, which has a large number of LUTs with output set containing POs 2 and 5 (1 is the LSB).

(c) *Circuits with large number of LUTs with fanout equal to one for some or all primary outputs:* This will result in a large pool of marked LUTs whenever a single primary output is diagnosed as corrupt. This can also be improved by adequate circuit synthesis at design time. On a small scale, this was observed for the 5xp1 benchmark as it has the highest percentage of LUTs with unary fanout as shown in Fig. 17. Similarly, on a large scale, this was also observed for the apex4 benchmark, which has large number of LUTs with unary fanout for some POs.

Scalability of Evaluation: Recovery time has a direct correspondence with the number of test patterns utilized during the fitness evaluation of a candidate solution. NDER attempts to reduce the required number of test patterns by pruning circuit elements to form its pool of marked LUTs. In particular, NDER assessment would at its worst perform identical to a conventional GA whereby all PIs have been marked for evaluation. The conventional GA approach requires repair evaluation having the maximal cardinality of 2^n tests. On the other hand, Fig. 19 shows the expected value of this same metric for the NDER technique. Here the benchmark-specific probability distribution of the number of PIs expected for fitness evaluation is based on a combinatorial analysis of each circuit output line. For example, the probability of NDER incurring exhaustive evaluation for the alu2 benchmark is only 0.55, even

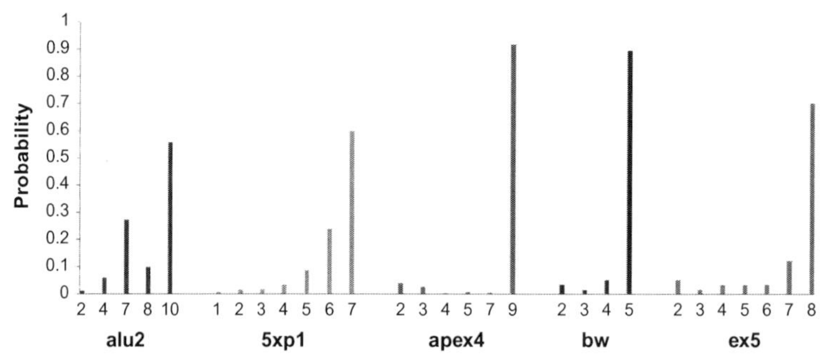

FIG. 19

Probability distributions of the size of subset of PI(s) required for evaluation over 500 random faults.

if fault location is equiprobable among the resource pool. Furthermore, the outcome that the cardinality of test pattern set is 2^7 has probability 0.27. In other words, Fig. 19 quantifies the expectation of reduced evaluation time for NDER based on the structure of the circuit at hand. In particular, it shows the trend that the probability of exhaustive evaluation decreases as the fanout of the CUT is increased, thus improving refurbishment scalability for circuits with larger fanout.

Summary of Results: NDER leverages circuit characteristics to realize self-healing capability that can accommodate to larger circuits than those feasible with conventional EH techniques. It takes into account the output discrepancy conditions, formulated as spanning from exhaustive down through aggressive pruning heuristics, to reduce the search space and maintain the functionality of viable components in the subsystems which use them. For instance, the failure syndrome expressed as the bitwise output discrepancy is utilized to form a concise set of suspect resources at failure time. Thus, the individuals comprising the population of the GA are dynamically determined, which balances the metrics of fault coverage, size of the search space, and implementation complexity. Its capability to preserve pristine LUTs promotes gainful search independent of the underlying device technology characteristics and failure mechanisms.

2.3 SUMMARY

Survivable embedded systems that achieve resiliency without the use of detailed fault models are ideal for the challenges of the nanoscale era. Namely, reliability challenges such as process variation follow a probabilistic distribution which limits the utility of models. Also many interacting variables, such as temperature and aging, become critical to correct operation. While measurement of such parameters at a sufficient granularity becomes prohibitive, on the other hand the reaction to their intrinsic operation in terms of a suitable health metric induces low overhead and remains model-independent.

3 TRADEOFFS OF RESILIENCE, QUALITY, AND ENERGY IN EMBEDDED REAL-TIME COMPUTATION

The reconfiguration techniques of FaDReS and PURE provide a continuum of output quality via online exploration of the search space for tradeoffs between *energy consumption* and *application throughput* in real-time to realize resilience. Both approaches efficiently utilize available resources based upon input signal characteristics to maintain user-desired quality requirements. The flexibility of reconfigurable fabrics to time-multiplex resources with different functions at runtime is utilized for FD with minimal overhead to energy consumed by the recovery operation. In both cases, healthy resources can be identified in bounded number of reconfigurations. In FaDReS, healthy resources are utilized to achieve diagnosis of all resources in the datapath, thus focusing on the timely completion of FI phase. Whereas, PURE utilizes healthy resources immediately for computation of highest-priority functions, thus focusing on availability and quality of throughput during FR phase. The continuum of availability vs. diagnosis latency tradeoff is presented in this section for these two approaches. While these two approaches focus on DCT core in H.263 video encoder. Interesting tradeoffs for ME are plausible where the number of resources allocated to computation can be varied based on input signal characteristics and the vacated resources can be utilized for FH or for lowering of power consumption. Finally, the ability to sustain multiple faults using spatial redundancy schemes is demonstrated while operating at near-threshold voltages (NTVs).

3.1 PERFORMANCE, POWER, AND RESILIENCE CHARACTERIZATION FOR FaDReS AND PURE ALGORITHMS

To assess the performance improvement, power consumption and resilience of the FaDReS and PURE algorithms, case studies were evaluated with various benchmarks, using either PSNR or output discrepancy as a health metric.

In the case of the H.263 video encoder's DCT module, the 8 × 8 DCT is computed by using eight PEs. Each PE performs a 1D-DCT on a row of input pixels to produce an output coefficient. In the current prototype, the video encoder application is run on the on-chip processor. All of the subblocks except the DCT block are implemented in software, the latter being implemented in hardware. The image data from the video sequences is written by the processor to the frame buffer. In order to facilitate 2-D DCT operations, the frame buffer also serves as transposition memory and is implemented by Virtex-4 dual port Block-RAM. Upon completion of the DCT operation, it is read back from the frame buffer to the PowerPC through the Xilinx General Purpose Input-Output (GPIO) core. By the pipeline design of the DCT core, the effective throughput of the DCT core is one pixel per clock. Internally, the PEs utilize DSP48 blocks available in Virtex-4 FPGAs. A 100 MHz core can provide a maximum throughput of 100 M-pixels/s in order to meet the real-time throughput requirement for 176 × 144 resolution video frames at 30 fps, the minimum computational rate should be 760K pixels/s. The PSNR computation time is much longer than that consumed by PEs processing data stream in parallel, i.e., 0.25 ms/frame. It is worth mentioning, however, that a failure to meet real-time deadline in PSNR computation due to a slow processor will only impact the fault-detection/handling latency rather than the computational throughput of the concurrently operating PE-array implemented in the reconfigurable fabric.

The priority of functions is naturally in descending order as the DC coefficient contains the most content information of a natural image. These 8 PEs in the processing throughput datapath are covered by the resilience scheme. For this purpose, depending upon the area/power margins available, RSs are created at design time or generated at runtime considering the priority of functions. As shown in Fig. 20, FH is performed at runtime with a small quality degradation during the diagnosis process. The diagnosis time of divide-and-conquer approach is very short, however, this greedy approach incurs quality degradation during FH process. The quality

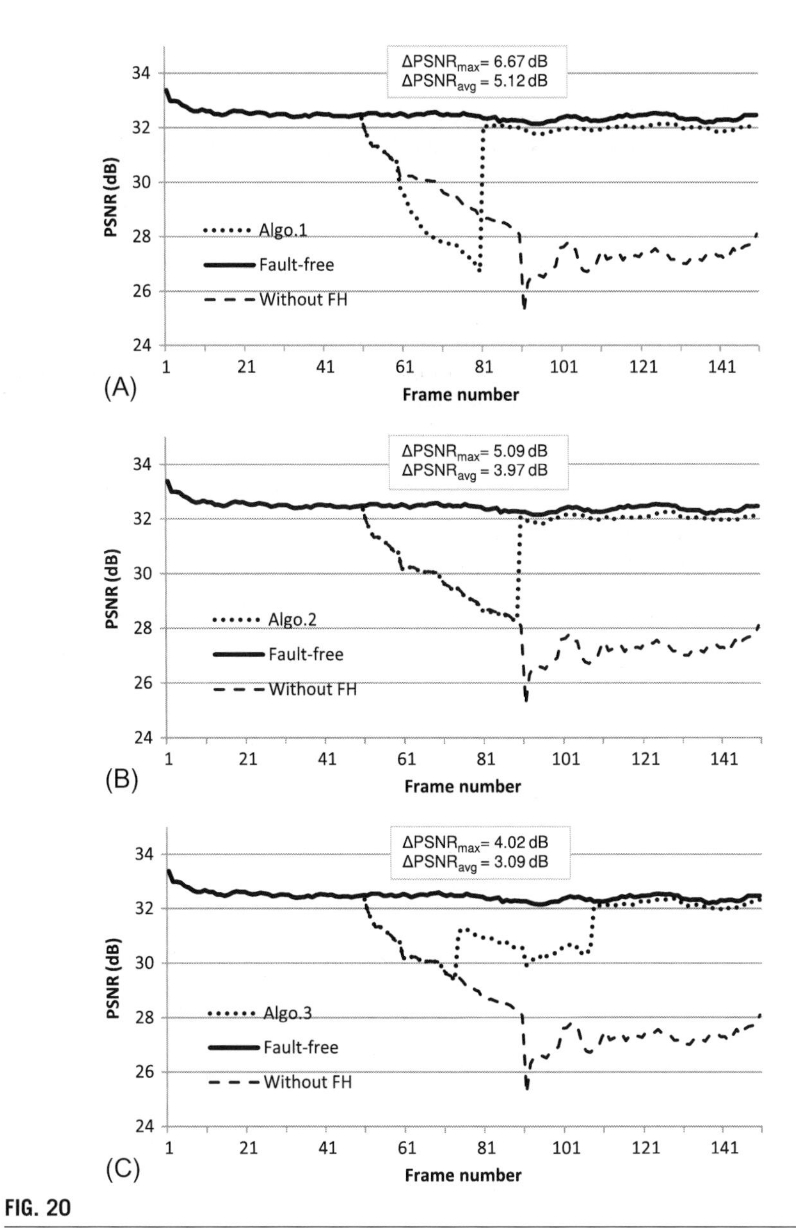

FIG. 20

Operational examples of the three algorithms. (A) Fault handling using Algorithm 1: divide-and-conquer. (B) Fault handling using Algorithm 2: FaDReS. (C) Fault handling using Algorithm 3: PURE.

degradation during FH process is improved in FaDReS approach at the cost of some diagnosis latency. PURE approach provides the best availability of the system during FH and the PSNR is maintained above 29.5 dB. Fig. 21 shows the PSNR and bit-rate of the video stream from an operational encoder before, during, and after FH using the PURE approach.

During diagnosis, PURE can achieve higher useful throughput than alternative approaches due to the escalation of healthy resources to the top priority functional assignments. FH results with a video encoder show that average PSNR in PURE's case is only 3.09 dB below that of a fault-free encoder, compared to a divide-and-conquer approach which incurs average PSNR loss of 5.12 dB during the FD phase. This metric provides a useful indication of quality during refurbishment. Fig. 22 illustrates the qualitative results of the approach. The recovered image in

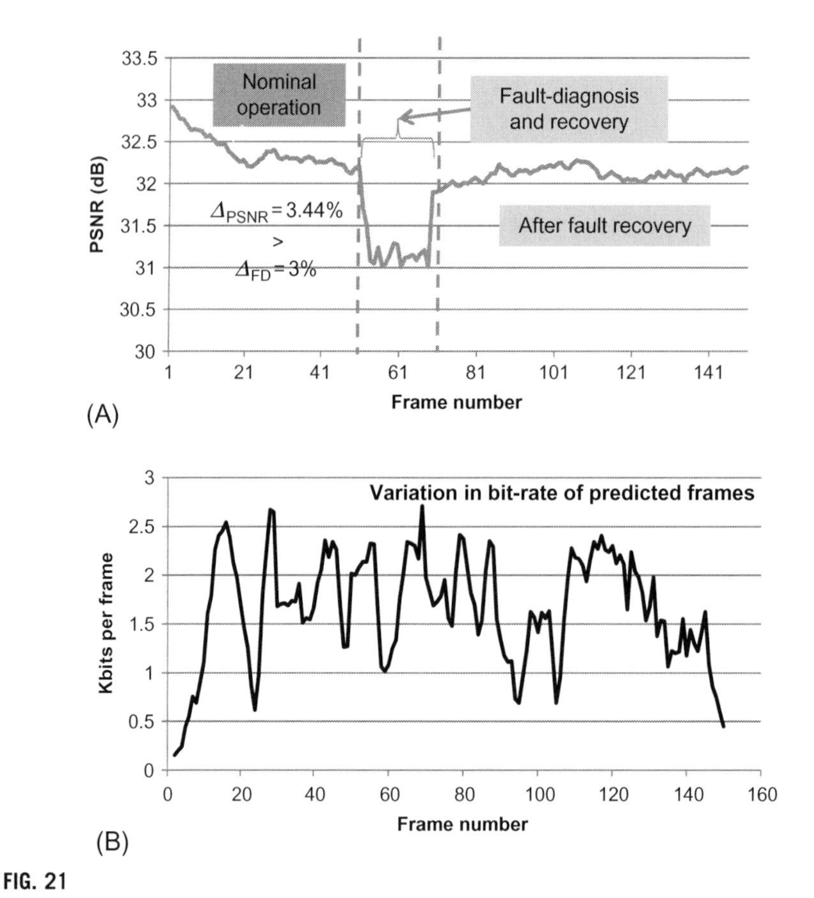

FIG. 21

PSNR and bit-rate of the encoder employing PURE. (A) PSNR of `silent.qcif` video sequence. (B) Bit-rate of `silent.qcif` video sequence.

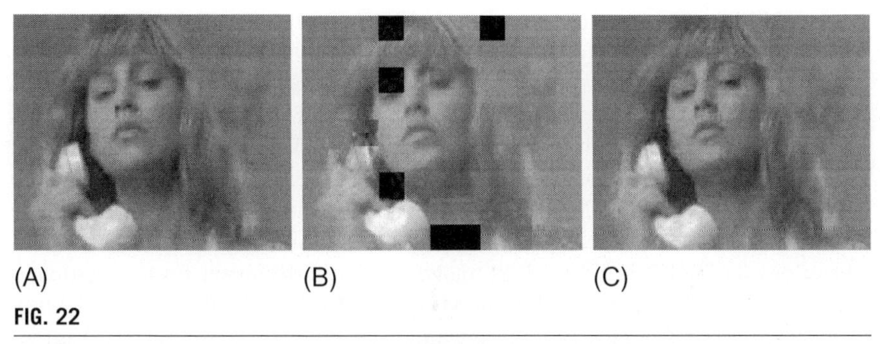

FIG. 22

Qualitative results on sequence from ASU video library [44]. (A) Reconstructed frame (without faults). (B) The effect of a faulty PE. (C) Recovered frame (after fault handling).

Fig. 22C after applying the FH scheme is visually much better than Fig. 22B where a faulty PE in the DCT block disrupts the image in the frame buffer.

The PURE approach maintains throughput by retaining viable modules in the datapath while divide-and-conquer does not take them into account. Moreover, latency of the diagnosis phase can be reduced by employing multiple dynamic RSs. For instance, 90% of diagnosable conditions can be identified in a single reconfiguration using $N_s = 3$ slacks while $N_s = 1$ slack identifies only 50% diagnosable conditions. A 90% CPDC is achieved in more than 3 testing arrangement instances when using a single slack. Compared to a static topology scheme where PE arrangement is fixed at design time, diagnosability can be increased from a single defective node to six defective nodes using as few as $r = 4$ reconfigurations and $N_s = 2$ slacks. In general, the diagnosability at the completion of PURE's algorithm is $N - 2$ after every possible pair combination is evaluated since at least one healthy pair is necessary to eliminate suspect status.

Time-dependent dielectric breakdown (TDDB) and EM are two significant causes of permanent faults over the device lifetime [45]. To quantify the *survivability* of the system employing the PURE fault-handing flow, the anomaly detection, diagnosis, and recovery times are considered here. The *availability* is generally defined in terms of mean-time-to-failure (MTTF) and mean-time-to-repair (MTTR). The impact of radiation and aging-induced degradation on reliability of FPGA-based circuits has been analyzed by authors in [45–47].

In this analysis, we use a TDDB failure rate of 10% LUT per year and an EM failure rate of 0.2% per year as demonstrated in [45] for MCNC benchmark circuits simulating their 12-year behavior. Considering a DCT core, 312 utilized LUTs in a PE spanning one partial reconfiguration region (PRR) exhibiting a 10.2% failure rate means 32 LUTs fail per year. If the failure rate is uniformly distributed over time, then a worst case scenario would correspond to a MTTF of 11 days between LUT failures.

The MTTR is the sum of times required for anomaly detection, diagnosis, and recovery. To assess the detection latency, faults are injected into the DCT module at frame number 50 of the `news.qcif` video sequence [44]. As a result, the PSNR drops at frame number 59. Thus, the detection time is 0.3 s for a 30 fps frame rate. Using one slack, the maximum cost is 196 frames or 6.5 s for a frame rate of 30 fps. Given the diagnosis data, the time to identify faulty nodes is negligible as the on-chip processor operating at 100 MHz clock rate can mark faulty nodes in very short time once the syndrome matrix is computed. Similarly, the time required for eight reconfigurations during FR is 1.6 s. Thus, total time to refurbishment for this particular example is 8.4 s. With these values of MTTF and MTTR, the PURE's availability is 99.999%. Moreover, significant throughput is maintained during FD phase as evident by the minimum values of PSNR in Fig. 20. Thus, the impact on signal quality even during the period of unavailability is minimal.

To analyze the power overhead of the proposed scheme, Xilinx XPower Estimator (XPE) 11.1 has been utilized. The static nonreconfigurable design power requirement is 1541 mW which consists of 591 mW Quiescent and 950 mW dynamic power. The individual components of the total dynamic power are listed in Table 12. In addition, the power estimates of *partial reconfiguration modules (PRMs)* are listed in Table 13. An estimate of power consumption for reconfiguration is based upon the resource count of HWICAP core reported in the Xilinx datasheet [48]. The measurement and estimation of power consumption in Virtex FPGAs is thoroughly demonstrated in [49]. The PEs employ a power-efficient design since the utilized DSP48 multipliers generally consume only 2.3 mW operating at 100 MHz at a typical toggle rate of 38% [50]. As illustrated by Table 13, considering the example of 8×8 DCT module, the fault-tolerant version consumes more power (i.e., 142 mW) than that required by a simple DCT module (i.e., 72 mW). However, the extra cost of power is considerably less than that required by TMR which would consume a factor of three times the power of a single module. Table 14 lists the *dynamic energy consumption, E* of the *partial reconfigurable (PR)* design throughout the isolation phases for various fault-rates described in terms of number of faulty modules, N_f. Thus, operating in uniplex mode with a single RS, the FaDReS power

Table 12 Dynamic Power Consumption of the Static Design

Resource	Utilization Count	Dynamic Power (W)
Clock	2	0.113
Logic	8812 LUTs [a]	0.328
I/Os	144	0.361
BRAM	1	0.002
DCM	1	0.082
PowerPC processor	1	0.064
Total		0.950

[a] *Includes dynamic power of 5656 FFs, 445 shift registers, and 71 select RAMs.*

Table 13 Dynamic Power Consumption of the Reconfigurable Design

Module	Resource Count	Dynamic Power (W)
PRM_1	82 LUTs, 44 FFs, 1 DSP48	0.021
PRM_2	117 LUTs, 78 FFs, 1 DSP48	0.023
PRM_3	103 LUTs, 64 FFs, 1 DSP48	0.021
PRM_4	120 LUTs, 81 FFs, 1 DSP48	0.023
PRM_5	94 LUTs, 56 FFs, 1 DSP48	0.021
PRM_6	122 LUTs, 83 FFs, 1 DSP48	0.023
PRM_7	107 LUTs, 68 FFs, 1 DSP48	0.023
PRM_8	113 LUTs, 74 FFs, 1 DSP48	0.023
HWICAP	2115 LUTs, 728 FFs	0.061
4 × 4 DCT		0.044
6 × 6 DCT		0.059
8 × 8 DCT		0.072
8 × 8 DCT with one RS		0.081

Table 14 Dynamic Energy Consumption of the PR Design During FI Phase

N_f	1	2	3	4	5	6	7
E (Joules)	1.52	2.89	4.37	5.82	7.08	8.04	8.64

consumption including the HWICAP totals only 65.7% of TMR during FH and roughly only 33% of TMR during fault-free operation.

3.2 ENERGY SAVINGS AND FAULT-HANDLING CAPABALITY OF FHME

An architecture proof-of-concept for a ME engine which adapts the throughput datapath based on the anticipation of computational demand in dynamic environments was demonstrated earlier. The input signal characteristics were exploited to anticipate the time-varying computational complexity as well as to instantiate DRs to realize fault resilience. The scheme employed APEs which either perform as AEs to maintain quality/throughput, serve as DRs to increase reliability levels, or hibernate passively as RS available to other tasks.

Experimental results from a hardware platform for FPGAs-based video encoding demonstrates power efficiency and fault tolerance of the ME engine. A significant reduction in power consumption is achieved ranging from 83% for low-motion-activity scenes to 12.5% for high motion activity video scenes. The scenes' motion activity are utilized to improve redundancy for the purpose of priority-based diagnosis of the computing modules. In addition, a graceful degradation strategy is developed to recover from hard errors by adapting the search range of candidate motion vectors. This adaptive hardware scheme is shown to automatically demote the faulty resources in FPGA devices based on online performance evaluation.

3.2.1 Energy saving in reconfigurable design

Given some tolerance of bitrate variation, the number of APEs that can be vacated depends upon a scene's motion activity as illustrated by results from videos in Table 15. Thus, a significant number of RSs can be created dynamically for low motion activity video scenes. On the other hand, disabling the AEs in high motion activity video scenes causes an increase in bitrate. One way to examine these interacting effects is to calculate power savings of the ME architecture as the PEs are configured with blank bitstreams. The number of inactive PEs influences the bitrate overhead which can be measured for a given video sequence. For example, Fig. 23 shows the trend of saving between 20 and 120 mW as N_{RS} is varied.

Fig. 24A illustrates the effectiveness of a greedy algorithm which adapts the number of active APEs in the datapath according to the runtime conditions. As it is evident from various frames of a video sequence, the number of predicted computational resources (i.e., N_{AE}) dynamically adapts according to the average

Table 15 Number of Vacated APEs While Bitrate Is Within a 3% Tolerance

Video Sequence	Motion Activity	Baseline ME PSNR (dB)	Baseline ME Bitrate	FHME PSNR (dB)	FHME Bitrate	N_{RS}
Soccer	High	32.32	8.43	32.31	8.62	1
Football	High	31.44	14.22	31.45	14.61	2
Ice	Medium	33.56	6.29	33.51	6.38	4
Suzie	Low	34.29	2.05	34.22	2.07	5

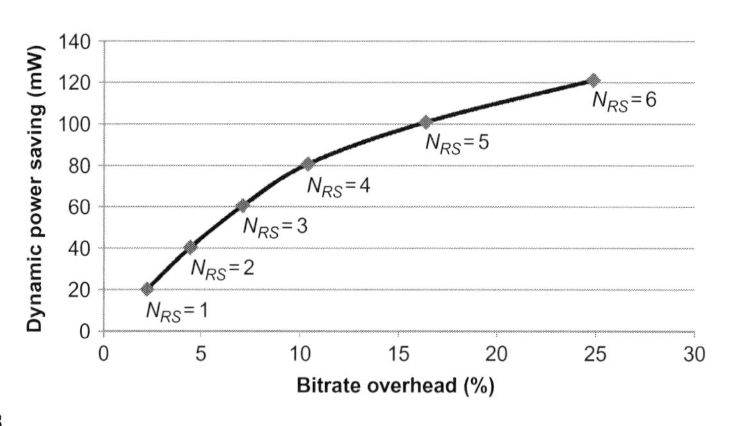

FIG. 23

Power saving at the cost of increased bitrate for Soccer video sequence.

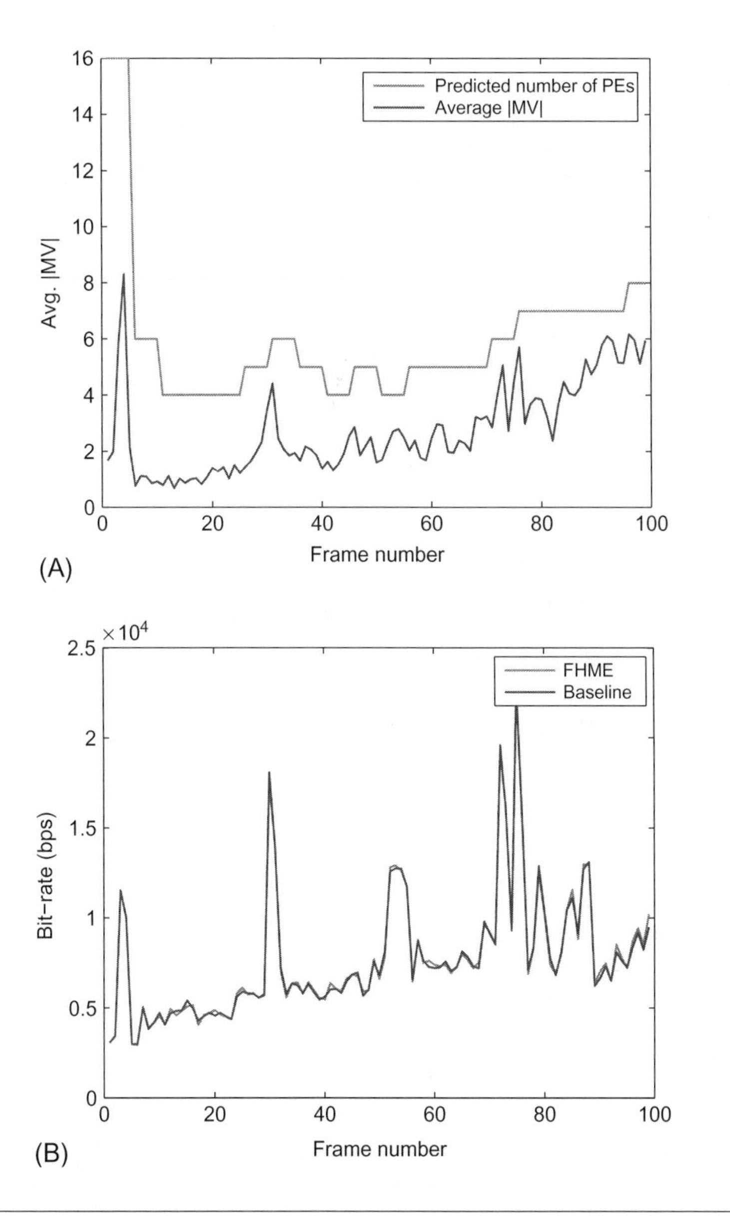

FIG. 24

Dynamic computational resource prediction for `crew` video sequence. (A) Number of active APEs predicted by a greedy resource allocation algorithm. (B) Bitrate variation.

MV value. A higher than the minimum required search range leads to power overhead while a lower search range increases the SAD values and bitrate increases consequently. Furthermore, as demonstrated via Fig. 24B, the bitrate increase is small implying that the computational resource prediction algorithm works well. Fig. 25 shows power savings and consequently the bitrate overhead of FHME for various video sequences. The average power consumption is reduced from 320 to 280 mW thus saving 12.5% for a high-motion-activity video sequence (`football`) while the power savings in case of low-motion-activity sequences is significantly

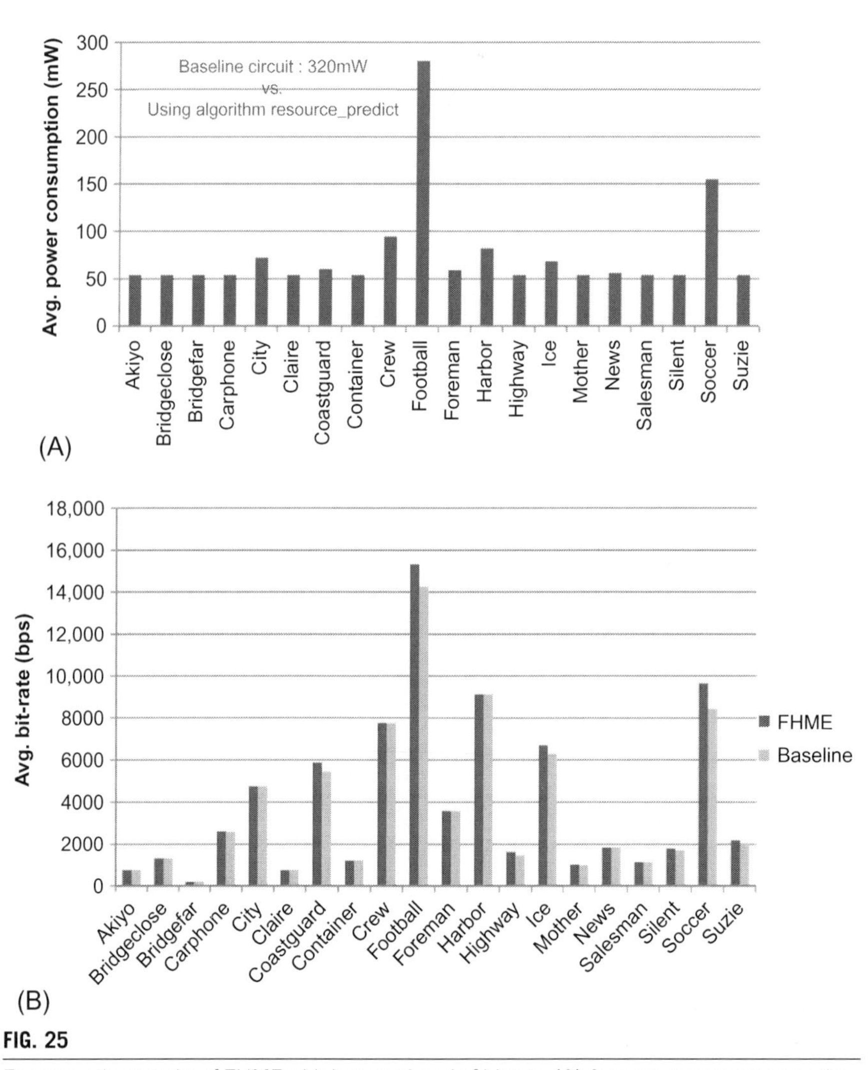

FIG. 25

Energy saving results of FHME with low overhead of bitrate. (A) Average power consumption. (B) Bitrate.

as high as 83% in the case of bridgefar. Also, the bitrate overhead due to prediction errors is negligibly small as it can be seen from Fig. 25B where columns for FHME and the baseline circuit are comparable.

Fig. 26 illustrates the qualitative results of a 704 × 576, 30 fps, city.4cif video sequence [51]. The FHME architecture provides considerable reduction in power consumption with only a slight degradation in image quality as evident from the PSNR measure as well.

3.2.2 Online recovery results of FHME core

We evaluate the effectiveness of the FHME approach by simulating fault injections in the post place-and-route simulation model of the ME circuit generated by the Xilinx software flow. For this purpose, Verilog HDL (.v) and Xilinx User Constraints File (.ucf) source files are modified to simulate stuck-at faults at look-up tables (LUTs) inputs, FF-inputs, and routing resource. A stuck-at "one" or "zero" fault manifests itself in the form of some output deviation from the actual output.

Fig. 27A illustrates an operational example from a video encoder containing a FHME core for the foreman video sequence. For this simulation, random faults are injected in AE_1 at frame 30 which is highlighted in Fig. 27A as Event ①. Without any fault information available, the Algorithm 4 adjusts the search range; however, bitrate does not get improved because of the faulty AE. Fault detection is triggered at the instant shown by Event ②. Then, the Algorithm 4 is triggered for diagnosis purposes in which an evaluation period of 10 frames is selected for a discrepancy check. The reconfiguration of a healthy module from the RS set to the AE set completes the recovery process at an instant labeled by Event ③. Fig. 27B shows another scenario in which two DRs are employed for FD purposes, and hence, fault-isolation latency is improved from a latency of 51 frames down to 40 frames. Table 16 illustrates the average reduction in compression efficiency in terms of bitrate increase to compare fault-free, baseline, and FHME modules. This shows a significant reduction in bitrate overhead from 117% to 24% depending upon the number of DRs available.

3.2.3 Comparisons and tradeoffs for TMR vs. DRFI

An operational comparison of two FH techniques to mitigate local permanent damage, namely TMR and Distance-ranked fault identification (DRFI) [52] is illustrated in Fig. 28. While the benefit of TMR is instantaneous fault-masking capability, its FH capacity is limited to failure of only a single module in the triplicated arrangement. On the other hand, DRFI's diagnosis and recovery phase involves multiple reconfigurations, and thus the FH latency is significantly longer than that of TMR. On the positive side, DRFI can sustain multiple failures in the design. After fault detection, the diagnosis and repair mechanism is triggered which selects configurations from the pool while healthy resources are promoted to higher preference which results in improved throughput quality. Our results show that the recovery latency is 1.2 s for the video encoder case study. If the mission cannot tolerate recovery delay of this interval then TMR is preferable for the first fault, however multiple faults impacting distinct TMR modules lead to an indeterminate result.

(A)

(B)

FIG. 26

Power and quality tradeoff results for `city.4cif` video sequence. (A) Image in the baseline's frame buffer, PSNR = 33.02 dB, Power = 320 mW. (B) Image in the FHME's frame buffer, PSNR = 32.09 dB, Power = 120 mW.

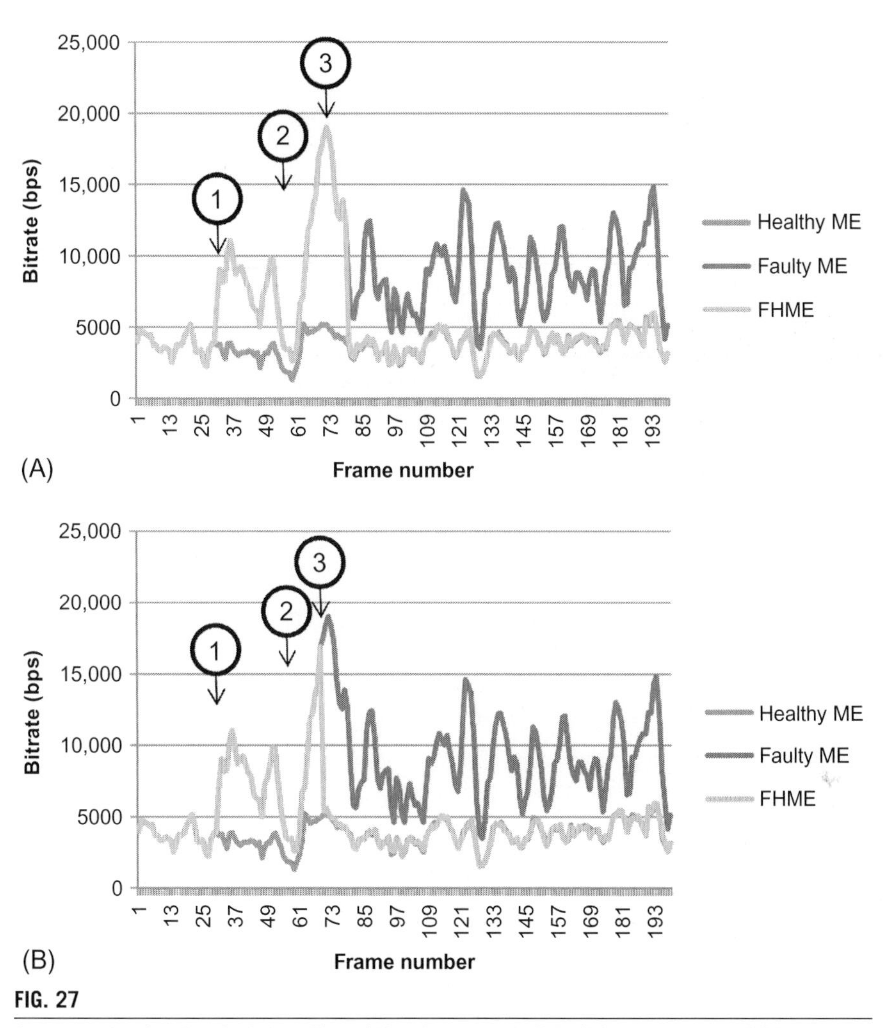

FIG. 27

An example of online fault handling. (A) using a single DR. (B) using a pair of DRs.

Fig. 28 illustrates the operational differences of TMR and DRFI techniques to mitigate permanent faults. Although DRFI incurs a recovery latency of approximately 1 s, it can sustain recovery after second, third, or even subsequent faults while TMR is only able to recover from a single fault per module over the entire mission. If the mission cannot tolerate a recovery delay of this duration then TMR is preferable for the first fault, yet DRFI may be preferred for handling multiple faults and for provision of graceful degradation. However, an important insight is that TMR and DRFI need not be mutually exclusive. If device area is not at a premium, then each configuration could be realized in a TMR arrangement for low-latency initial FR, along with DRFI being applied to a pool of TMR configurations at the next higher

Table 16 Bitrate of Encoded Bitstream for Foreman Video Sequence

Condition	Average Bitrate (bps)	Average Increase in Bitrate (%)
Fault-free ME	3755	0.0 (ref.)
Faulty baseline ME	8166	117.4
FHME with a single DR	5246	39.7
FHME with a pair of DRs	4678	24.6

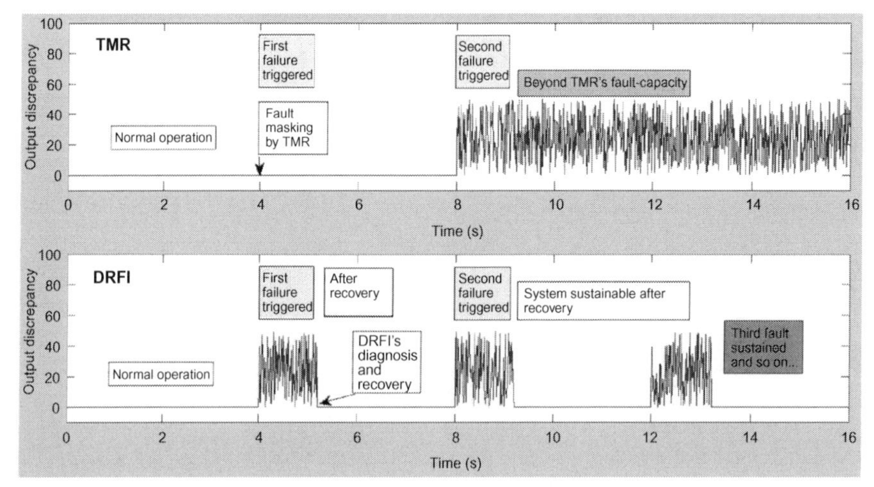

FIG. 28

A comparison of DRFI [52] with TMR in terms of system's capability to sustain multiple failures.

layer. Thus, DRFI need not necessarily be exclusive of TMR, but orthogonal to it if sustainability and graceful degradation are also sought.

In order to assess the resource overhead of TMR and DRFI over a uniplex scheme, we implemented a DCT core on Virtex-4 FPGAs in two different arrangements. The TMR arrangement involves a triplicated design while the DRFI arrangement involves a duplicated design along with the configuration memory required to store multiple bitstreams in order to mitigate hardware faults. While a simplex arrangement uses 33% of the area of TMR, the DRFI technique uses approximately 67% of the area of TMR when considering the logic resource count. In addition, DRFI contains provision of reconfiguration thereby utilizes a reconfiguration controller and peripherals present on the chip.

For DRFI, configuration ranking is invoked to leverage the priority inherent in the computation to mitigate performance-impacting phenomena such as extrinsic fault sources, aging-induced degradations, or manufacturing process variations.

Such reconfigurations are only initiated periodically, for example when adverse events such as discrepancies occur. Fault recovery is performed by fetching alternative partial Configuration Bitstreams which are stored in a CompactFlash external memory device. Alternative designs can be implemented using multiple methods described in this chapter. A configuration port, such as the ICAP on Xilinx FPGAs, provides an interface for the reconfiguration controller to reconfigure the alternative bitstreams. Together these components support the reconfiguration flows needed to realize a runtime adaptive approach to FH architectures.

3.3 RELIABILITY AND ENERGY TRADEOFFS AT NTV

NTV operation offers reduced energy consumption with an acceptable increase in delay for targeted applications. However, increased *soft error rate* (SER) has been identified as a significant concern at NTV. In this section, tradeoffs are evaluated regarding use of spatial redundancy to mask increased SER at NTV.

While NTV offers an attractive approach to balance energy consumption versus delay for power-constrained applications such as high-performance computing [53], it may also introduce reliability implications. In particular, radiation-induced *single event upsets* (SEUs) which cause soft errors are expected to increase at this operating region [54, 55]. These errors may manifest as a random bit flip in a memory element or a transient charge within a logic path which is ultimately latched by a flip-flop. While soft errors in memory elements are feasible to detect and correct using error correcting codes [53], their resolution in logic paths typically involves the use of spatial or temporal redundancy which allow area versus performance tradeoffs [56]. In particular, the cost of redundancy at NTV, expressing the delay cost of NMR systems at NTV for common values of N within a given energy budget at nominal supply voltage is investigated in this section.

3.3.1 Soft errors in logic paths

The contribution of soft errors in logic paths as opposed to memory elements is becoming significant as the supply voltage is scaled down to the threshold region. It was predicted in [57] that the SER of logic circuits per die will become comparable to the SER for unprotected memory elements, which was later verified through experimental data for a recent microprocessor [55]. Operation at NTV is expected to exaggerate these trends. For instance, [55] states that SER increases by approximately 30% per each 0.1 V decade as V_{DD} is decreased from 1.25 to 0.5 V. In the NTV region, it is shown through both simulation and experiment at the 40 and 28 nm nodes, that SER doubles when V_{DD} is decreased from 0.7 to 0.5 V. Primarily, the critical charge needed to cause a failure decreases as V_{DD} is scaled and SER has an exponential dependence on critical charge [55]. Such trends are consistent with decreasing feature sizes due to technology scaling [58].

The SER in logic paths can be reduced by schemes such as gate sizing [59] or dual-domain supply voltage assignments [60] to harden components which are more susceptible to soft errors. These techniques tradeoff increased area and/or power to reduce SER of the logic circuit, but may not be able to provide comprehensive coverage. For instance, SER reduction of only 33.45% is demonstrated in [60] using multiple voltage assignments. One option identified for masking soft errors is spatial redundancy, and in particular the readily-accepted use of TMR, as being effective for mitigating soft errors. TMR is considered to be appropriate for applications which demand immunity to soft errors and also are able to accommodate its inherent overheads.

Spatial redundancy is often employed in mission-critical applications to ensure system operation even in unforeseen circumstances, such as autonomous vehicles, satellites, and deep space systems [61–63]. It is also been employed in commercial systems such as high-performance computing applications [64] where significant increase in compute node availability is sought. Here, use of compute-node level redundancy at the processor, memory module, and network interface can improve reliability by a factor of 100- to 100,000-fold.

3.3.2 NMR systems at near-threshold voltage

Nanoscale devices are susceptible to process variations created by precision limitations of the manufacturing process. Phenomena such as *random dopant fluctuations* (RDF) and *line-edge roughness* (LER) are major causes of such variations in CMOS devices [65]. The increased occurrence of PV results in a distribution of threshold voltage V_{th}. As the V_{th} increases, the increase in switching time affects the delay performance of the circuit. Such variability is observed to become magnified by continued scaling of process technology node [65]. For example, the effect of RDF is magnified as number of dopant atoms is fewer in scaled devices such that the addition or deletion of just a few dopant atoms significantly alters transistor properties. In addition, a large impact in circuit performance occurs as the transistor on-current is highly variable near the threshold region [54]. Recent approaches for dealing with increased PV at NTV in multicore devices include leveraging the application's inherent tolerance for faults through performance-aware task-to-core assignment based on problem size [66].

While operation at NTV can be seen to increase PV by approximately 5-fold as quantified in [53] for simplex arrangements, the effect in *N*MR systems has not been previously investigated. In the case of an *N*MR arrangement, it can be expected that the worst-case delay will exceed that of any single module. Generally, if the worst-case delay of an instance *i* of *N*MR system is τ_i, then the overall delay of *N*MR system τ_{NMR} is given by:

$$\tau_{NMR} = \max_{1 \leq i \leq N} (\tau_i) + \delta \tag{2}$$

where δ represents the delay of the voting logic, which contributes directly to the critical delay. Furthermore, the chance of having an instance with higher than average delay increases with *N*, which has been validated through experimental results. Overall, these results are in agreement with distributions of 128-wide SIMD architectures demonstrated in [67], whereby the speed of the overall architecture is also determined by the slowest SIMD lane. Herein, we focus on the performance of *N*MR systems as compared to simplex systems, with $3 \leq N \leq 5$.

Intradie variations for both 22 and 45 nm technology nodes are simulated using the Monte-Carlo method in HSPICE. A viable alternative approach for simulating PV at NTV is also proposed in [68], which captures both the systematic effects due to lithographic irregularities and the localized variations due to RDF. The random effects are modeled through the variation in V_{th} caused due to RDF and LER effects. The standard deviation σV_{th} values are adopted from [65] which range from 25.9 to 59.9 mV for 45 nm process and 22 nm process, respectively.

3.3.3 Energy cost of mitigating variability in NMR arrangements

A fundamental approach to alleviate observed delay variations in the near-threshold region is to add one-time *timing guard bands* [67]. Guard bands can be realized by operating at reduced frequency via a longer clock period, or by operating at a slightly elevated voltage to compensate for the increased delay variations. To achieve an expected yield of approximately 99% for *N*MR systems while achieving the same worst-case performance at a fixed NTV of simplex system; the voltage for *N*MR system is increased such that the respective delay distributions have the following statistical characteristics for $N \geq 3$:

$$(\mu_{NMR} + 3 * \sigma_{NMR}) \leq (\mu_{Simplex} + 3 * \sigma_{Simplex}) \tag{3}$$

where μ_{NMR}, $\mu_{Simplex}$ represent the mean delays for *N*MR and simplex systems, respectively, and σ_{NMR}, $\sigma_{Simplex}$ represent the respective standard deviations. The three sigma rule has the property that nearly all (99.7%) of the instances have the delay less than (μ_{NMR} + 3 * σ_{NMR}). Thus, this results in a high expectation for an *N*MR system to have same worst-case delay as simplex system, i.e., the same throughput performance.

3.3.4 Cost of increased reliability at NTV

Operation of *N*MR systems in the near-threshold region allows for consideration of interesting tradeoffs. For example, increasing *N* from 1 to 3 can be evaluated as a means to increase reliability within the same energy budget as a simplex system operating at nominal voltage. This is valid provided that the increase in delay, and thus corresponding drop in performance, is acceptable. Note that this pursuit of increased reliability is predicated upon the assumption that the source of variability in the near-threshold region is due to variation in V_{th} for which these results are restricted. Further study needs to be performed to determine the reliability levels provided by *N*MR systems in this operating region due to other noise sources such as variation in V_{DD} [69], inductive noise and temperature [70].

Based on these assumptions, we point out these tradeoffs in Fig. 29. For instance, it shows the feasibility of TMR operation at approximately 0.69*V* (operating point B on plot) on average given an identical energy budget of a simplex system operating at nominal voltage of 1.1*V* (operating point A on plot) while incurring a delay difference of 2.58*X*. Similarly, it is possible to achieve 5MR with approximately 0.545*V*

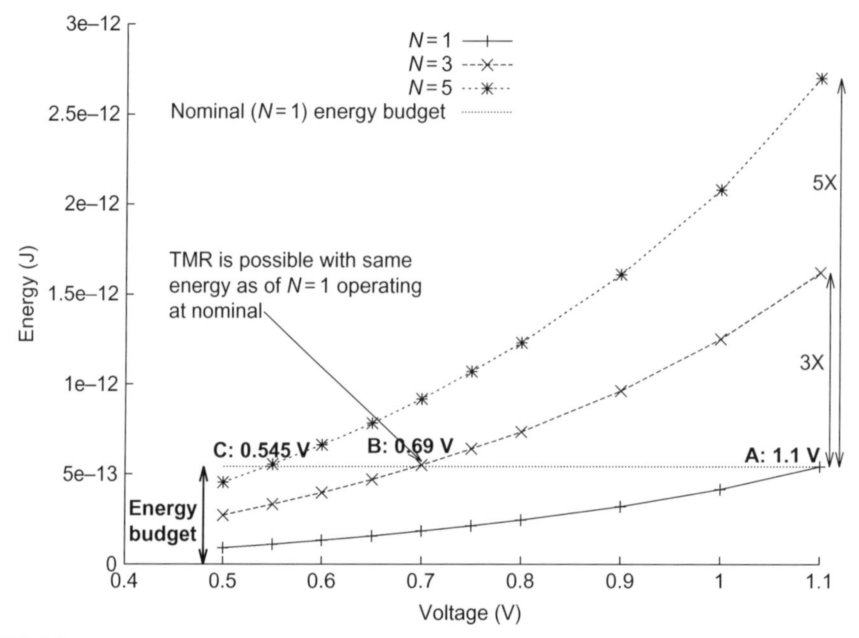

FIG. 29

Mean energy (each point is obtained by averaging at least 1000 samples) of *N*MR systems at 45 nm technology node. Operation at NTV corresponds to tradeoffs between performance and energy consumption required for soft error masking.

(operating point C on plot) operation on average while incurring a performance impact of 7.15*X*. This represents a greater-than-linear increase in delay as a function of *N*, as compared to TMR operation. Thus for mission-critical applications, this offers insights into tuning the degree of redundancy facilitated by a near-threshold computing paradigm.

To consider the feasibility of reducing energy consumption while simultaneously providing soft error masking, we again observe Fig. 29. The energy requirement of a simplex arrangement at 1.1 V is about 0.541 pJ, while the TMR curve at 550 mV is only about 0.330 pJ. Thus, a TMR arrangement at 550 mV results in an energy savings of 38.4% as compared to simplex system at 1.1 V. This means compared to a simplex arrangement at nominal voltage, selecting a supply voltage of 550 mV allows for provision of TMR for soft error masking in the presence of technology scaling while still reducing the energy requirement significantly.

3.4 SUMMARY

The presented approaches provide graceful degradation over the FD and FH periods, which spans a few frames for video encoding applications. During this time a partial throughput is available as an intrinsic provision of degraded mode. Thus, priority-aware FR is successfully demonstrated for the video encoding application. This work can be easily extended to other applications which can provide a health metric for uniplex assessment for faulty behavior. Then, the reconfigurability of FPGAs can be utilized for hard-error mitigation and thus the survivability of lengthy missions operating in harsh environments can be targeted. In these environments, energy consumption is a strict constraint, to which effect, operation in near-threshold region allows for consideration of energy and delay tradeoffs by varying the degree of redundancy. Thus, redundancy can be seen as a degree of freedom enabled by decreasing supply voltage. Furthermore, augmenting the use of temporal redundancy in the form of timing-speculation circuits to reduce variation effects and spatial redundancy would also be worthwhile to investigate.

REFERENCES

[1] J. Sloan, D. Kesler, R. Kumar, A. Rahimi, A numerical optimization-based methodology for application robustification: transforming applications for error tolerance, in: Proceedings of the IEEE/IFIP International Conference on Dependable Systems and Networks (DSN), 2010, pp. 161–170.

[2] M. Gao, H.M.S. Chang, P. Lisherness, K.T.T. Cheng, Time-multiplexed online checking, IEEE Trans. Comput. 60 (9) (2011) 1300–1312.

[3] E. Stott, P. Sedcole, P. Cheung, Fault tolerant methods for reliability in FPGAs, in: Proceedings of the International Conference on Field Programmable Logic and Applications (FPL), 2008, pp. 415–420.

[4] M. Garvie, A. Thompson, Scrubbing away transients and jiggling around the permanent: long survival of FPGA systems through evolutionary self-repair, in: Proceedings of the IEEE International On-Line Testing Symposium (IOLTS), 2004, pp. 155–160.

[5] D. Keymeulen, R.S. Zebulum, Y. Jin, A. Stoica, Fault-tolerant evolvable hardware using field-programmable transistor arrays, IEEE Trans. Reliab. 49 (3) (2000) 305–316.

[6] J.C. Gallagher, The once and future analog alternative: evolvable hardware and analog computation, in: Proceedings of the NASA/DoD Conference on Evolvable Hardware, 2003, pp. 43–49.

[7] W. Barker, D.M. Halliday, Y. Thoma, E. Sanchez, G. Tempesti, A.M. Tyrrell, Fault tolerance using dynamic reconfiguration on the POEtic tissue, IEEE Trans. Evol. Comput. 11 (5) (2007) 666–684.

[8] R.F. DeMara, K. Zhang, C.A. Sharma, Autonomic fault-handling and refurbishment using throughput-driven assessment, Appl. Soft Comput. 11 (2011) 1588–1599.

[9] J.M. Emmert, C.E. Stroud, M. Abramovici, Online fault tolerance for FPGA logic blocks, IEEE Trans. Very Large Scale Integr. VLSI Syst. 15 (2) (2007) 216–226.

[10] S. Dutt, V. Verma, V. Suthar, Built-in-self-test of FPGAs with provable diagnosabilities and high diagnostic coverage with application to online testing, IEEE Trans. Comput. Aided Des. Integr. Circuits Syst. 27 (2) (2008) 309–326.

[11] M. Abramovici, C.E. Stroud, J.M. Emmert, Online BIST and BIST-based diagnosis of FPGA logic blocks, IEEE Trans. Very Large Scale Integr. VLSI Syst. 12 (12) (2004) 1284–1294.

[12] M.G. Gericota, G.R. Alves, M.L. Silva, J.M. Ferreira, Reliability and availability in reconfigurable computing: a basis for a common solution, IEEE Trans. Very Large Scale Integr. VLSI Syst. 16 (11) (2008) 1545–1558.

[13] N. Imran, J. Lee, R.F. DeMara, Fault demotion using reconfigurable slack (FaDReS), IEEE Trans. Very Large Scale Integr. VLSI Syst. 21 (7) (2013) 1364–1368.
[14] I. Naveed, R.F. DeMara, J. Lee, J. Huang, Self-adapting resource escalation for resilient signal processing architectures, J. Signal Process. Syst. (2013) 1–24.
[15] N. Imran, R. Ashraf, R.F. DeMara, Evaluating quality and resilience of an embedded video encoder against a continuum of energy consumption, in: Invited submission to 2014 Workshop on Suite of Embedded Applications and Kernels (SEAK-2014), San Francisco, CA, June 1, 2014.
[16] R. Jain, The Art of Computer Systems Performance Analysis, John Wiley & Sons, New York, 2008.
[17] K.V. Palem, L.N. Chakrapani, Z.M. Kedem, A. Lingamneni, K.K. Muntimadugu, Sustaining Moore's law in embedded computing through probabilistic and approximate design: retrospects and prospects, in: Proceedings of the International Conference on Compilers, Architecture, and Synthesis for Embedded Systems, CASES, ACM, Grenoble, France, 2009, pp. 1–10.
[18] V.K. Chippa, D. Mohapatra, A. Raghunathan, K. Roy, S.T. Chakradhar, Scalable effort hardware design: exploiting algorithmic resilience for energy efficiency, in: Proceedings of the 47th ACM/IEEE Design Automation Conference (DAC), 2010, pp. 555–560.
[19] D. Mohapatra, G. Karakonstantis, K. Roy, Significance driven computation: a voltage-scalable, variation-aware, quality-tuning motion estimator, in: Proceedings of the 14th ACM/IEEE International Symposium on Low Power Electronics and Design (ISLPED), ACM, San Fancisco, CA, 2009, pp. 195–200.
[20] P.S. Ostler, M.P. Caffrey, D.S. Gibelyou, P.S. Graham, K.S. Morgan, B.H. Pratt, H. M. Quinn, M.J. Wirthlin, SRAM FPGA Reliability Analysis for Harsh Radiation Environments, IEEE Trans. Nucl. Sci. 56 (6) (2009) 3519–3526.
[21] N. Imran, R. Ashraf, R.F. DeMara, On-demand fault scrubbing using adaptive modular redundancy, in: Proceedings of the International Conference on Engineering of Reconfigurable Systems and Algorithms (ERSA 2013), Las Vegas, NV, July 22–25, 2013.
[22] J. Heiner, N. Collins, M. Wirthlin, Fault tolerant ICAP controller for high-reliable internal scrubbing, in: Proceedings of the 2008 IEEE Aerospace Conference, 2008, pp. 1–10.
[23] E. Mizan, T. Amimeur, M.F. Jacome, Self-imposed temporal redundancy: an efficient technique to enhance the reliability of pipelined functional units, in: Proceedings of the 19th International Symposium on Computer Architecture and High Performance Computing, 2007, pp. 45–53.
[24] K. Paulsson, M. Hubner, J. Becker, Strategies to on-line failure recovery in self-adaptive systems based on dynamic and partial reconfiguration, in: Proceedings of the First NASA/ESA Conference on Adaptive Hardware and Systems (AHS), 2006.
[25] S. Mitra, E.J. McCluskey, Which concurrent error detection scheme to choose? in: Proceedings of the International Test Conference, 2000, pp. 985–994.
[26] C.A.B. Smith, The counterfeit coin problem, Math. Gaz. 31 (293) (1947) 31–39.
[27] R. Dorfman, The detection of defective members of large populations, Ann. Math. Stat. 14 (4) (1943) 436–440.
[28] E. Litvak, X.M. Tu, M. Pagano, Screening for the presence of a disease by pooling sera samples, Journal of the American Statistical Association 89 (426) (1994) 424–434.
[29] S. Mitra, W.-J. Huang, N.R. Saxena, S.Y. Yu, E.J. McCluskey, Reconfigurable architecture for autonomous self-repair, IEEE Des. Test Comput. 21 (3) (2004) 228–240.

[30] R.C. Gonzalez, R.E. Woods, Digital image processing, third ed., Prentice Hall, Upper Saddle River, NJ, 2008.

[31] G.V. Varatkar, N.R. Shanbhag, Error-resilient motion estimation architecture, IEEE Trans. Very Large Scale Integr. VLSI Syst. 16 (10) (2008) 1399–1412.

[32] T. Wiegand, H. Schwarz, A. Joch, F. Kossentini, G.J. Sullivan, Rate-constrained coder control and comparison of video coding standards, IEEE Trans. Circuits Syst. Video Technol. 13 (7) (2003) 688–703.

[33] M. Bucciero, J.P. Walters, M. French, Software fault tolerance methodology and testing for the embedded PowerPC, in: Proceedings of the IEEE Aerospace Conference, Big Sky, MT, 2011, pp. 1–9.

[34] N. Imran, R.A. Ashraf, J. Lee, R.F. DeMara, Activity-based resource allocation for motion estimation engines, J. Circ. Syst. Comput. 24 (1) (2015) [Online] http://dx.doi.org/10.1142/S0218126615500048.

[35] S. Vigander, Evolutionary fault repair of electronics in space applications, Dissertation, University of Sussex, UK, (2001).

[36] M. Garvie, A. Thompson, Scrubbing away transients and jiggling around the permanent: long survival of FPGA systems through evolutionary self-repair, in: Proceedings of the 10th IEEE International On-Line Testing Symposium (IOLTS 2004), 2004, pp. 155–160.

[37] R.F. DeMara, K. Zhang, Autonomous FPGA fault handling through competitive runtime reconfiguration, in: Proceedings of the NASA/DoD Conference on Evolvable Hardware, 2005, pp. 109–116.

[38] R.N. Al-Haddad, R.S. Oreifej, R. Ashraf, R.F. DeMara, Sustainable modular adaptive redundancy technique emphasizing partial reconfiguration for reduced power consumption, Int. J. Reconfig. Comp. 2011 (2011) [Online] http://dx.doi.org/10.1155/2011/430808.

[39] R. Salvador, A. Otero, J. Mora, E. de la Torre, L. Sekanina, T. Riesgo, Fault tolerance analysis and self-healing strategy of autonomous, evolvable hardware systems, in: Proceedings of the 2011 International Conference on Reconfigurable Computing and FPGAs (ReConFig), 2011, pp. 164–169.

[40] F. Cancare, S. Bhandari, D.B. Bartolini, M. Carminati, M.D. Santambrogio, A bird's eye view of FPGA-based Evolvable Hardware, in: Proceedings of the 2011 NASA/ESA Conference on Adaptive Hardware and Systems (AHS), 2011, pp. 169–175.

[41] S. Yang, Logic synthesis and optimization benchmarks version 3, Tech. Rep., Microelectronics Center of North Carolina, 1991.

[42] B.L. Synthesis, V. Group, ABC: a system for sequential synthesis and verification, release 10216, Tech. Rep., University of California Berkeley, 2011. http://www.eecs.berkeley.edu/alanmi/abc.

[43] R.S. Oreifej, R.N. Al-Haddad, H. Tan, R.F. DeMara, Layered approach to intrinsic evolvable hardware using direct bitstream manipulation of virtex II pro devices, in: Proceedings of the International Conference on Field Programmable Logic and Applications (FPL 2007), 2007, pp. 299–304.

[44] Trace, Video trace library: YUV 4:2:0 video sequences, retrieved on January 20, 2012 [Online], http://trace.eas.asu.edu/yuv/.

[45] S. Srinivasan, R. Krishnan, P. Mangalagiri, Y. Xie, V. Narayanan, M.J. Irwin, K. Sarpatwari, Toward increasing FPGA lifetime, IEEE Trans. Dependable Secure Comput. 5 (2) (2008) 115–127.

[46] C. Poivey, M. Berg, S. Stansberry, M. Friendlich, H. Kim, D. Petrick, K. LaBel, Heavy ion SEE test of Virtex-4 FPGA XC4VFX60 from Xilinx, test report, June 2007.

[47] C. Bolchini, C. Sandionigi, Fault classification for SRAM-Based FPGAs in the space environment for fault mitigation, IEEE Embed. Syst. Lett. 2 (4) (2010) 107–110.
[48] Xilinx, LogiCORE IP XPS HWICAP datasheet, retrieved on January 11, 2012 [Online], http://www.xilinx.com/.../xps_hwicap.pdf.
[49] J. Becker, M. Huebner, M. Ullmann, Power estimation and power measurement of Xilinx Virtex FPGAs: trade-offs and limitations, in: Proceedings of the 16th Symposium on Integrated Circuits and Systems Design (SBCCI 2003), 2003, pp. 283–288.
[50] Xilinx, XtremeDSP 48 Slice, retrieved on January 11, 2012 [Online], http://www.xilinx.com/technology/dsp/xtremedsp.htm.
[51] TNT, Video test sequences, Institute for Information Processing, Leibniz University of Hannover, retrieved on May 26, 2013 [Online], ftp://ftp.tnt.uni-hannover.de/pub/svc/testsequences/.
[52] N. Imran, R.F. DeMara, Distance-ranked fault identification of reconfigurable hardware bitstreams via functional input, Int. J. Reconfig. Comput. (2014), Article ID 279673, 21 pages.
[53] R.G. Dreslinski, M. Wieckowski, D. Blaauw, D. Sylvester, T. Mudge, Near-threshold computing: reclaiming Moore's law through energy efficient integrated circuits, Proc. IEEE 98 (2) (2010) 253–266.
[54] H. Kaul, M. Anders, S. Hsu, A. Agarwal, R. Krishnamurthy, S. Borkar, Near-threshold voltage (NTV) design: opportunities and challenges, in: Proceedings of the 49th ACM/EDAC/IEEE Annual Design Automation Conference (DAC), San Francisco, CA, 2012, pp. 1153–1158.
[55] A. Dixit, A. Wood, The impact of new technology on soft error rates, in: Proceedings of the IEEE International Reliability Physics Symposium (IRPS), 2011, pp. 5B.4.1–5B.4.7.
[56] L.G. Szafaryn, B.H. Meyer, K. Skadron, Evaluating overheads of multibit soft-error protection in the processor core, IEEE Micro 33 (4) (2013) 56–65.
[57] P. Shivakumar, M. Kistler, S.W. Keckler, D. Burger, L. Alvisi, Modeling the effect of technology trends on the soft error rate of combinational logic, in: Proceedings of the International Conference on Dependable Systems and Networks (DSN), 2002, pp. 389–398.
[58] M. Snir, R.W. Wisniewski, J.A. Abraham, S.V. Adve, S. Bagchi, P. Balaji, J. Belak, P. Bose, F. Cappello, B. Carlson, A.A. Chien, P. Coteus, N.A. DeBardeleben, P. C. Diniz, C. Engelmann, M. Erez, S. Fazzari, A. Geist, R. Gupta, F. Johnson, S. Krishnamoorthy, S. Leyffer, D. Liberty, S. Mitra, T. Munson, R. Schreiber, J. Stearley, E.V. Hensbergen, Addressing failures in exascale computing, Int. J. High Perform. Comput. Appl. 28 (2) (2014) 129–173.
[59] Q. Zhou, K. Mohanram, Gate sizing to radiation harden combinational logic, IEEE Trans. Comput. Aided Des. Integr. Circuits Syst. 25 (1) (2006) 155–166.
[60] K.-C. Wu, D. Marculescu, Power-aware soft error hardening via selective voltage scaling, in: Proceedings of the IEEE International Conference on Computer Design (ICCD), 2008, pp. 301–306.
[61] J.A.P. Celis, S. De La Rosa Nieves, C.R. Fuentes, S.D.S. Gutierrez, A. Saenz-Otero, Methodology for designing highly reliable Fault Tolerance Space Systems based on COTS devices, in: Proceedings of the IEEE International Systems Conference (SysCon), 2013, pp. 591–594.

[62] M. Pignol, How to cope with SEU/SET at system level? in: Proceedings of the 11th IEEE International On-Line Testing Symposium (IOLTS), 2005, pp. 315–318.
[63] R. Al-Haddad, R. Oreifej, R.A. Ashraf, R.F. DeMara, Sustainable modular adaptive redundancy technique emphasizing partial reconfiguration for reduced power consumption, Int. J. Reconfig. Comput. 2011 (430808) (2011) 1–25.
[64] C. Engelmann, H. Ong, S.L. Scott, The case for modular redundancy in large-scale high performance computing systems, in: Proceedings of the IASTED International Conference on Parallel and Distributed Computing and Networks (PDCN), vol. 641, 2009, pp. 189–194.
[65] Y. Ye, T. Liu, M. Chen, S. Nassif, Y. Cao, Statistical modeling and simulation of threshold variation under random dopant fluctuations and line-edge roughness, IEEE Trans. Very Large Scale Integr. VLSI Syst. 19 (6) (2011) 987–996.
[66] U.R. Karpuzcu, I. Akturk, N.S. Kim, Accordion: toward soft near-threshold computing, in: Proceedings of the 20th IEEE International Symposium on High Performance Computer Architecture (HPCA), 2014.
[67] S. Seo, R.G. Dreslinski, M. Woh, Y. Park, C. Charkrabari, S. Mahlke, D. Blaauw, T. Mudge, Process variation in near-threshold wide SIMD architectures, in: Proceedings of the 49th ACM/EDAC/IEEE Design Automation Conference (DAC), 2012, pp. 980–987.
[68] U.R. Karpuzcu, K.B. Kolluru, N.S. Kim, J. Torrellas, VARIUS-NTV: a microarchitectural model to capture the increased sensitivity of manycores to process variations at near-threshold voltages, in: Proceedings of the 42nd Annual IEEE/IFIP International Conference on Dependable Systems and Networks (DSN), IEEE Computer Society, Washington, DC, 2012, pp. 1–11.
[69] T.N. Miller, R. Thomas, X. Pan, R. Teodorescu, VRSync: characterizing and eliminating synchronization-induced voltage emergencies in many-core processors, in: Proceedings of the 39th Annual International Symposium on Computer Architecture (ISCA), 2012, pp. 249–260.
[70] Y. Kim, L.K. John, I. Paul, S. Manne, S. Michael, Performance boosting under reliability and power constraints, in: Proceedings of the International Conference on Computer-Aided Design (ICCAD), IEEE Press, San Jose, CA, 2013, pp. 334–341.
[71] J. Huang, M.B. Tahoori, F. Lombardi, Routability and fault tolerance of FPGA interconnect architectures, in: Proceedings of IEEE International Test Conference (ITC 2004), October, 2004, pp. 479–488.

CHAPTER

Resilience for extreme scale computing

5

R. Gioiosa

Pacific Northwest National Laboratory, Richland, WA, United States

1 INTRODUCTION

Supercomputers are large systems that are specifically designed to solve complex scientific and industrial challenges. Such applications span a wide range of computational intensive tasks, including quantum mechanics, weather forecasting, climate research, oil and gas exploration, molecular dynamics, and physical simulations, and require an amount of computing power and resources that go beyond what is available in general-purpose computer servers or workstations. Until the late 1990s, supercomputers were based on high-end, special-purpose components that underwent extensive testing. This approach was very expensive and limited the use of supercomputers in a few research centers around the world. To reduce the acquisition and operational costs, researchers started to build supercomputers out of "common-off-the-shelf" (COTS) components, such as those used in general-purpose desktop and laptop. This technology shift greatly reduced the cost of each supercomputer and allowed many research centers and universities to acquire mid- and large-scale systems.

Current supercomputers consist of a multitude of compute nodes interconnected through a high-speed network. Each compute node features one or more multicore/multithreaded processor chips, several memory modules, one/two network adapters, and, possibly, some local storage disks. More recently, accelerators, in the form of graphical vector units or field-programmable gate arrays (FPGA), have also made their way to mainstream supercomputing.

On the other hand, however, COTS components brought a new set of problems that makes HPC applications run in a harsh environment. First, COTS components are intrinsically more vulnerable than special-purpose components and, because of cost reasons, follow different production process and verification/validation paths than military or industry mission-critical embedded systems or server mainframes. This means that the individual probability that every single component may fail is higher than in other domains. Second, COTS components are not specifically design to solve scientific applications and their single-processor performance was lower

Rugged Embedded Systems. http://dx.doi.org/10.1016/B978-0-12-802459-1.00005-1

than the vector processor, thus a larger number of processors are usually required to achieve the desired performance. Combining together an extremely large number of system components greatly increase the combined probability that at least one breaks, experience a soft error, or stop functioning completely or partially. Third, HPC workloads have been shown to be extremely heterogeneous and can possibly stress different parts of the supercomputers, such as memory, processors, or network. This heterogeneous behavior, in turn, increases the thermal and mechanical stress, effectively reducing the life span of each component and the entire system. The location and the facility where the supercomputer is installed play an important role in the resilience of the system. The cooling system of the machine room must be appropriate to reduce the temperature of densely-packet compute nodes with limited space for air flow. Modern supercomputers take cooling a step forward and provide liquid cooling solution embedded in the compute node racks. Moreover, recent studies confirm that supercomputer installed at higher altitudes, such as Cielo at Los Alamos National Laboratory (LANL), are more exposed to radiation and show higher soft error rates [1]. Finally, the mere size of current supercomputers makes it impossible to employ resilience solutions commonly used in other domains, both because of cost and practical reasons. For example, traditional resilience techniques, such double- or triple-module redundancy (DMR, TMR), are prohibitive in the context or large supercomputers because of the large number of compute nodes and compute nodes' components. Moreover, additional redundant or hardening hardware might increase the overall cost considerably. Even a 10% increase in energy consumption might results in a large increase in the electricity bill if the system consumes a large amount of power.

Without specific and efficient resilience solutions, modern HPC systems would be extremely unstable, to the point that it is utopistic to assume that a production run of a scientific application will not incur in any error or will be able to terminate correctly. This chapter will review some of the most common resilience techniques adopted in the HPC domains, both application-specific and system-oriented. The chapter will also provide initial insights and research directions for future exascale systems.

2 RESILIENCE IN SCIENTIFIC APPLICATIONS

Research scientists rely on supercomputers to solve computation intensive problems that cannot be solved on single workstation because of the limited computing power or memory storage. Scientific computations usually employ double-precision floating point operations to represent physics quantities and to compute velocity and position of particles in a magnetic field, numerical approximations, temperature of fluids or material under thermal solicitation, etc. The most common representation of double-precision floating point values follows the IEEE Standard for Floating-Point Arithmetic (IEEE 754) [2]. With a finite number of bits, this format is necessarily an approximate representation of real numbers. As a consequence, rounding up errors are common during floating-point computations and must be taken into account when writing scientific code. Additionally, many scientific problems do not have a closed form that can be solved by using a direct method. In these cases, iterative methods are used to approximate the solution of the problem. In contrast to direct methods, which provide an exact solution to the problem, iterative methods produce a sequence of improving approximate solution for the problem under study. Iterative methods are said to be "convergent" if the corresponding sequence converges for given initial approximations based on a measurement of the error in the result (the residual). During the execution, iterative methods form a "correction equation" and compute the residual error. The approximation process is repeated until the residual error is small enough, which implies that the computed solution is acceptable, or until a maximum number of iterations have been reached, in which case the solution might not be acceptable.

The uncertainty introduced by the floating-point representation and the iterative approximation require scientific applications to assess the *quality of the solution*, often in the form of a residual error. The quality of the solution is typically reported in addition to the solution itself: for example, a molecular dynamics application might produce a final state in which the molecules have evolved and a residual error to assess the feasibility of the dynamics. By tuning the residual errors, scientists can increase the quality of the computed solution, generally at the cost of higher computing time. For example, a larger number of iteration steps might be necessary to reach an approximation with a lower residual error, i.e., a better solution.

When running in faulty environments, computation errors can also occur because of bit-flips in processor registers and functional units, or memory. Using the residual error, HPC applications can generally assess whether a fault has produced an error or if it has vanished. However, the very approximate nature of both floating-point representation and iterative, indirect mathematical methods, makes most HPC applications "naturally tolerant" to a small number of bit-flips. For example, consider the case of a bit-flip in a double-precision floating-point register. In the IEEE 754 representation, one bit indicates the sign of the number, 11 bits store the exponent, and 52 bits store the mantissa. Because of the lossy representation, a bit-flip in the mantissa may introduce only a small perturbation and, thus, has much lower chances to affect the final result than a bit flip in the exponent or the sign [3]. In the case of iterative methods, a small perturbation caused by a fault may delay the convergence of the algorithm. The application could still converge to an acceptable solution with a few more iterations or terminate with a less accurate solution. This example exposes a trade-off between performance and accuracy: given enough time, iterative methods could converge to an acceptable solution even in the presence of faults that introduce small perturbations. On the other hand, because of the fault, the application may never converge, thus it is necessary to estimate or statically determine how many extra iterations should be performed before aborting the execution. This is a difficult problem with high-risk/high-reward impact. On one hand, the applications might have run for weeks and a few more hours of work could provide a solution and avoid complete re-execution. On the other hand, the application may never converge. In the absence of mathematical foundations to estimate if and when the application will converge, the practical solution is to statically set a fixed number of maximum iterations after which the application is aborted. As the number of soft errors increases, the choice of whether to invest time and resources in the hope that the application will converge will become even more important, thus methods to estimate the probability that the application will converge are required.

For large perturbations, there is a possibility that the algorithm will never converge. In this case, the application will reach the maximum number of iterations and terminate without providing an acceptable solution. Normally, however, these impossible convergence conditions can be detected by the application itself. In these cases, the application will typically abort as soon as the condition is detected and will report an error, saving the remaining execution time. This technique is used, for example, in LULESH, a shock hydrodynamics proxy application developed by the ASCR ExMatEx Exascale Co-Design Center [4]. In particular, LULESH periodically checks the consistency of the solution and aborts if either the volume of the fluid or any element have negative values.

Compared to other classes of applications, HPC applications can be classified in the following categories:

Vanished (V): Faults are masked at the processor-level and do not propagate to memory, thus the application produces correct outputs (COs) and the entire internal memory's state is correct.

Output not affected (ONA): Faults propagate to the application's memory state, but the final results of the computation are still within the acceptable error margins and the application terminates within the number of iterations executed in fault-free runs.

Wrong output (WO): Faults propagate through the application's state and corrupt the final output. The application may take longer to terminate.

Prolonged execution (PEX): Some applications may be able to tolerate corrupted memory states and still produce correct results by performing extra work to refine the current solution. These applications provide some form of inherent fault tolerance in their algorithms, though at the cost of delaying the output.

Crashed (C): Finally, faults can induce application crashes. We consider "hangs," i.e., the cases in which the application does not terminate, in this class.

This classification extends the one previously proposed in [5–8] to account for specific HPC features. Compared to the classification presented in Chapter 2, SDCs are identified as ONA, WO, and PEX, depending on their effects on the application's state and output. Analysis based on output variation, such as fault injection analysis, cannot distinguish V from ONA, thus we introduce the class CO to indicate the sum of V and ONA when the application produces correct results within the expected execution time. Correct results with longer executions are in PEX.

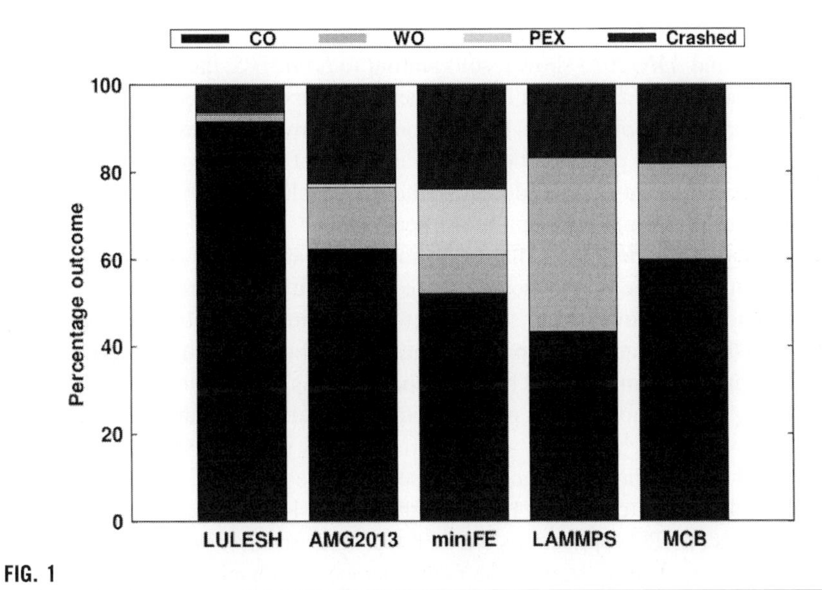

FIG. 1

Resilience analysis based on statistical fault-injection outcomes.

A recent study [9] shows the outcomes of injecting faults into several HPC applications from the U.S. DOE Exascale Co-Design centers,[1] DOE Office of Science, and the CORAL benchmark suite.[2] Fig. 1 shows the statistical result outcomes of injecting single-bit fault at architectural register level. In this experiments, each application was executed 5000 times and one single-bit fault was injected in each run into a random MPI process. Faults are injected into input or output registers of arithmetic operations, both integer and floating-point using a modified version LLFI [10], an application-level fault injector based on LLVM [11], for MPI applications. The application output is considered corrupted (WO) if it differs significantly from the fault-free execution or if the application itself reports results outside of the acceptable error boundaries. A similar analysis is presented, for a different set of applications, in [10, 12, 13].

The results in the figures show that each application present different vulnerability. Single-bit faults injected in *LULESH* appear to result in correct execution (CO) in over 90% of the cases, i.e., the computed solution is within the user-specified error margins. Moreover, generally, no additional iteration is performed (PEX) to achieve the result. For this application, only less than 10% of the executions result in crashes and less than 5% of the experiments produce wrong results. *LAMMPS* sits at the other extreme of the vulnerability spectrum, with only 40% of the experiments producing correct results within the expected number of iterations. In over 40% of the cases, the

[1]http://science.energy.gov/ascr/research/scidac/co-design/.
[2]https://asc.llnl.gov/CORAL/.

results is wrong (WO) and, in about 20% of the experiments the application crashes. While *MCB* and *AMG2013* show results similar to *LAMMPS*, though they appear to be more robust, *miniFE* shows a large percentage of experiments (over 15% of the cases) that produce correct results but take longer to converge (PEX). These are interesting cases, as they expose a particular characteristic of scientific applications that is not necessarily present in other domains: the user could trade-off the accuracy and correctness of the computed solution for performance.

Generally, the experiments that result in crashes are mainly due to bit flips in pointers that cause the applications to access a part of the address space that has not been allocated. However, in some cases, applications may abort autonomously because the injected fault either produced an "unfeasible" value or the convergence of the algorithm is compromised beyond repair. This is the case for *LULESH*: if the energy computed at time step i is outside of the acceptable boundary, the application aborts the execution calling `MPI_Abort()` routine. As potentially wrong execution are detected earlier, the number of WO cases is limited.

The results presented in Fig. 1 and in other works [10, 12, 13] show interesting characteristics of HPC applications but do not provide the user or an intelligent runtime with actionable tools to tune the fault-tolerance mechanism during the execution of a parallel application. Moreover, statistical analysis based on output variation do not necessarily assess that a fault has not propagated into the application memory state. In fact, output analysis might be based on the residual reported by the application, which may also be a part of the output.

Fig. 2 shows an analysis of fault propagation for *LULESH* under the same circumstances of the experiments reported in Fig. 1. Although the results in Fig. 1 show that *LULESH* appears to be very robust, the plot in Fig. 2 shows that many faults

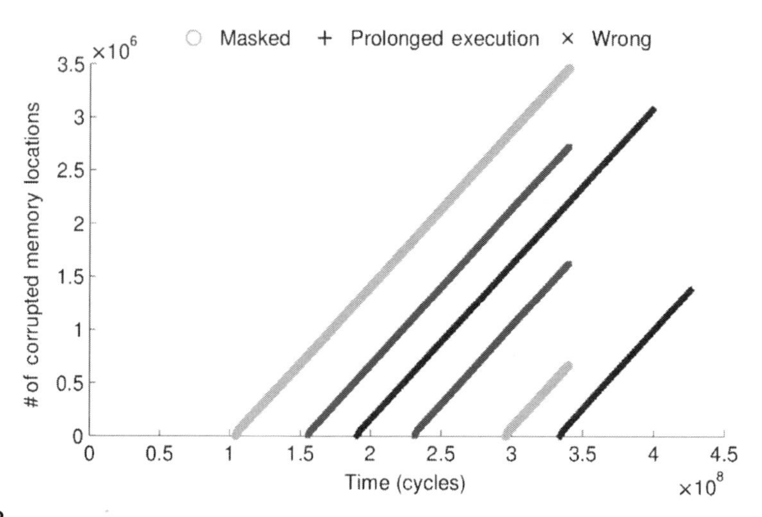

FIG. 2

Fault propagation analysis for *LULESH*.

propagate through the application state, even when the final outcome appears correct. The fact that the result appears to be correct, thus, does not mean that faults have not propagated but only that effects of a bit-flip may be confused with the approximation due to the use of floating point operations. Nevertheless, faults propagate in *LULESH* with each time step.

Fig. 2 provides another interesting information: faults propagate linearly in time. This information can be used to build a fault propagation model and to estimate the number of memory location corrupted by a fault detected at time t_f. By using linear regression or machine learning techniques, it is possible to derive a family of linear fault propagation models (one per experiment) to estimate the number of corrupted memory locations (CML) as function of the time t:

$$CML(t) = a \cdot t + b \tag{1}$$

where a expresses the propagation speed and b is the intercept with the time axis.[3] The value of b for a specific experiment can be determined as:

$$b = -a \cdot t_f \tag{2}$$

By averaging the propagation speed factor of all experiments, it is possible to derive the *fault propagation speed* (FPS) factor for an application. FPS expresses how quickly a fault propagates in the application memory state and represent an intrinsic characteristic of an HPC application. FPS is an actionable metric that can be used to estimate the number of memory locations contaminated by a fault. For example, assuming that a fault is detected at time t_f, then the maximum number of contaminated memory locations in an time interval $[t_1, t_2]$ is:

$$max(CML(t_1, t_2)) = FPS \cdot (t_2 - t_1) \tag{3}$$

which assumes that t_f is close to the lower extreme of the interval, t_1. This information can be used, for example, to decide when and if a rollback should be triggered.

Table 1 shows the FPS values for all the application reported in Fig. 1. The table show that, when considering the speed at which faults propagates, *LULESH* is much more vulnerable than *LAMMPS*, as faults propagate at a rate of 0.0147 CML/s in the former and 0.0023 CML/s in the latter. *MCB* is the most vulnerable application among the ones tested. For this application faults propagate at a rate of 0.0531 CML/s.

Table 1 Fault Propagation Speed Factors

App.	LULESH	LAMMPS	MCB	AMG2013	miniFE
FPS	0.0147	0.0025	0.0562	0.0144	0.0035
SDev	1.48E−4	0.96E−4	26.7E−4	6.82E−4	2.89E−4

[3]This formula assumes that the fault is detected when it occurs. If there is a delay between the occurrence and the detection of the fault, Δt needs to be taken into account in the computation of b.

3 SYSTEM-LEVEL RESILIENCE

The most appealing resilience solutions, from the user prospective, are those that automatically make applications and systems tolerant to faults without or with minimal user intervention. Compared to application-specific fault tolerance solutions presented in Section 4, system-level resilience techniques provide fault-tolerance support without modifications of the application's algorithm or structure. Applications and fault-tolerance solutions are mostly decoupled, thus the user can chose the most appropriate system-level solutions, i.e., a dynamic library, for each application or can simply use the one already provided by the supercomputer system administrator.

Among the system-level fault-tolerance techniques, checkpoint/restart solutions are the most widely used, especially in the context of recovering from hard faults, such as compute node crashes [14, 15]. *Checkpointing* refers to the action of recording the current total or partial state of a computation process for future use. Checkpoint/restart solutions periodically save the state of a computational application to stable/safe storage, such as a local disk or a remote filesystem storage media, so that, in the event of a failure, the application can be restarted from the most recent checkpoint. Generally, checkpoint/restart solutions work well under the following assumptions:

Fault detection: Faults can always be detected, either using hardware or software techniques, or can be safely assumed, such as in the case of a nonresponding compute node. This model is often referred to as *fail-stop semantics* [16].
Consistency: The state of the entire application, including all compute nodes, processes, and threads, can be stored at the same time and all in-flight communication events must be processed before the checkpoint is taken or delayed until the checkpoint is completed.
Ubiquity: In the case of a compute node failure, a checkpointed process can restart on a different node and the logical communication layer can safely be adapted to the new physical mapping.

Many properties of a checkpoint/restart solution affect the execution of parallel applications. First, the direct *overhead* introduced by periodically saving the state of the application takes resources that would have been used to make progress in the computation. An application that is checkpointed during the execution runs in a *lock-step* fashion, where all processes compute for a certain amount of time and then stop so that a consistent checkpoint can be taken. The *checkpointing interval* is the amount of time that intercurs between two consecutive checkpoints. Generally, the checkpointing interval is constant, especially for transparent solutions that periodically save the state of the application without user intervention. However, adaptive checkpointing intervals can be used if the application present different execution phases or if the vulnerability of the system increases. Adaptive checkpoint/restart solutions present higher complexity, thus solutions with fixed checkpointing interval are more common. The *size* of the checkpoint refers to the amount of data that is saved to stable storage to properly restart the application in case of failure. The size of the checkpoint strongly depends on the application's memory footprint but is also affected by the meta-data used by the checkpoint/restart system to describe the state of the application on storage. To reduce the checkpointing size, researchers have investigated alternatives that avoid saving the entire application state (*full checkpointing*) but rather save only the data that has changed since the last checkpoint (*incremental checkpointing*). Incremental checkpointing might reduce the amount of data that needs to be saved at every checkpoint but generally increases the restarting time after a failure. In fact, it might be necessary to use more than a single checkpoint to rebuild the full state of the application, possibly starting from a full checkpoint. Moreover, whether or not incremental checkpointing is beneficial for an application depends on the memory access patterns. Several applications, such as streaming or stencil computations, tend to rewrite the entire memory space within a short amount of time, possibly shorter than the checkpointing interval. Finally, a checkpoint/restart solution is considered *transparent* if it does not require user intervention (except, perhaps, to set some system-level parameters, such as the checkpointing interval), or *semitransparent* if the user needs to provide guidance about which data structures need to be saved or when.

A general objective of any resilience solution, including checkpoint/restart systems, is to minimize the impact on the application performance and power consumption. In other terms, the amount of resources that can be dedicated to resilience, either temporally or spatially, is limited. For checkpoint/restart solutions, the two major objectives are to minimize the runtime overhead and the disk space. The former is affected by the checkpointing interval, the amount of data to be saved and the location of the storage media. Saving the checkpoint in memory or on local disk reduces the transfer time but also increases the risk of not being able to restart the application, if the entire node fails. The disk space used is mainly a function of the checkpoint size. As discussed above, incremental checkpointing can be used to minimize the size of a single checkpoint. However, more than one checkpoint needs to be stored on disk to ensure proper restart in case of failure. In the other hand, in case of full checkpoint, it could be possible to discard all the other checkpoints taken before the last.

3.1 USER-LEVEL CHECKPOINTING

Checkpoint/restart solutions are designed and implemented at different levels of the hardware/software stack. The choice of where to implement the solution affects most of its properties, including transparency to the application, size of the state to be saved, and comprehensiveness of the state saved. Fig. 3 shows possible implementations within the hardware/software stack. At higher level, a checkpoint/restart solution can be implemented within the applications. This approach is generally expensive in terms of user effort and is seldom adopted, as many checkpoint/restart

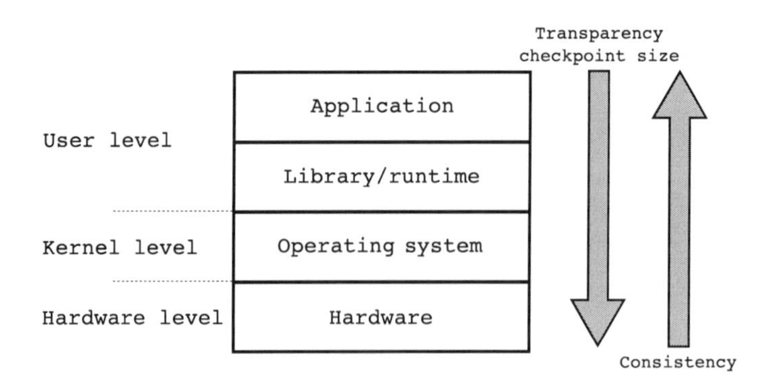

FIG. 3

Checkpointing/restart solution landscape.

operations are common to many applications. The next step is to group together common checkpoint/restart functionalities within a user-level library. These libraries provide specific APIs to the user to instrument which data structures should be saved and when should a checkpoint be taken to guarantee consistency. Solutions implemented at this level are considered intrusive, as they require application code modifications, but are also the ones that provide the user with APIs to express which data structures are really important for the recovery of the state of the application in case of failure. The checkpoint size is generally smaller as compared to other solutions implemented at lower level. However, the user has to explicitly indicate which data structures should be saved during checkpointing and when the checkpoint should be taken to guarantee consistency. Compiler extensions can be used to ease the burden on the programmer by automatically identify loop boundaries, global/local data structure, and perform pointer analysis. The compiler can, then, either automatically instrument the application or provide a report to the user that will guide the manual instrumentation. At restart time, the user may need to provide a routine to redistribute data among the remaining processes or to fast-forward the execution of the application to the point where the most recent checkpoint was taken.

At lower level, a runtime system can be used to periodically and automatically checkpoint a parallel application. These solutions typically come in the form of a runtime library that can be preloaded before the application is started, as an extension of the programming model runtime, such as MPI [17, 18], or as compiler modules that automatically instrument the application code and link to the checkpoint/restart library for runtime support. The main advantage of this approach is the higher level of transparency and the lower level of involvement of the user, thus higher productivity. A single checkpoint/restart solution can generally be used with many applications without major effort from the programmer. It usually suffices to recompile or relink the applications against the checkpoint/restart library and set some configuration parameters, such as the checkpointing interval. The primary disadvantage is the lost of application knowledge that can be used to disambiguate which data structures are actually essential for the correct representation of the state of the application and which can be omitted from the checkpoint. The checkpoint size generally increases and, for long-running application, the space complexity may be a limiting factor.

Completely transparent runtime checkpoint/restart solutions do not require user intervention and periodically save the application state based on a timer or the occurrence of an event, such as at loop boundaries identified by the compiler. The flexibility and transparency provided by automatic checkpoint/restart runtime solutions come at the cost of a generally larger checkpoint size and increased complexity to guarantee consistency. In fact, without user intervention, it becomes difficult to distinguish the data structures that contain the application state from those that are temporal or redundant. Typically, local variables stored on the process' stacks and read-only memory regions are omitted from the checkpoint. However, even within the heap and read-write memory regions there might be redundant or temporal data structures that need not to be saved. Consistency is also difficult to guarantee. For tightly coupled applications that often synchronize with global barriers, the checkpoint can be easily taken when the last application's process reaches the barrier and before all processes leave it. Several automatic checkpoint/restart solutions intercepted the `pMPI` barrier symbol in the MPI communication library to ensure that all processes where at the same consistent state when the checkpoint was taken. Applications that often synchronize with global barriers, however, do not tend to scale well, especially on large supercomputer installations. Applications that exchange asynchronous messages and that synchronize with local barriers scale more effectively but also pose problems regarding the consistency of checkpointing. In particular, the automatic checkpointing runtime needs to ensure that there are no in-flights messages. If the checkpoint is taken when the message has left the source process but has not yet reached the destination, the message's content would be neither in the context of the source process nor the context of the destination and it becomes lost. At restart time, there is no information about the lost message: the source process will not resend the message, as it appears already sent in its context, and the destination process will wait for a message that will never arrive, thereby possibly blocking the progress of the entire application.

3.2 PRIVILEGED-LEVEL CHECKPOINTING

In some cases, the application state does not contain sufficient information to seamlessly restart the system after a failure. This happens when some hardware or system software states must be restored before restarting the application. Solutions implemented at supervisor/hypervisor level, i.e., operating system (OS) or virtual machine manager (VMM) level, provide the necessary support to save the state of the entire compute node as part of the checkpoint. These solutions are capable of restoring the entire system, both hardware and application, at the point when the failure occurred. With this approach, the application does not need to be fast-forwarded to the time when the fault occurred just to set the underlying hardware and OS to the proper state.

Privileged access to hardware and software compute node information may also facilitate the gathering of the state information. The OS has unrestricted access to the state information of all processes running in a compute node and to the underlying hardware. Implementing incremental checkpoint/restart solutions based on page frame *dirty bit* is usually easier at this level because the OS can directly manipulate the page table entries. The OS can mark a page as dirty and as to-be-saved at the next checkpointing interval without incurring in the high overhead cost of using the `mprotect()` system call. A similar operation performed from a user-level library or application would involve a two-step procedure: First, the process address space needs to be protected from write access through the `mprotect()` system call. On attempt to modify a page, the OS will send a `SIGSEV` signal to the process. This signal must be intercepted by the process or checkpointing library signal handler and the page must be marked as to-be-saved. In this endeavor, the user/kernel boundary is passed twice, once from the `mprotect()` system call for restricting access to a page, and once from the signal handler to record that the page has been modified since the last checkpoint.

One of the OS-level checkpoint/restart solution is the Transparent Incremental Checkpointing at Kernel level, TICK [19] TICK is a fully transparent solution implemented as a Linux kernel module that provides the user with functionalities to perform both full and incremental checkpointing with extremely low overhead. As explained above, tracking dirty pages at kernel level is more efficient than at user level. TICK uses the dirty bits stored in the page table entry of each page to track which page has been modified since the last checkpoint. On the first access to a protected page, the OS triggers a page fault exception (similar to the case in which a page is protected through `mprotect()` system call) and TICK stores the address of the page is a bitmask so that the page can be saved at the next checkpoint.

Fig. 4 shows the runtime overhead introduced by TICK to save the full application state for two applications, SAICs Adaptive Grid Eulerian (SAGE) and Sweep3D. SAGE hydrocode is a multidimensional (1D, 2D, and 3D), multimaterial, Eulerian hydrodynamics code with adaptive mesh refinement. The code uses second order accurate numerical techniques, and is used to investigate continuous adaptive

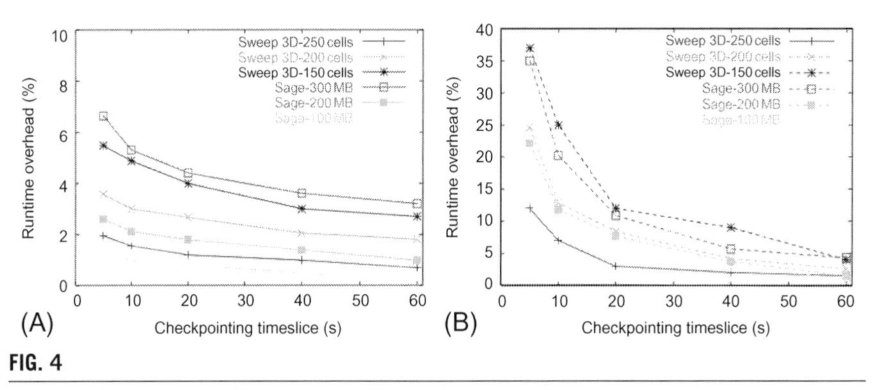

FIG. 4

TICK runtime overhead for memory and disk checkpointing as the checkpointing interval increases. (A) Memory and (B) disk.

Eulerian techniques to stockpile stewardship problems. SAGE has also been applied to a variety of problems in many areas of science and engineering including water shock, energy coupling, cratering and ground shock, stemming and containment, early time front end design, explosively generated air blast, and hydrodynamic instability problems [20]. The ASCI Sweep3D benchmark is designed to replicate the behavior of particle transport codes which are executed on production machines [21]. The benchmark solves a one-group time-independent discrete ordinates neutron transport problem, calculating the flux of neutrons through each cell of a three-dimensional grid along several directions of travel. Both SAGE and Sweep3D were developed at the LANL.

As expected, the runtime overhead decreases with the increased checkpointing interval, both for in-memory and on-disk checkpointing. However, Fig. 4A shows that, even with extremely small checkpointing intervals (5 s) the runtime overhead introduced to save the full application state in memory is less than 8%. Saving the checkpoint to disk introduces larger overhead, especially for small checkpointing intervals. For more realistic checkpointing intervals, 60 s or more, the runtime overhead is less than 4%, when saving the checkpoint in memory, and 5% when saving the checkpoint to disk. It is worth noticing that, for on-disk checkpointing, the overhead greatly reduces for checkpointing intervals larger than 10 s. This is because 10 s coincides with the normal dirty pages synchronization time interval, thus the pages containing the application checkpoint can be piggy-backed to the other dirty pages with a unique stream of data to disk.

Fig. 5 shows the TICK runtime overhead for SAGE when using full and incremental checkpointing and saving the application state to disk. The plot shows that for small checkpointing intervals, incremental checkpoint is more efficient than full checkpointing, with a runtime overhead of 14% instead of 26% with a checkpointing

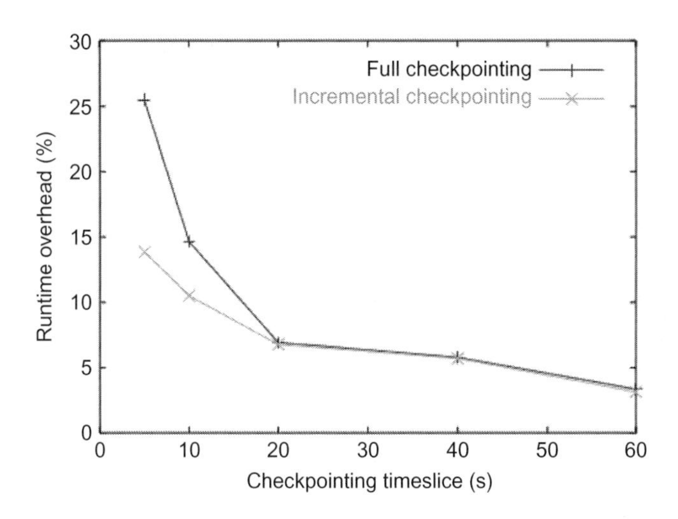

FIG. 5

TICK runtime overhead of full versus incremental checkpointing for SAGE.

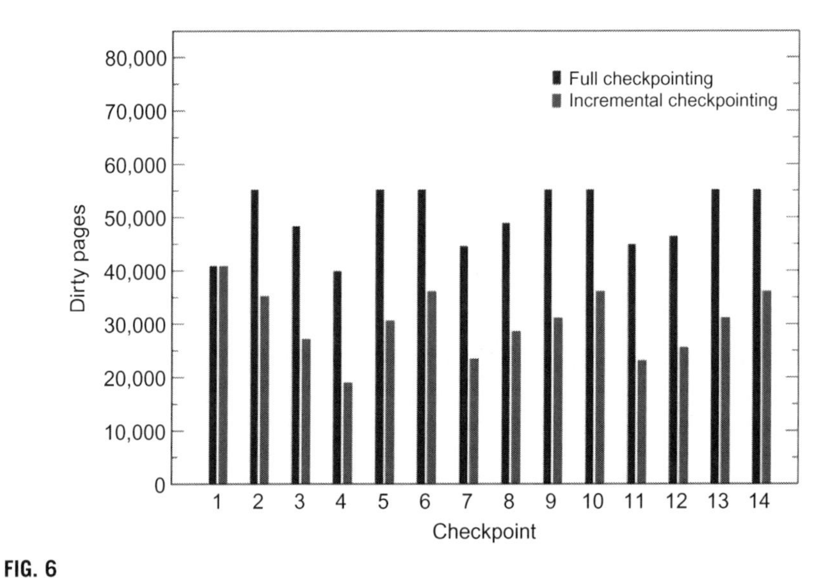

FIG. 6

Number of pages saved in full versus incremental checkpoint mode (checkpointing interval 20 s.

interval of 5 s. However, as the size of the checkpointing interval increases, the relative advantage of incremental checkpoint decreases and eventually disappears with checkpointing intervals larger than 20 s. With such checkpointing intervals, in fact, the application rewrite most of its data structures, thus the number of dirty pages to be save in each incremental checkpointing interval approximates the full set of pages. Fig. 6 shows that, indeed, the number of pages to be saved with incremental and full checkpoint is fairly similar with a checkpointing interval of 20 s.

An OS-level implementation also provides access to information that are not generally available in user-mode, such as anonymous and file memory mappings, pending signals and I/O transfers, and hardware state. This information can help restore the state of the system and to understand which data has been actually transferred to I/O. At every checkpointing interval, TICK also saves the state of the system (both hardware and OS). At restart time, TICK restore the state of the system as it was before the failure, hence avoiding that the application be fast-forwarded in time just to restore the correct hardware and software state.

Despite their efficiency, OS-level implementations are not widely common, mainly because of the difficulty in porting and maintaining code among various OSs and even among different versions of the same OS. Virtual machines (VM) solutions have been proposed as an easier and more portable way to provide supervisor privileges to checkpoint/restart solutions. The general idea is that the application runs within a VM rather than directly on the host compute node. The VMM periodically saves the entire state of the VM on safe storage. For checkpoint/restart, the VMM leverage VM migration functionalities usually present in modern VMM to migrate VMs from one compute node to another in data-center installation. Upon restart, the checkpointed VM is migrated to an available compute node and restarted. Since there interaction between VM and host node is managed by the VMM and the guest OS is not aware of the actual hardware on top of which the VM is running, restarting a VM is a pretty seamless procedure.

Although privileged-level solutions provide easy access to system software and hardware information and provide faster restart, they incur in larger checkpoint sizes and increased consistency complexity. Privileged-level solutions do not reason in terms of data structures, but rather manage memory pages. A page is considered dirty even if a single byte in the page has been modified since the last checkpoint. Saving the entire page even for a minimal modification could results in an unnecessary space overhead. On the other hand, as discussed earlier for incremental checkpointing, parallel applications tend to rewrite their address space fairly quickly. If the checkpointing interval is large enough, there are good chances that most of the data contained in a page has been updated since the last checkpoint.

4 APPLICATION-SPECIFIC FAULT TOLERANCE TECHNIQUES

System-level resilience techniques attempt to provide transparent solutions to guarantee the correct execution of parallel applications. Although general, these solutions may incur in extra-overhead caused by the lack of knowledge regarding the application's algorithm and data structures. For example, full-checkpoint/restart systems save the state of the application without distinguishing between temporal and permanent data. As such, system-level techniques may save more data that it is actually necessary. In contrast, application-specific resilience and fault-tolerance solutions (also called application-based fault tolerance) leverage application knowledge and internal information to selectively protect or save data structures that contain the fundamental application state. Many scientific application's algorithms have mathematical *invariant properties* that can be leveraged to understand the correctness of partial solutions. For example, the user could assert that a certain quantity be positive, thus errors are easily detectable if the sign of the quantity turns negative. In other cases, redundant information can be added to detect or detect/correct a finite number of faults.

On the positive side, application-specific resilience techniques generally introduce lower runtime and space overhead, are highly portable because they do not depend on the underlying architecture, and are generally more salable. Computing the residual error falls into the application-specific resilience techniques category, among the error detection solutions. The major drawback of application-specific techniques is the high-effort of maintainability and the lower productivity. Compared to system-level solutions, application-specific techniques have to be implemented for each algorithm and algorithm's variant, as they are generally part of the application itself. Moreover, modifications to the algorithm (e.g., changing the solver of a linear algebra system) may require a reimplementation of the resilience part as well, as the original algorithm may have different mathematical properties.

Fortunately, many applications use fundamental operations that can be embedded in mathematical libraries together with their corresponding resilience solutions. One of such algorithm is the matrix-matrix multiplication used in many scientific codes. Let $A_{n,\ m}$ be a n-by-m matrix and $B_{m,\ n}$ be a m-by-n matrix. The result of multiplying A by B is a matrix $C_{n,\ n} = A_{n,\ m} \times B_{n,\ m}$. In this case, the problem is determining whether matrix $C_{n,\ n}$ is correct, given that the input matrices $A_{n,\ m}$ and $B_{n,\ m}$ are correct, and possibly correcting errors due to bit-flips. Check-summing is a well-know techniques used to guarantee the integrity of data in scientific computations, communication and data analytic, among others.

Check-summing techniques are based on adding redundant information to the original data. Once the computation is completed, the redundant information is used to determine the correctness of the results, in this case the matrix C. Let's consider a generic matrix $A_{n,\ m}$:

$$A_{n,m} = \begin{bmatrix} a_{1,1} & a_{1,2} & a_{1,3} & \dots & a_{1,m} \\ a_{2,1} & a_{2,2} & a_{2,3} & \dots & a_{2,m} \\ \dots & \dots & \dots & \dots & \dots \\ a_{n,1} & a_{n,2} & a_{n,3} & \dots & a_{n,m} \end{bmatrix}$$

The column checksum matrix A^c of matrix A is an $(n+1)$-by-m matrix which consists of matrix A's first n rows plus an additional $(n+1)^{th}$ row that contains column-wise sums of the A's elements. The elements of the additional row are defined as:

$$a_{n+1,j} = \sum_{i=1}^{n} a_{i,j} \text{ for } 1 \le j \le m$$

Similarly, the row checksum matrix A^r of matrix A is a n-by-$(m+1$ matrix, which consists of the first m columns of A and a row followed by a $(m+1)^{th}$ column containing the sum of each row. The elements of A^r are defined as:

$$a_{i,m+1} = \sum_{j=1}^{m} a_{i,j} \text{ for } 1 \le i \le n$$

Finally, the full checksum matrix A^f of matrix A is an $(n+1)$-by-$(m+1)$ matrix composed by the column checksum matrix A^c and the row checksum matrix A^r and whose elements are defined as:

$$a_{n+1,m+1} = \sum_{j=1}^{m} \sum_{i=1}^{n} a_{i,j}$$

A^c and A^r can also be expressed as:

$$A^c = \begin{bmatrix} A \\ e^T A \end{bmatrix},\ e^T = [1\ 1\ \dots\ 1]^T$$

$$A^r = [A\ \ Ae],\ e^T = \left[e^T\right]^T$$

thus,

$$A^f = \begin{bmatrix} A & Ae \\ e^T A & e^T Ae \end{bmatrix}$$

The computation of the original matrix C is extended to account for the redundant checksum information. In the case of $C = A \times B$, the column checksum matrix A^c and the row checksum matrix B^r are used to compute full checksum matrix C^f:

$$C = A \times B \Rightarrow C^f = A^c \times B^r = \begin{bmatrix} A \\ e^T A \end{bmatrix} \times [B \ \ Be] = \begin{bmatrix} AB & ABe \\ e^T AB & e^T ABe \end{bmatrix} = \begin{bmatrix} C & Ce \\ e^T C & e^T Ce \end{bmatrix}$$

Obviously, computing C^f requires extra operations as compared to computing the original matrix C. In this case, thus, resilience is achieved through lower performance and higher energy consumption. The redundant information in C^f can be used to assess the correctness of C. More in details, the column and row checksum of the computed matrix C can be compared to the ones produces by multiplying A^c and B^r. Let k_j and r_i be the column and row checksum of matrix C, respectively, such that:

$$k_j = \sum_{i=1}^{n} c_{i,j}, \quad r_i = \sum_{j=1}^{m} c_{i,j}$$

then, k_j and r_i can be compared to $c_{n+1,\, j}$ and $c_{i,\, m+1}$, respectively, to detect errors in each column j or row i, respectively. In this particular case, one single-bit error per column or row can be detected. If we assume that one error occurred during the computation, then the error can be located at the intersection between the inconsistent column $j*$ and inconsistent row $r*$. With this technique, a single error can also be corrected by adding the difference between the checksum computed during the computation of C^f and the ones computed after the computation, k_j and r_i.

$$\hat{c}_{i^*,j^*} = c_{i^*,j^*} + |k_{j^*} - r_{i^*}|$$

Although checksum is a fairly well-know and wide-spread solution, many other application-specific techniques have been developed for scientific algorithms [22–27].

5 RESILIENCE FOR EXASCALE SUPERCOMPUTERS

HPC systems have historically been vulnerable to hard errors caused by the extremely large number of system components. Mechanical parts, such as storage disks, and memory DIMM failures were, and still are, among the most common reasons for system shutdown and require manual intervention or replacement. While this is still true in current-generation supercomputers and expected to worsen for future exascale systems because of the even larger number of system components, the stringent power constrains expected in the exascale era will change the balance between hard and soft errors. New low-power and near-threshold voltage (NTV) technologies, along with higher temperature and voltage tolerance techniques, are expected to be deployed on exascale system to address the power challenge [28].

Without additional mitigation actions, these factors, combined with the sheer number of components, will considerably increase the number of faults experienced during the execution of parallel applications, thereby reducing the mean time to failure (MTTF) and the productivity of these systems. Two additional factors further complicate the exascale resilience problem. First, power constraints will limit the amount of energy and resources that can be dedicated to detect and correct errors, which will increase the number of undetected errors and silent data corruptions (SDCs) [29]. Second, transistor device aging effects will amplify the problem. Considering that supercomputers stay in production for about 5 years, aging effects can dramatically diminish the productivity and return of investment of large-scale installations. The net effect is that new resilience solutions will have to address a much larger problem with fewer resources.

To address the exascale resilience problem, it is expected that new radical solutions be developed. Such solutions are currently under investigation and are not yet ready for deployment. This section will list the desirable properties and the main challenges that exascale resilience solutions will have to address.

5.1 CHECKPOINT/RESTART AT EXASCALE

Global checkpoint/restart will be prohibitive for exascale systems because of both space and time complexity. Given the massive number of parallel processes and threads running on an exascale system, saving the entire or even the incremental state of each thread on safe storage would require a prohibitively large amount of disk space. Since the I/O subsystem is not expected to scale as much as the number of cores, saving the states of each thread on disk would create severe I/O bottlenecks and pressure on the filesystem and disks. The serialization in the I/O subsystem, together with the large I/O transfer time necessary to save all thread states, will also increase the checkpoint time. Projections show that the checkpointing time might closely match or exceed the checkpointing interval, for representative HPC applications, thus the application may spend more time taking checkpoints than performing computation.

The second, but not less important, problem is how to guarantee checkpoint consistency for an application that consists of hundreds millions threads. Waiting for all the threads to reach a consistency point might considerably reduce the system utilization and throughput. Moreover, researchers are looking at novel programming models that avoid the use of global barriers, natural consistency points, as they have been identified as a major scalability limiting factor. Reintroducing these barriers, in the form of checkpoint consistency points, would bring back the problem and is, thus, not desirable.

5.2 FLAT I/O BANDWIDTH

Many parameters are expected to scale on the road to exascale. Table 2 shows the past, current, and expected typical values of the most important system parameters, including the core count, memory size, power consumption, and I/O bandwidth.[4]

[4]Source http://www.scidacreview.org/1001/html/hardware.html.

Table 2 Past, Current, and Expected Supercomputer Main Parameters

Systems	2009	2011	2015	2018
System peak FLOPS	2 Peta	20 Peta	100–200 Peta	1 Exa
System memory	0.3 PB	1 PB	5 PB	10 PB
Node performance	125 GF	200 GF	400 GF	1–10 TF
Node memory BW	25 BG/s	40 GB/s	100 GB/s	200–400 GB/s
Node concurrency	12	32	O(100)	O(1000)
Interconnect BW	1.5 GB/s	10 GB/s	25 GB/s	50 GB/s
System size (nodes)	18,700	100,000	500,000	O(million)
Total concurrency	225,000	3 million	50 million	O(billion)
Storage	15 PB	30 PB	150 PB	300 PB
I/O	0.2 TB/s	2 TB/s	10 TB/s	20 TB/s
MTTI	O(days)	O(days)	O(days)	O(1 day)
Power	6 MW	10 MW	10 MW	20 MW

Most of the system parameters see an increase, albeit not always linear, from petascale to exascale systems. For example, the total number of processor core increases from 225,000 in 2009, to 3 and 5 million in 2011 and 2015, and it is expected to be in the order of billion in exascale systems. The I/O bandwidth per compute node, on the other hand, remains practically flat. While the overall I/O bandwidth is expected to double from 2015 to 2018 and to reach about 20 TB/s, the number of compute nodes per system is expected to increase much more, hence the I/O bandwidth available per compute node may decrease. This means that, on one side the number of cores and process threads will increase by two or three orders of magnitude, while, on the other side, the I/O bandwidth used to transfer checkpoints from local memory to stable storage will remain constant or slightly decrease. Moving checkpoints to stable disk is one of the aspects of the exascale data movement problem [30]. With memory systems becoming more hierarchical and deeper and with the introduction of new memory technologies, such as nonvolatile RAM or 3D-stacked on-chip RAM, the amount of data moved among the memory hierarchy might increase to the point of dominating the computation performance. Moreover, the energy cost of moving data across the memory hierarchy, currently up to 40% of the total energy of executing a parallel application [30], may also become the primary contribution to the application energy profile.

Considering the expected projections, extending current checkpointing solutions to exascale will considerably increase the pressure on data movement. New resilience solutions for exascale will necessarily need to reduce data movement, especially I/O transfers. Local and in-memory checkpointing are possible solutions, especially for systems that features nonvolatile memory (NVRAM) [31]. In this scenario, uncoordinated checkpoints are saved to local NVRAM. In case of failure, the failed compute node will restart and resume its state from the checkpoint store in the local NVRAM memory, while the other nodes will either wait at a consistency point or restore their status at a time before the failure occurred. This solution has its own limitations: First, if the failed node cannot be restarted, the local checkpoint is lost. To avoid such situations, the each local checkpoint can be saved to remote, safe storage with a time interval much larger than the local checkpointing interval. Second, the checkpoint/restart system needs to guarantee consistency when recovering the application. As explained in the next section, novel parallel programming models might help to alleviate the need of full consistency across the application treads.

5.3 TASK-BASED PROGRAMMING MODELS

Synchronizing billions of threads through global barriers might severely limit the scalability of HPC applications on future exascale systems and the system productivity. Novel programming models are emerging, which provide asynchronous synchronization among subset of application threads, generally much smaller than the total number of threads. Task-based programming models are among the most promising programming models for exascale systems [32–34]. Task-based programming models follow a *divide-and-conquer* approach in which the problem is decomposed in several subproblems recursively until a small unit of work (*task*) is identified. The efficacy of task-based approaches is based on the idea of overprovision the system with a myriad of tasks that will be managed and assigned to worker threads by an intelligent runtime scheduler. The number of instantiated tasks is much larger than the available computing units, i.e., CPU and GPU cores, so that there are usually tasks available to run on idle cores. The runtime scheduler takes care of mapping a task to a computing element.

Fig. 7 shows a possible task-based implementation of the Fibonacci algorithm. The program in Fig. 7A follows a fork-join semantics in which tasks are *spawn* at each level of the recursion and synchronize after completing their computation. The

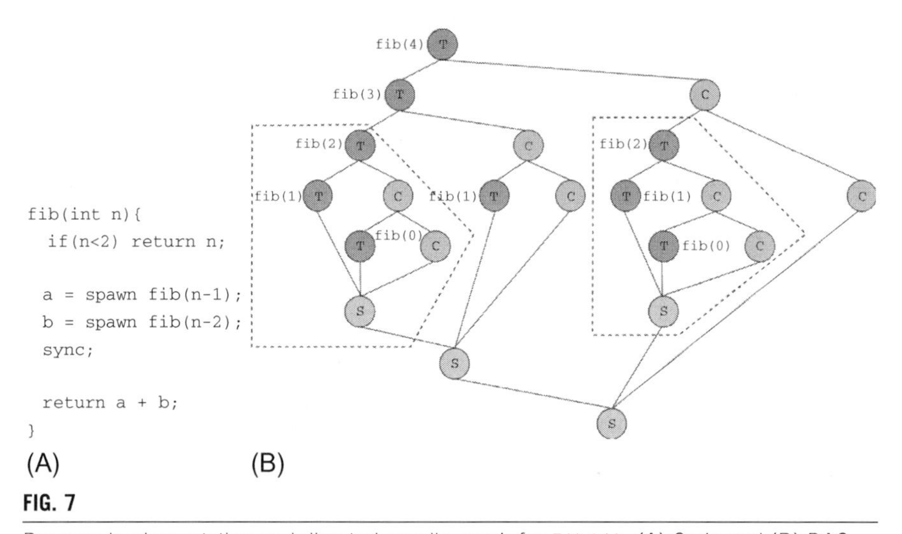

fib(int n){
if(n<2) return n;
a = spawn fib(n-1);
b = spawn fib(n-2);
sync;
return a + b;
}
fib(4)
fib(3)
fib(2)
fib(1)
fib(0)
T
C
S
(A)
(B)

FIG. 7

Program implementation and directed acyclic graph for `Fib(4)`. (A) Code and (B) DAG.

When a task *spawns* another task (*fork* point), the runtime scheduler decides whether the current worker thread will execute the newly spawn tasks (blue nodes in Fig. 7B) or their *continuation* (yellow nodes in Fig. 7B). The suspended task or continuation will be placed on a local stack. Idle worker threads will be able to steal from busy worker threads' stacks thereby providing automatic load balancing. After completing their computation, each task will reach a synchronization point (gray nodes in Fig. 7B). Only after all the spawn tasks reach the synchronization point, a task can continue its computation. The synchronization points (or *join* point), thus, act as local barriers.

Task-based programming models are attractive for two main reasons: their adaptivity and the decoupling of computation and data. The adaptive and dynamic nature of the task-based approach allows for adapting the computation to a given set of computing resource whose availability varies in time and for performing automatic load balancing. One of the possible causes that may reduce the amount of computing resources available is the occurrence of failures. In case of failure, the work performed by the failed node can be re-executed by another node without forcing all the other compute nodes to roll-back to a previous checkpoint. For example, if the compute nodes that executed the tasks within red, dashed lines in Fig. 7B fail, other compute nodes can re-execute the lost computation without rolling back the other live worker threads.

Dividing the execution of an application in tasks also provides an inherent form of soft errors containment. In fact, if a fault is detected before a new task is spawn and if the current task as only modified local variables or local copies of variables, it is possible to re-execute only the current task without even notifying the other worker threads of the detected failure. This model is particularly suitable for *data-flow* programming models, in which tasks exchange data only through input and output parameters. In case tasks also modify global data structure, it might still be required to perform a form of checkpointing. However, since task-based programming models decouple computation and data structure, it is possible to restart the computation with a different set of worker threads without explicitly redistributing data. This means that the checkpoint consistency does not need to be guaranteed for all worker threads in the systems but only for the ones that are somehow impacted by the failure, i.e., the worker threads that spawn the failed tasks. This greatly reduces the complexity of taking checkpoint and avoid the use of global barriers as synchronization points.

5.4 PERFORMANCE ANOMALIES

An interesting new dimension for exascale systems is represented by the fact that hardware components *gracefully* degrade performance in the presence of contingent problems or because of aging. This behavior are not necessarily caused by hardware faults but are still considered *performance anomalies* and undesirable. For example, a processor core may reduce the operating frequency if the temperature raises over the safe threshold. Similarly, a network links may reduce the transferring bandwidth if the number of packets corrupted reduces the quality of service. In these cases the system continues to operate "correctly" but the performance provided does not match

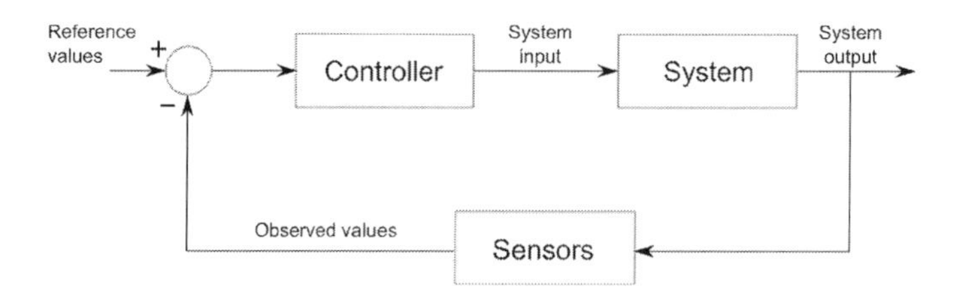

FIG. 8

Closed control loop.

the expectation of the users. Although these examples are not generally considered under the scope of resilience techniques, the symptoms they expose are often similar to the ones generated by hard and soft errors. Moreover, the effects of these performance faults can be assimilated to the PEX cases analyzed in Section 2. It is thus necessary to distinguish performance anomalies caused by hard or soft errors from those caused by contingent conditions. In most cases, it is frustrating to see that an application does not perform as expected even if everything looks OK and performances of previous runs, in the "exact same conditions," were higher.

To be able to disambiguate actual hardware faults from performance anomalies or, more in general, be able to understand whether the performance provided by an application in a particular run are satisfactory, researchers have proposed lightweight runtime monitoring that analyze the performance of a parallel application and compare current performance against previous executions or application performance models. In case a discrepancy between observed and expected performance or in case a fault is detected, the monitor may activate a controller module that will take the necessary actions to guarantee the correct and satisfactory execution of the application. Fig. 8 shows a graphical representation of how monitor, controller and applications are placed in a closed feedback loop.

The feedback system depicted in Fig. 8 can be used to detect performance anomalies caused by the thermal management system, which may reduce the processor frequency or skip processor cycles to reduce the temperature of the system. Fig. 9 shows the outcome of this test case: the plot reports the performance in FLOPS of NEKBone during the execution of the application. NEKBone is a thermal hydraulic proxy application developed by the ASCR CESAR Exascale Co-Design Center to capture the basic structure of NEK5000, a high order, incompressible Navier-Stokes solver based on the spectral element method.[5] NEK5000 is an MPI-based code written in Fortran and C that employs efficient multilevel iterative solvers and designed for large eddy simulation (LES) and direct numerical simulation (DNS) of turbolence in complex domains. NEKBone mimics the computationally intense linear solvers that accounts for a large percentage of the more complicated NEK5000 computation, and the communication costs required for nearest-neighbor data exchange and vector

[5]https://cesar.mcs.anl.gov/content/software/thermal_hydraulics.

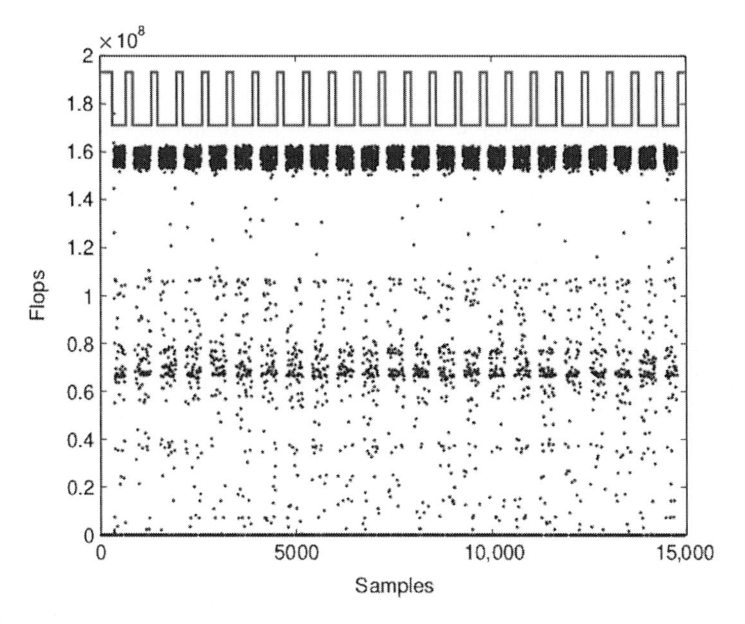

FIG. 9

Performance anomalies detection for NEKBone.

reduction. In particular, the NEKBone kernel solves a standard 3D Poisson equation using the spectral element method with an iterative conjugate gradient solver with an orthogonal scaling preconditioner.

From the plot it is evident how some processor cycles are periodically skipped and the application performance decreases. A monitoring system can detect the sudden lost of performance and activate the controller, which can identify the reason behind the performance drop (e.g., high temperature) and take the proper corrective actions to maintain satisfactory performance and lower the temperature.

Although the example in Fig. 9 is just a test case, it highlights a general trend for exascale systems. Given the extreme level of parallelism, the large number of system components, the thigh power budget, and the high probability of soft and hard fault, exascale systems will be dynamic and applications will be forced to run in changing execution environments. Self-aware/self-adaptive systems become essential to be able to run in such environment and still obtain the desired performance and correctness while maintaining the target power budget.

6 CONCLUSIONS

Although not generally deployed in hostile or unsafe environments, supercomputers have historically be particularly susceptible to soft and hard faults because of their extremely large number of system components, the general complexity of these systems, and the environmental conditions in which they operate. Without appropriate resilience solutions, it would be impractical to run a large-scale application on modern supercomputers without incurring in some form or errors.

Many fault-tolerance solutions have been developed throughout the past years. Some of these solutions, such as checkpoint/restart, are transparent to the user and can be employed with different application without or with minimal code modification. Other solutions are more intrusive and strongly depends on the mathematical characteristics of a particular algorithm or algorithm's implementation. Such application-level solution allow the user to specify which data structures are essential for the correct recovery of the application in case of failure and which are only temporal.

As systems become larger and include more components, the probability of failures increases. Moreover, the strict power limits enforced on exascale systems will force the use of novel low-power and NTV technologies, which, together with the massive number of system components, will negatively impact system resilience. Projections show that the number of soft error that may result in SDC will become dominant in the exascale era and that novel resilience solutions are necessary to guarantee the correct execution of HPC applications.

Current solutions based on global checkpoint seem impractical at the scale of future generation supercomputers, both because of the limited I/O bandwidth and because of the difficulties in guaranteeing consistency upon restart. Novel programming models, such as task-based programming models, are emerging as new programming models for exascale systems. These new programming models incorporate, directly or indirectly, primitives or properties that can be leverage to increase system resilience to the point that is needed by exascale supercomputers.

REFERENCES

[1] V. Sridharan, N. DeBardeleben, S. Blanchard, K.B. Ferreira, J. Stearley, J. Shalf, S. Gurumurthi, Memory errors in modern systems: the good, the bad, and the ugly, SIGARCH Comput. Archit. News 43 (1) (2015) 297–310.

[2] IEEE Standard for Floating-Point Arithmetic, IEEE Std 754-2008 (2008) 1–70.

[3] J. Elliott, F. Mueller, M. Stoyanov, C. Webster, Quantifying the impact of single bit flips on floating point arithmetic, ORNL Technical Report number ORNL/TM-2013/282, Oak Ridge (TN), (August 2013).

[4] I. Karlin, A. Bhatele, J. Keasler, B.L. Chamberlain, J. Cohen, Z. Devito, R. Haque, D. Laney, E. Luke, F. Wang, D. Richards, M. Schulz, C.H. Still, Exploring traditional and emerging parallel programming models using a proxy application, in: Proceedings of the 2013 IEEE 27th International Symposium on Parallel and Distributed Processing (IPDPS '13), IEEE Computer Society, Washington, DC, 2013, pp. 919–932.

[5] C. Bender, P.N. Sanda, P. Kudva, R. Mata, V. Pokala, R. Haraden, M. Schallhorn, Soft-error resilience of the IBM POWER6 processor input/output subsystem, IBM J. Res. Dev. 52 (3) (2008) 285–292.

[6] H. Cho, S. Mirkhani, C.Y. Cher, J.A. Abraham, S. Mitra, Quantitative evaluation of soft error injection techniques for robust system design, in: Proceedings of the 50th ACM/EDAC/IEEE Design Automation Conference (DAC), 2013.

[7] N.J. Wang, A. Mahesri, S.J. Patel, Examining ACE analysis reliability estimates using fault-injection, in: Proceedings of the 34th Annual International Symposium on Computer Architecture (ISCA '07), San Diego, CA, 2007, pp. 460–469.
[8] N.J. Wang, J. Quek, T.M. Rafacz, S.J. patel, Characterizing the effects of transient faults on a high-performance processor pipeline, in: Proceedings of the International Conference on Dependable Systems and Networks (DSN), 2004.
[9] R. Ashraf, R. Gioiosa, G. Kestor, R. DeMara, C. Cher, P. Bose, Understanding the propagation of transient errors in HPC applications, in: Proceedings of the ACM/IEEE International Conference on High-Performance Computing, Networking, Storage and Analysis (SC), 2015.
[10] J. Wei, A. Thomas, G. Li, K. Pattabiraman, Quantifying the accuracy of high-level fault injection techniques for hardware faults, in: Proceedings of the International Conference on Dependable Systems and Networks (DSN), 2014.
[11] C. Lattner, V. Adve, LLVM: a compilation framework for lifelong program analysis & transformation, in: Proceedings of the 2004 International Symposium on Code Generation and Optimization (CGO'04), Palo Alto, CA, 2004.
[12] D. Li, J.S. Vetter, W. Yu, Classifying soft error vulnerabilities in extreme-scale scientific applications using a binary instrumentation tool, in: Proceedings of the International Conference on High Performance Computing, Networking, Storage and Analysis (SC), Salt Lake City, UT, 2012.
[13] V.C. Sharma, A. Haran, Z. Rakamarić, G. Gopalakrishnan, Towards formal approaches to system resilience, in: Proceedings of the 19th IEEE Pacific Rim International Symposium on Dependable Computing (PRDC), 2013.
[14] J.C. Sancho, F. Petrini, K. Davis, R. Gioiosa, S. Jiang, Current practice and a direction forward in checkpoint/restart implementations for fault tolerance, in: Proceedings of the 19th IEEE International Parallel and Distributed Processing Symposium, 2005, 8 pp.
[15] I.P. Egwutuoha, D. Levy, B. Selic, S. Chen, A survey of fault tolerance mechanisms and checkpoint/restart implementations for high performance computing systems, J. Supercomput. 65 (3) (2013) 1302–1326.
[16] R.D. Schlichting, F.B. Schneider, Fail-stop processors: an approach to designing fault-tolerant computing systems, ACM Trans. Comput. Syst. 1 (3) (1983) 222–238.
[17] G.E. Fagg, J.J. Dongarra, FT-MPI: fault tolerant MPI, supporting dynamic applications in a dynamic world, in: Recent Advances in Parallel Virtual Machine and Message Passing InterfaceSpringer, New York, 2000, pp. 346–353.
[18] G. Bosilca, A. Bouteiller, F. Cappello, S. Djilali, G. Fedak, C. Germain, T. Herault, P. Lemarinier, O. Lodygensky, F. Magniette, et al., MPICH-V: toward a scalable fault tolerant MPI for volatile nodes, in: Proceedings of the ACM/IEEE 2002 Supercomputing Conference, IEEE, 2002, pp. 29.
[19] R. Gioiosa, J.C. Sancho, S. Jiang, F. Petrini, K. Devis, Transparent incremental checkpoint at kernel level: a foundation for fault tolerance for parallel computers, Proceedings of the 17th the ACM/IEEE International Conference on High-Performance Computing, Networking, Storage and Analysis (SC), (2005).
[20] D.J. Kerbyson, H.J. Alme, A. Hoisie, F. Petrini, H.J. Wasserman, M. Gittings, Predictive performance and scalability modeling of a large-scale application, in: Proceedings of the 2001 ACM/IEEE Conference on Supercomputing (SC '01), ACM, Denver, CO, 2001, pp. 37.
[21] O. Lubeck, M. Lang, R. Srinivasan, G. Johnson, Implementation and performance modeling of deterministic particle transport (Sweep3D) on the IBM Cell/B.E., Sci. Program. 17 (1–2) (2009) 199–208.

[22] G. Bosilca, R. Delmas, J. Dongarra, J. Langou, Algorithm-based fault tolerance applied to high performance computing, J. Parallel Distrib. Comput. 69 (4) (2009) 410–416.
[23] Z. Chen, Algorithm-based recovery for iterative methods without checkpointing, in: Proceedings of the 20th International Symposium on High Performance Distributed Computing (HPDC '11), ACM, San Jose, CA, 2011, pp. 73–84.
[24] Z. Chen, J. Dongarra, Algorithm-based checkpoint-free fault tolerance for parallel matrix computations on volatile resources, in: Proceedings of the 20th International Parallel and Distributed Processing Symposium, IPDPS 2006, 2006, 10 pp.
[25] K.H. Huang, J.A. Abraham, Algorithm-based fault tolerance for matrix operations, IEEE Trans. Comput. C-33 (6) (1984) 518–528.
[26] P. Du, A. Bouteiller, G. Bosilca, T. Herault, J. Dongarra, Algorithm-based fault tolerance for dense matrix factorizations, in: Proceedings of the 17th ACM SIGPLAN Symposium on Principles and Practice of Parallel Programming (PPoPP '12), ACM, New Orleans, LA, 2012, , pp. 225–234.
[27] P. Wu, C. Ding, L. Chen, F. Gao, T. Davies, C. Karlsson, Z. Chen, Fault tolerant matrix-matrix multiplication: correcting soft errors on-line, in: Proceedings of the Second Workshop on Scalable Algorithms for Large-Scale Systems (ScalA '11), ACM, Seattle, WA, 2011, pp. 25–28.
[28] R.G. Dreslinski, M. Wieckowski, D. Blaauw, D. Sylvester, T. Mudge, Near-threshold computing: reclaiming Moore's law through energy efficient integrated circuits, Proc. IEEE 98 (2) (2010) 253–266.
[29] P. Kogge, K. Bergman, S. Borkar, D. Campbell, W. Carlson, W. Dally, M. Denneau, P. Franzon, W. Harrod, K. Hill, J. Hiller, S. Karp, S. Keckler, D. Klein, R. Lucas, M. Richards, A. Scarpelli, S. Scott, A. Snavely, T. Sterling, R.S. Williams, K.A. Yelick, ExaScale computing study: technology challenges in achieving exascale systems, Tech. Rep., DARPA-2008-13, DARPA IPTO, 2008.
[30] G. Kestor, R. Gioiosa, D.J. Kerbyson, A. Hoisie, Quantifying the energy cost of data movement in scientific applications, in: Proceedings of the 2013 IEEE International Symposium on Workload Characterization (IISWC), 2013, pp. 56–65.
[31] S. Gao, B. He, J. Xu, Real-time in-memory checkpointing for future hybrid memory systems, in: Proceedings of the 29th ACM on International Conference on Supercomputing (ICS '15), ACM, Newport Beach, CA, 2015, pp. 263–272.
[32] J. Dongarra, et al., The international exascale software project roadmap, Int. J. High Perform. Comput. Appl. (2011) 1094342010391989.
[33] K. Wang, K. Brandstatter, I. Raicu, Simmatrix: simulator for many-task computing execution fabric at exascale, in: Proceedings of the High Performance Computing Symposium, Society for Computer Simulation International, 2013, p. 9.
[34] J. Dongarra, Impact of architecture and technology for extreme scale on software and algorithm design, in: Department of Energy Workshop on Cross-cutting Technologies for Computing at the Exascale, 2010.

CHAPTER

Security in embedded systems*

J. Rosenberg
Draper Laboratory, Cambridge, MA, United States

1 NOT COVERED IN THIS CHAPTER

This chapter is about logical and network security (or lack thereof) of embedded devices. The physical security of the environs of the embedded device is not touched on here. In many situations, such as when the embedded devices are part of an enterprise—used in manufacturing or part of some sort of infrastructure or part of the financial operations (e.g., point of sale (POS) systems)—or when used in a contained military vehicle like a submarine or bomber, the physical security of the environment takes care of the embedded device. In situations where the embedded device is just one of many Internet of things (IoT), physical security of one device may not be vital but security of the whole array of devices is tied to the network that connects them, we will be discussing network security for embedded devices in this chapter. While the concept of denial of service (DoS) will be covered here, it is difficult if not impossible for the embedded device itself to locally defend itself against DoS attacks. Finally, where physical security of the embedded device is of critical importance, but the device cannot be part of any large organizational security infrastructure, an antitamper approach must be taken. Antitamper technology equips the device with special hardware and software that makes sure the device is, first, nonoperational when not in the hands of those authorized to possess it, and second, refuses to give up any secrets it contains if the unauthorized parties in possession try to tamper with the device in an attempt to steal its secrets. Perhaps even more important, if a device is considered *inherently secure*, antitamper is necessary to maintain that designation. We will only briefly cover antitamper as it is a large topic unto itself but we will discuss how an embedded device can become inherently secure in this chapter.

*Please visit the companion website http://booksite.elsevier.com/9780128024591 for part two of this chapter which covers the following topics in detail: Important Security Concepts, Security And Network Architecture, Software Vulnerability And Cyber Attacks, Security And Operating System Architecture.

Rugged Embedded Systems. http://dx.doi.org/10.1016/B978-0-12-802459-1.00006-3

2 MOTIVATION

2.1 WHAT IS SECURITY?

In the English language, security is defined as the state of being free from danger or threat.

Consequently we lock our car. We like to walk along well-lit streets. Many of us have alarm systems on our houses. Thanks to important banking laws enacted after the depression, we trust the government to back up losses at our bank. We use credit cards knowing that if they are stolen we are not responsible for expenditures on them. We are (mostly) free from danger or threat in our physical being and in our financial dealings.

But what happens when your house or car is broken into and your things are stolen, or you hear footsteps behind you on a dark street, or your bank is hacked and your identity is stolen? Understandably you have a visceral reaction. It feels like you have been personally violated.

The early days of computer hacking was simply vandalism usually designed to disable a computer. The first significant computer worm was The Morris worm of Nov. 2, 1988. It was the first to be distributed via the Internet. And it was the first to gain significant mainstream media attention. It also resulted in the first conviction in the United States under the 1986 Computer Fraud and Abuse Act. It was written by a graduate student at Cornell University, Robert Morris, and launched from MIT (where he is now a full professor).

But as the stakes got higher and higher and people who wanted things that were not theirs got more sophisticated in their programming skills, much bigger targets were taken on including big corporations, public infrastructure, and the military.

We are talking about security of embedded systems in this chapter so let's begin with the fundamental principles on which computer security is based.

2.2 FUNDAMENTAL PRINCIPLES

Computer security is defined in terms of what is called the CIA triad: Confidentiality, Integrity, and Availability [1]. Let's take each concept in turn and list out the traits of that principle.

2.2.1 Confidentiality

- Assurance that information is not disclosed to unauthorized individuals, programs, or processes.
- Ensures that the necessary level of secrecy is enforced at each junction of data processing and prevents unauthorized disclosure.
- Some information is more sensitive than other information and requires a higher level of confidentiality.
- The military and intelligence organizations classify information according to multilevel security designations such as Secret, Top Secret, and so forth that require protection from disclosure.
- Individuals have personally identifiable information such as social security numbers that also require protection from disclosure to unauthorized individuals.

- Attacks against confidentiality perpetrated by individuals tend to be about personal gain, so they go after credit card numbers that they use themselves or sell on the black market.
- On the other hand, Nation States may be after corporate confidential information like plans for a new fighter jet or national or corporate secrets.
- Either individuals, corporations, or government information may be the target of insider attacks, who have a variety of motives for accessing information they are not authorized to.

2.2.2 Integrity

- Assures that the accuracy and reliability of information and systems are maintained and any unauthorized modification is prevented.
- Must make sure data, resources, are not altered in an unauthorized fashion.
- Modification of data on disk, or in transit would violate integrity.
- Modification of any programs, computer systems, and network connections would also violate integrity.
- Vandals may try to change or destroy data just to create confusion or fear.
- Nation states may try to change data to affect outcomes beneficial to them. The military would fear an attack that changes battlefield recognizance information or command directives. These would be integrity attacks.

2.2.3 Availability

- Ensures reliability and timely access to data and resources to authorized individuals.
- Information systems to be useful at all must be available for use in a timely manner so productivity is not affected.
- A system that loses connectivity to its database would become useless to most users.
- A system that runs so slowly that users cannot get their work done becomes useless as well.
- Denial of service attacks have a wide variety of forms but all tend to keep some aspect of a system so busy responding to the DoS traffic that it no longer functions for its legitimate users.

Computer security is necessary because there are threats. The best way to think about what types of security is needed is to establish a model for those threats. This will get us into some basic vocabulary that is used to describe threats.

2.3 THREAT MODEL

Key vocabulary used in security discussions includes "vulnerability," "threat," "risk," and "exposure," which would create a lot of confusion if they are really the same thing; which they are not. We will define them here and, more importantly, we will define a threat model in terms of how these concepts interact with each other.

2.3.1 Vulnerability

A vulnerability is a software, hardware, procedural, or human weakness that provides a hacker an open door to a network or computer system to gain unauthorized access to resources in the environment. Weak passwords, lax physical security, unmatched applications, an open port on a firewall, or bugs in software enable an attacker to leverage that flaw to gain access to the computer that software is running on. The Heartbleed bug was a vulnerability deployed on 66% of servers on the Internet [2] (not counting email, chat or VPN servers, or any embedded devices) created by a programmer's failure to verify the bounds of a buffer that allowed unlimited external access to a server's internal memory. For years, the vast majority of home wireless access points were shipped with the administration password of "password" and most users never changed that creating a huge vulnerability.

2.3.2 Threat

A threat is a potential danger to information or systems. The danger is that someone or something will identify a specific vulnerability and use it against the company or individual. A hacker or attacker coming into a network from across the internet is a threat agent. The entity that takes advantage of a vulnerability is a threat agent. A threat agent might be an intruder accessing the system through an open port on firewall. Or it could be a process accessing data in a way that violates security policy. An employee making a mistake that exposes confidential information is an (inadvertent) threat agent. The insider who, for their own reasons, wants to cause harm or steal information taking advantage of their insider access privileges is a threat agent.

2.3.3 Risk

Risk is the likelihood of a threat agent being able to take advantage of a vulnerability and the operational impact if they do. Risk ties together the vulnerability, threat and likelihood of exploitation to the resulting operational loss. If firewall ports are open, if users are not educated on proper procedures, if an intrusion detection system is not installed or configured correctly, risk goes up. If a known vulnerability is not addressed, risk goes up because bad guys pay attention to discovered vulnerabilities, they did not discover themselves.

2.3.4 Asset

Assets can be physical, information, monetary, or reputation. A physical asset includes computers, network equipment, or attached peripherals. Information assets are things such as customer data, proprietary information, and secret information. Monetary assets include the fines or direct expenses from a breach or loss of stock value in the stock market. And reputation asset is what is lost due to the poor perceptions of a company that has suffered a high visibility breach.

2.3.5 Exposure

Exposure is an instance of being subjected to losses or asset damage from a threat agent. A vulnerability exposes an organization to possible damages. Just as failure to have working fire sprinklers exposes an organization not only to fire danger but also losses of physical assets and potential liabilities, poor password management exposes an organization to password capture by threat agents who would then gain unauthorized access to systems and information.

2.3.6 Safeguard

A safeguard (or countermeasure) mitigates potential risk. It could be software, hardware, configurations, or procedures that eliminate a vulnerability or reduce likelihood a threat agent will be able to exploit a vulnerability.

The relationships between all of these concepts form a threat model: A threat agent creates a threat, which exploits a vulnerability that leads to a risk that can damage an asset and causes an exposure, which can be remedied through a safeguard or countermeasure.

If no one and nothing had access to our computers and networks we would be done with this discussion and this would be a very short chapter. Threats exist because there is access (usually through a network a system is connected to but also even when a system is air-gapped), so the next thing we need to include, when thinking about what security is, has to be access control. Readers who are aware of embedded systems that are not connected to any network or are on a network that is air-gapped from any network, where threat agents are active and think those devices are safe from threat just need to remember how Stuxnet was able to destroy 2000 uranium processing centrifuges that were air-gapped from the Internet.

2.4 ACCESS CONTROL

Fundamental to security is controlling how resources are accessed. A subject is an active entity that requests interaction with an object. An object is a passive entity that contains information. Access is the flow of information between a subject and an object. Access controls are security features that control how subjects and objects communicate and interact with each other.

2.4.1 Identification

All access control, and all of security for that matter, hinges on making sure with high confidence we know who is requesting access to a resource. If everyone can claim to be Barack Obama and gains access as such, there is little hope we will have much security in our systems. Identification is the method for ensuring that a subject is the entity it claims to be. Once an authoritative body has accepted the entity is who they say they are, that information is mapped to something that can be used to claim that identity such as a login user name or account number.

2.4.2 Authentication

A weakness in most systems is that identification is quite easy to spoof. Stronger defense against what are called social engineering attacks requires a second piece of the credential beyond identification such as password, passphrase, cryptographic key, personal identification number, anatomical attribute (biometric), token, or answers to a set of shared secrets.

2.4.3 Authorization

Authorization asks the question: Does the subject have the necessary rights and privileges to carry out the requested actions? If so, the subject is authorized to proceed.

2.4.4 Accountability

Keeping track by identity of what each subject did in the system is called accountability. It is used for forensics, to detect attacks, and to support auditing.

Clearly, different organizations have different needs when it comes to security. The national intelligence and defense organizations need to protect confidential, secret, and top secret information in separate buckets and people without sufficient clearance must never be allowed to see higher level information. A bank must not allow customer social security numbers to get to unauthorized individuals. These are examples of security policies.

2.5 SECURITY POLICY

A security policy is an overall statement of intent that dictates what role security plays within the organization. Security policies can be organizational policies, issue-specific policies, or system-specific policies, or a combination of all of these.

Security policies might:

- identify assets the organization considers valuable;
- state company goals and objectives pertaining to security;
- outline personal responsibility of individuals within the organization;
- define the scope and function of a security team;
- outline the organization's planned response to an incident including public relations, customer relations, or government relations; and
- outline the company's response to legal, regulatory, and standards of due care.

2.6 WHY CYBER?

Cyber is short for cybernetics. It is used as an adjective defined as: of, relating to, or characteristic of the culture of computers, information technology, and virtual reality. That doesn't really explain why it started to be used so heavily because it was a simple substitution for *Information Technology* or *Computers* or even *Internet*. The best answer might just be that it sounds cool and the media tend to drive the popular terminology. Whatever the reason, our subject is now described by the terms *cyber-security*, *cyber-attacks*, and *cyber-threats*.

2.7 WHY IS SECURITY IMPORTANT?

Security has become vitally important because of the critical nature of what our computers and networks do now compared to two or even one generation ago. For instance, they run the electric grid. On Aug. 14, 2003, shortly after 2 p.m. Eastern Daylight Time, a high-voltage power line in northern Ohio brushed against some overgrown trees and shut down—a fault, as it's known in the power industry. The line had softened under the heat of the high current coursing through it. Normally, the problem would have tripped an alarm in the control room of FirstEnergy Corporation, an Ohio-based utility company, but the alarm system failed due to a software bug—a vulnerability. (Here the threat agent was Mother Nature not a hacker.) All told, 50 million people lost power for up to 2 days in the biggest blackout in North American history. The event contributed to at least 11 deaths and cost an estimated $6 billion. The entire power grid is vulnerable to determined hackers and a nationwide blackout—according to the Wall Street Journal quoting US intelligence sources, nation-states have penetrated the US power grid [3].

Similarly, computers control our water supply. EPA advisories have outlined the risk of cyber attack and its consequences. Our drinking water and wastewater utilities depend heavily on computer networks and automated control systems to operate and monitor processes such as treatment, testing, and movement of water. These industrial control systems (ICSs) have improved drinking water and wastewater services and increased their reliability. However, this reliance on ICSs, such as supervisory control and data acquisition (SCADA) embedded devices, has left the Water Sector like other interdependent critical infrastructures, including energy, transportation, and food and agriculture, vulnerable to targeted cyber attacks or accidental cyber events. A cyber attack causing an interruption to drinking water and wastewater services could erode public confidence, or worse, produce significant panic or public health and economic consequences. In factories producing chemicals, processing food, and manufacturing products, every valve, pump, and motor has the same sort of ICS processor control and monitoring embedded system each one of which is highly susceptible to cyber attack.

Transportation systems, air traffic control (ATC), airplanes themselves, trains, ships, even our cars have dozens of processors in them and many of those are in embedded systems with little or no physical protection from cyber attack. Havoc of disruption to our air, train, ship, or road-based transportation would have a significant impact on our economy and society. Transportation represents 12% of the GDP, it is highly visible and most citizens are dependent on it. Just if GPS went down it would shut down most of the above systems. You just have to see what happens when one stoplight is out and multiply that by tens of thousands just for autotransportation.

The stock market and commodity trading systems use thousands of embedded systems to function. A major disruption to the stock market stops the flow of investment capital and thus could shut down the economy.

Besides these major systems, other segments of the economy that are heavily dependent on highly vulnerable embedded systems include financial systems including POS, e-commerce, and inventory. Twenty percent of the economy is health-related including hospitals, medical devices, and even insurance companies. While not life-critical, the entertainment segment is big business and an important aspect of life including movies, TV, cable systems, satellite TV, video games, music, audio, and e-books.

Increasingly, our homes and all the devices in them such as thermostats, washers, dryers, stoves, microwaves, and even newer toasters are all controlled by embedded systems; they are all part of the trend toward an IoT.

Perhaps the heaviest user of embedded systems to date is our military. Military systems from satellites, to weapons systems, to command & control (C&C) are heavily dependent on embedded processors. Recent hacks against Central Command, the Office of Personnel Management (not an embedded device attack), and hundreds not reported, point to the danger.

And of course, the Internet and World Wide Web themselves are controlled by millions of embedded devices. Designed to withstand a direct nuclear hit, the Internet is indeed very resilient but never was the cyber attacking envisioned or planned for and no mechanisms to help defend against them were ever designed in. In fact, the entire Web is designed to help the attackers with ways to hide their tracks, dynamically change IP addresses, and to anonymize their location and attack paths.

Let's talk some more about the IoT because this is the future of embedded devices and represents a huge potential source of vulnerabilities. What is the IoT and what is the vision? Currently there are 9 billion interconnected devices, which is expected to reach 24 billion devices by 2020. This next major expansion of our computing base will be outside the realm of the traditional desktop and will sit squarely in the embedded device space. In the IoT paradigm, many of the objects that surround us will be on the network in one form or another. Radio frequency identification (RFID) and sensor network technologies will represent a major portion of the sensors that will drive the IoT expansion. Many new types of sensors will be part of the IoT including smart healthcare sensors, smart antennas, enhanced RFID sensors, RFID applied to retail, smart grid, and household metering sensors, smart traffic sensors, and many others. IoT will have major implications for privacy concerns and impressive new levels of big data storage, management, analytics and, yes, security. IoT will be such a large market opportunity that it is the latest "gold rush" and companies are rushing new products to market as fast as possible with very little attention being paid to innate security. The hackers must be rubbing their hands in anticipation of their coming opportunities.

2.8 WHY ARE CYBER ATTACKS SO PREVALENT AND GROWING?

In 2014, companies reported 42.8 million detected attacks (the number unreported is probably at least as much) worldwide, a 48% year-over-year increase [4]. Some nation-states who are sometimes considered "adversaries" of the US have acknowledged they have large teams of cyber warriors [5]. But there are a lot of factors that have contributed to the exponential growth of cyber attacks. We identify a few of those here. The rapid growth and severity of incidents is stunning. Over 220,000 new malware variants appear every day according to AV-TEST. The number of new malware programs has rapidly grown over the last 10 years to exceed 140 million per year.

Source: AV-TEST, https://www.av-test.org.

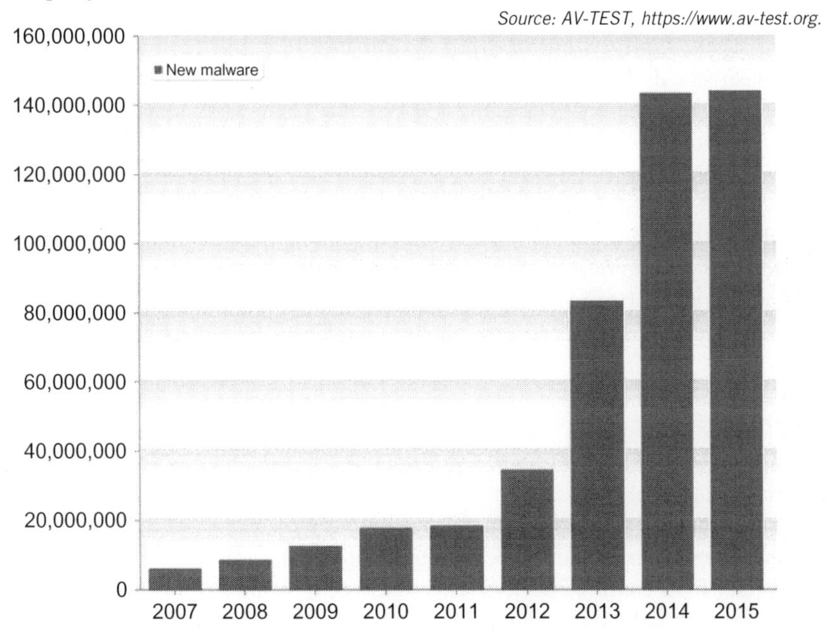

2.8.1 Mistakes in software

Mistakes are a big part of how cyber attackers ply their trade. Programmers make a mistake, that becomes a bug in an application and the cyber attackers exploit that bug as a cyber vulnerability. Many mistakes in the software in the SCADA controllers, and the Windows-based systems commonly used to program them, were leveraged to destroy 2000 centrifuges in the 2008 Stuxnet attack. Similarly, it was mistakes in various aspects of the systems at Sony, Target, Home Depot, and hundreds of others that allowed those famous cyber attacks to proceed. Stuxnet, however, is in a new class of dangerous attacks designed to destroy physical assets from across the Internet but this category also includes the Shamoon attack on Saudi Aramco that destroyed 30,000 desktop computers and in late 2014 hackers struck an unnamed steel mill in Europe. They did so by manipulating and disrupting control systems to such a degree that a blast furnace could not be properly shut down, resulting in "massive"—though unspecified—damage.

According to Lloyd's of London, businesses lose \$400B per year due to cyber attacks but due to potential lawsuits and liability issues, much cybercrime goes unreported and so is presumably much higher than \$400B. According to the US Department of Homeland Security, about 90% of security breaches originate from defects in software. The embedded processors in mobile devices and the applications built for them are becoming the most common vector of attack on companies. These mobile apps typically have poor security because developers rush to bring them to market before properly implementing security protocols. More on software bugs and vulnerabilities to cyber attack is discussed later in this chapter.

2.8.2 Opportunity scale created by the Internet

The sheer size of the Internet, the existence of malcontents, determined nation-state adversaries, and the financial gains now possible from cyber attacks, are all reasons for growth of cyber crime.

Size of the Internet as measured by domains is 271 M, [6] servers, and as measured by users is 3.17B users.[1] There has been an exponential explosion of smart phones that are Internet connected. In 2015, there were 2B smart phone users worldwide and that is expected to grow to over 6B by 2020.[1] GE estimates the "Industrial Internet" has the potential to add $10 to $15 trillion[2] to global GDP over the next 20 years. Cisco states that its forecast for the economic value of the IoT is $19 trillion[3] in the year 2020.

2.8.3 Changing nature of the adversaries

The first cyber-criminals were purely vandals. The previous generation's graffiti is hacking into a site and scribbling on their home page. There is an aspect of Haves vs. Have-nots on the Internet where the have-nots can be across borders and oceans and still attack the rich people in the rich countries. Thus phase two of cyber-crime became identity theft and wholesale thefts of credit cards which are re-sold as in the Target POS credit card theft. This phase also included a rash of cyber-ransom attacks against individuals and small businesses. Meanwhile increasing steadily are the actions of nation-states, where the aim is to steal secrets and to obtain an advantageous posture for a future action that will include cyber-war as part of an overall strategy. These nation states steal identities of government employees, and contractors in order to blackmail individuals into turning over secrets, they are behind Flame—an attack against the embedded devices in personal computers (e.g., Keyboard, microphone, camera)—whose mission is espionage and is considered by many to be the most complex malware ever found. The US's main adversaries have penetrated critical infrastructure and have, it is believed, left behind mechanisms allowing them to take action against that infrastructure when it suits their overall aims.

2.8.4 Financial gain opportunities

The Target attack, enabled by the HVAC-contractor vulnerability, ultimately compromised the POS systems where consumer's credit cards are swiped. The foreign organized crime hackers who perpetrated the attack were highly enriched by their ability to resell those credit cards on the black market [7]. Those stolen cards were being offered at "card shops" starting at $20 each, up to more than $100. Between 1 and 3 million of those credit cards were ultimately sold on the black market, raising an estimated $53.7 million for the hackers.[4] The attack costs Target $148 million,

---

[1]Statistica.
[2]General Electric Corporation.
[3]World Economic Forum.
[4]Newsweek.

and costs credit card issuers institutions $200 million. The CEO lost his job and company profits fell 46% the quarter after the breach.

2.8.5 Ransomware

Ransomware is malware that locks your keyboard or computer to prevent you from accessing your data until you pay a ransom, usually demanded in Bitcoin. The digital extortion racket is not new—it's been around since about 2005, but attackers have greatly improved on the scheme with the development of ransom cryptoware, which encrypts your files using a private key that only the attacker possesses, instead of simply locking your keyboard or computer.

It is not the case that ransomware just affects desktop machines or laptops; it also targets mobile phones and if it became more lucrative, the IoT would be next (think: control of your Nest thermostat during very cold weather).

Symantec gained access to a C&CI server used by the CryptoDefense malware and got a glimpse of the hackers' haul based on transactions for two Bitcoin addresses the attackers used to receive ransoms. Out of 5700 computers infected with the malware in a single day, about 3% of victims appeared to shell out for the ransom. At an average of $200 per victim, Symantec estimated that the attackers hauled in at least $34,000 that day. Extrapolating from this, they would have earned more than $394,000 in a month. This was based on data from just one command server and two Bitcoin addresses; the attackers were likely using multiple servers and Bitcoin addresses for their operation.

Conservatively, at least $5 million is extorted from ransomware victims each year. But forking over funds to pay the ransom doesn't guarantee attackers will be true to their word and victims will be able to access their data again. In many cases, Symantec reports, this doesn't occur.

2.8.6 Industrial espionage

Worldwide, around 50,000 companies a day are thought to come under cyberattack with the rate estimated as doubling each year. One of the means perpetrators use to conduct industrial espionage is by exploiting vulnerabilities in computer software. Malware and spyware as a tool for industrial espionage, are designed to transmit digital copies of trade secrets, customer plans, future plans and contacts. Newer forms of malware include devices which surreptitiously switch on mobile phones camera and recording devices and in some cases, to monitor every keystroke at the keyboard.

Operation Aurora was a series of cyber attacks conducted by advanced persistent threats such as the Elderwood Group based in Asia [8]. The attack has been aimed at dozens of organizations, including Adobe Systems, Juniper Networks, Rackspace, Yahoo, Symantec, Northrop Grumman, Morgan Stanley, and Dow Chemical. The primary goal of the attack[5] was to gain access to and potentially modify source code repositories at these high tech, security and defense contractor companies. The Source Code Management systems were found to be wide open even though these

[5] McAfee.

were the crown jewels of these companies, in many ways more valuable than any financial or personally identifiable data that they may have and spend so much time and effort protecting.

2.8.7 Transformation into cyber warfare

Since 2010, when the cyberweapon Stuxnet was finally understood—and the damage a cyber attack could affect was fully absorbed—it has become clear that cyberwarfare was possible. At the very least, if not used as a weapon directly, cyber attacks were now part of any adversarial nation's foreign policy. The target most frequently cited is our critical infrastructure which in general is heavily dependent on embedded systems such as programmable logic controllers (PLCs) and SCADA controllers. This is exactly what Stuxnet targeted. And this is why this chapter is so important.

2.9 WHY ISN'T OUR SECURITY APPROACH WORKING?

Why hasn't what we've been doing been working to stop or even slow the relentless cyber attacking? We have really smart people; the inventiveness of our cyber security people is legend. We have created firewalls, intrusion detection systems, anomaly sleuthing, virus scanners, schemes to make an application morph itself into a different application to fool attackers, and every possible piece of software protection these brilliant people can invent, and still the attack frequency and seriousness goes up year after year.

We are living in a time of inherent in-security and there is nothing that seems to be working to slow things down much less fix the problem. One reason even experts seem to be throwing up their hands is that the assumption is that our computer systems are so complex that it is impossible to design them without vulnerabilities to cyber attacks.

Therefore, the best we can do, it seems, is to: build virtual walls around our networks using firewalls and intrusion detection systems, then constantly run virus scanning tools on every computer on those networks to look for known attack signatures, patch our operating systems, applications, firmware, and even those security systems (they have bugs too) once a week or so, and then what, pray?

Perimeters are known to be very porous. In fact, hackers like to joke that enterprise networks look like some kind of candy: crunchy on the outside but soft and chewy on the inside. That's because if they do penetrate the perimeter they are in and once in, they can pretty much move through the network and operate at will. In fact, one of the problems has become those perimeter systems themselves. They tend to run at a very high privilege level and because they are very large they have a corresponding number of bugs and those bugs are the entry point for the hackers. Patching is a particularly ineffective strategy for addressing cyber physical and other embedded systems because they can be much harder to reach, they tend to be long-lived and specialized, some are not on a network, and they substantially increase the attack surface of the environment with more patches to deliver and more opportunities for failure to protect.

It is important to remember that signature scanning and patching are only done after an attack has been identified and isolated. This is like waiting to lock your doors until after your neighbors have had a theft. What is most damning about patching's effectiveness is that we have found that common vulnerabilities and exposures (CVEs) [9] that were found between 1999 and 2011 represented 75% of the exploits reported in 2014. That's as much as 15 years since a vulnerability was reported without being patched [10].

2.9.1 Asymmetrical

A DARPA review of 9000 distinct pieces of malware in 2010 found that the average size of each independent attack component was only 125 lines of code. Meanwhile, the defensive systems enterprises use to protect their systems including intrusion detection continue to grow and some of these systems have reached 10 million lines of code. That is so complex—and these systems by nature have to have privileged access to the systems they are protecting—that they have become a desirable target for attack. Like all software, the larger it is the more bugs it has. Steve Maguire in his book *Writing Secure Code* found that across all deployed software regardless of application domain or programming language, one finds between 15 and 50 bugs per thousand lines of code (abbreviated KLOC). Rough estimates have been suggested that 10% of all bugs are potential cyber vulnerabilities. Given this, a 10-million lines of code defensive system has 150,000 bugs at best and potentially 15,000 security vulnerabilities that could enable an attacker entry into the network and all the systems the defensive system was designed to protect.

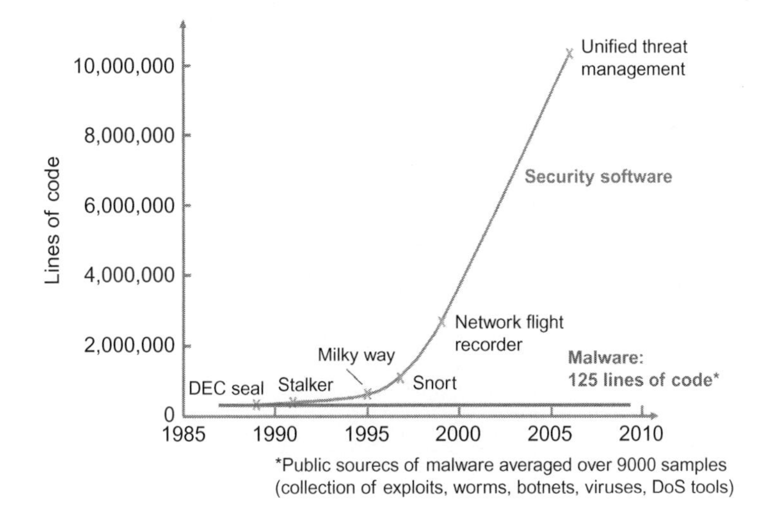

Size of malware compared to the increasing complexity of defensive software.

Malware over almost a 25-year span has remained at about 125 lines of code measured over 9000 samples. But in that same period, defensive systems have gotten increasing more complex to the point where they are over 10 million lines of code. Since studies have shown that there are consistently 15 bugs per KLOC, this means the threat and we are diverging. Figure credit DARPA.

2.9.2 *Architectural flaws*

The crux of the matter is that we are losing the cyber war mostly because the architecture of our computer processors has no support for cybersecurity, so it is practically child's play to attack them and have them do the bad guys bidding. No matter what virus protection, firewalls, or number of patches we apply, software is written by people and people cannot write perfect software, so vulnerabilities will always exist in any reasonably complex software. Bad guys are smart, they are patient, and they can win with very simple attack software.

Our legacy processor architectures are built around what many in the cyber security community call "Raw Seething Bits." That is their description of what it means to have a single undifferentiated memory, where there is no indication what each word in that memory is. There is no way to tell—and most importantly, no way for the processor to tell—if a word is an instruction or an integer or a pointer to a block of memory storing data. There is an important reason processors started out having a single undifferentiated memory: it was simple to build. Simple was important when logic was performed by vacuum tubes or, a bit later, by just a few really expensive transistors: it was easier and cheaper to build.

That architecture—still in use in all our computing devices whether servers, laptops, or embedded devices—is Von Neumann's 1945 stored-program computer (see Fig. 1). Having worked out the simplest architecture that worked, the industry started to perfect it, made transistors smaller and smaller, and ultimately started to see a trend

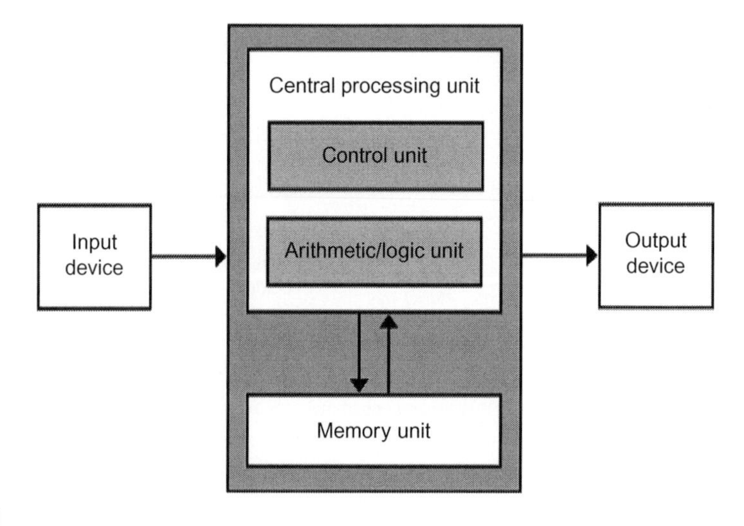

FIG. 1

The Von Neumann processor architecture.

that became known as Moore's Law take hold. Moore's Law stated that every 18th month the number of transistors occupying the same area would double. In fact the reason for the explosion of the Internet, mobile devices, the coming IoT, and information processing in general, is Moore's Law leveraging this simple architecture and driven by a mantra of smaller, cheaper, faster. Moore's Law has given us an iPhone more powerful than all computers NASA owned when it landed men on the moon. Transistors are now incredibly cheap, and while Moore's Law may have reached its limit, the 2000 transistors for the first microprocessor in 1971 has become 5.5 billion transistors in the current model. Ironically, it's the connectedness of everything enabled by Moore's Law that has made cyber threats so prevalent and so serious.

In 1945, Von Neumann and others described a simple but powerful processor architecture with a single internal memory. This architecture continues to dominate the architecture of processors in billions of devices today. This single memory where instructions, data, pointers, and all the data structures needed by an application are stored with no way to tell what is what. It is this memory sometimes called "raw, seething bits" that prevents the processor from cooperating with the program to enforce security. Figure under the Creative Commons Attribution-Share Alike 3.0 by Kapooht.

As our processors became much more powerful and sophisticated, people began to trust them to protect increasingly valuable things. As processor power made more things possible, people's expectations also grew and so software size has grown exponentially to keep up. All the while programming languages like C and C++ continued to provide direct access to the "raw, seething bits" of memory without manifest identity, types, boundaries, or permissions to help explain what each word in memory was for. Even Java—12 years newer than C++ and 23 years newer than C—which was designed to be safer and more helpful to the programmer, retains the C/C++ risks because it builds on libraries written in C and C++ and that unsound foundation leaves it vulnerable to bugs and attacks. This has left everything up to the programmer. When programming in C/C++/Java, the default is unsafe. As we have seen, a buffer when allocated does not protect itself from being overwritten; the programmer always has to do extra work (write more code) to make it safe. This is not so much the fault of the programming language as it is of things below the level of the language the programmer is writing in. It is this undefined behavior of the language, where most of the problems (which become cyber vulnerabilities) arise.

As cybersecurity started to become a concern, there were important constraints that had to be enforced in every program by every programmer to avoid cyber attack. But those vital constraints would only be enforced when every programmer got everything right on every single line of code. Any single mistake could become a vulnerability and sink the ship.

2.9.3 Software complexity—many vulnerabilities

NIST maintains a list of the unique software vulnerabilities (see https://nvd.nist.gov). Across all the world's software, whenever a vulnerability is found that has not been identified anywhere before, it is added to this list. As of this writing, that list was approaching 76,000 unique vulnerabilities. This means attackers have 76,000 things they can leverage to compromise the systems they attack. Vulnerabilities are how attackers get in. Bugs (or weaknesses) are how the software is written so that vulnerabilities are created.

The common weakness enumeration list contains a rank ordering of software errors (bugs) that can lead to a cyber vulnerability. Top 25 most dangerous software errors is a list of the most widespread and critical errors that can lead to serious vulnerabilities in software. They are often easy to find, and easy to exploit. They are dangerous because they will frequently allow attackers to completely take over the software, steal data, or prevent the software from working at all.

Rank	Score	ID	Name
[1]	93.8	CWE-89	Improper neutralization of special elements used in an SQL command ("SQL injection")
[2]	83.3	CWE-78	Improper neutralization of special elements used in an OS command ("OS command injection")
[3]	79.0	CWE-120	Buffer copy without checking size of input ("Classic buffer overflow")
[4]	77.7	CWE-79	Improper neutralization of input during web page generation ("Cross-site scripting")
[5]	76.9	CWE-306	Missing authentication for critical function
[6]	76.8	CWE-862	Missing authorization
[7]	75.0	CWE-798	Use of hard-coded credentials
[8]	75.0	CWE-311	Missing encryption of sensitive data
[9]	74.0	CWE-434	Unrestricted upload of file with dangerous type
[10]	73.8	CWE-807	Reliance on untrusted inputs in a security decision
[11]	73.1	CWE-250	Execution with unnecessary privileges
[12]	70.1	CWE-352	Cross-site request forgery
[13]	69.3	CWE-22	Improper limitation of a pathname to a restricted directory ("path traversal")
[14]	68.5	CWE-494	Download of code without integrity check
[15]	67.8	CWE-863	Incorrect authorization
[16]	66.0	CWE-829	Inclusion of functionality from untrusted control sphere
[17]	65.5	CWE-732	Incorrect permission assignment for critical resource
[18]	64.6	CWE-676	Use of potentially dangerous function
[19]	64.1	CWE-327	Use of broken or risky cryptographic algorithm
[20]	62.4	CWE-131	Incorrect calculation of buffer size
[21]	61.5	CWE-307	Improper restriction of excessive authentication attempts
[22]	61.1	CWE-601	URL redirection to untrusted site ("open redirect")
[23]	61.0	CWE-134	Uncontrolled format string
[24]	60.3	CWE-190	Integer overflow or wraparound
[25]	59.9	CWE-759	Use of a one-way hash without a salt

Rank ordered by a score based on factors including technical impact, attack surface, and environmental factors as determined by NIST and MITRE who maintain the list. Each item in the list has its ID listed with it for cross-reference to the list maintained at http://cwe.mitre.org.

Layer upon layer of software increases the "attack surfaces" that attackers can probe for weaknesses. Every layer is subject to the same 15–50 bugs per KLOC and the corresponding exploitable vulnerabilities. This has led many to conclude that while perimeters are absolutely necessary to keep the vandals and script-kiddies out, they are far from sufficient to keep out the determined and sophisticated hackers that nation-states now deploy.

2.9.4 Complacence, fear, no regulatory pressure to act

Another factor making today's security strategies not work is the simple fact that it is too easy to do nothing. There is the erroneous thinking that systems that are air-gapped are safe (the Stuxnet SCADA controllers were air-gapped). There is the fear and hoping that someone else gets hacked instead of you. In almost all situations for embedded systems there are few if any regulations that force one to act combined with a competitive environment that makes one not want to be first as it cuts into profit margins. The regulations we do have are mostly advisory in nature and have no teeth to require industry to act. It is a situation that is heading for a major catastrophe before serious action is taken.

2.9.5 Lack of expertise

Finally, security is not working for a huge number of organizations—especially smaller ones—simply because these organizations are not experts at IT, and expecting them to be on the cutting edge of IT security is totally unreasonable. Even if they outsource their IT function to a supposed expert, the IT outsourcing companies that service small organizations are themselves small businesses that cannot keep pace with the rapid change in the cybersecurity domain.

2.10 WHAT DOES THIS MEAN FOR THE IoT SECURITY?

The IoT is an extremely fast growing area with many established companies building products. But many of the companies are small startups trying to leverage this brand new market into a business. Projections by Cisco and others predict 50 billion connected IoT devices by 2020. There are lots of old products that are getting this "new" name IoT such as routers and home devices. New or old, companies participating in IoT have very little time to get a product to market so they do whatever they have to, to get their first products to market in the least amount of time. Products in the IoT space have to be inexpensive, so margins are thin and all participants—large and small companies—cut corners and build products that are powered by specialized computer chips made by companies such as Broadcom, Qualcomm, and Marvell. These chips are cheap, and the profit margins slim. Aside from price, the way processor manufacturers differentiate themselves from each other is by features and bandwidth. They typically put a version of the Linux operating system onto the chips, as well as a bunch of other open-source and proprietary components and drivers. They do as little engineering as possible before shipping, and there's little incentive to update their systems until absolutely necessary.

The system manufacturers—usually original device manufacturers (ODMs) who often don't get their brand name on the finished product—choose a chip based on price and features, and then build their IoT device such as a router or server, or something else. They tend not to do a lot of engineering either. The brand-name company on the box adds a user interface perhaps a few new features, they run batteries of tests to make sure everything works, and then they are done, too.

The problem with this process is that no one entity has any time, margin room (or maybe any cash at all), expertise, or even ability to patch the software once it is shipped. The chip manufacturer is busy shipping the next version of the chip. The ODM is busy upgrading its product to work with this next chip. Maintaining the older chips and products is just not a priority. But the situation with the software is worse.

Much of the software is old, even when the device is new. It is common in home routers that the software components are 4–5 years older than the device. The minimum age of the Linux operating system is around 4 years. The minimum age of the Samba file system software is 6 years. They may have had all the security patches applied, but most likely not. No one has that job or has made it a priority. Some IoT components are so old that they're no longer being patched. This patching is especially important because security vulnerabilities are found more easily as systems age.

To make matters worse, it's often impossible to patch the software or upgrade the components to the latest version. Often, the complete source code isn't available. They'll have the source code to Linux and any other open-source components but many of the device drivers and other components are just "binary blobs" with no source code at all. That's the most pernicious part of the problem: one can't patch code that is just binary.

Even when a patch is possible, it's rarely applied. Users usually have to manually download and install relevant patches. But since users never get alerted about security updates, and don't have the expertise to manually administer these devices, it doesn't happen. Sometimes the ISPs have the ability to remotely patch routers and modems, but this is also rare.

The result is hundreds of millions of devices that have been sitting on the Internet, unpatched and insecure, for the last 5–10 years. Devices that have not been on the Internet are being put on it and hackers are taking notice. When TrackingPoint enabled their self-aiming rifles for Wi-Fi, hackers wasted no time in hacking in and re-aiming the rifle and firing it remotely [11]. Similarly, when the skateboard company Boosted brought a remote-controlled powered skateboard to market and failed to encrypt the Bluetooth communications, hackers simply took control and could make a board going 20 miles an hour suddenly stop ejecting its rider [12]. Malware DNS Changer attacks home routers as well as computers. In a South American country, 4.5 million DSL routers were compromised for purposes of financial fraud.

Last month, Symantec reported on a Linux worm that targets routers, cameras, and other embedded devices.

This is only the beginning. What we will see soon are some easy-to-use hacker tools and once we do the script kiddies and vandals will get into the game.

All the new IoT devices will only make this problem worse, as the Internet—as well as our homes and bodies—becomes flooded with new embedded devices that will be equally poorly maintained and unpatchable. Still, routers and modems pose the biggest problem because they are: between users and the Internet so turning them off is usually not an option; more powerful and more general in function than other embedded devices; the one 24/7 computing device in the house, and therefore a natural place for lots of new features.

We were here before with personal computers, and we fixed the problem by disclosing vulnerabilities which forced vendors to fix the problem. But that approach won't work the same way with embedded systems. The scale is different today: more devices, more vulnerability, viruses spreading faster on the Internet, and less technical expertise on both the vendor and the user sides. Plus, as we have shown, we now have vulnerabilities that are impossible to patch. We have a formula for disaster: huge numbers of devices all connected to the Internet, more functionality in devices with a lack of updates, a pernicious market dynamic that has inhibited updates and prevented anyone else from updating, and a sophisticated hacker community chomping at the bit for this green field opportunity.

Fixing this has to become a priority *before* a disaster. We need better designs that start with security and don't build them in on the third or fourth revision, or never. Automatic and very secure update mechanisms—now that all these devices are going to be connected to the Internet—are essential as well.

2.11 ATTACKS AGAINST EMBEDDED SYSTEMS

Embedded systems face distinct challenges separate from networked IT systems. Conventional protective strategies are insufficient to mitigate current cyber vulnerabilities. Most organizations do not currently have sufficient embedded system expertise to provide long-term vulnerability mitigation against the adaptive threat we are seeing. While there is no silver-bullet solution, there are a broad-based set of immediate actions that can significantly mitigate embedded system cyber risk above and beyond basic hygiene:

1. Employ digital signatures and code signing to ensure software integrity of new applications and all updates. Require future systems to cryptographically verify all software and firmware as it is loaded onto embedded devices.
2. Mandate inclusion of software assurance tools/processes and independent verification and validation using appropriate standards as part of future system development. Use best commercial code tools and languages available.
3. Employ hardware/software isolation and randomization to reduce embedded cyber risk and improve software agility even for highly integrated systems.

4. Improve and build organizational cyber skills and capabilities for embedded systems.
5. Protect design/development information (e.g., Source code repositories and revision control systems). Implement security procedures sufficiently early that protection against exfiltration and exploitation is consistent with the eventual criticality of the fielded system.
6. Develop situational awareness hardware and analysis tools to establish a baseline for your embedded operational patterns such that this will inform the best mitigation strategies.
7. Develop and deploy continuously verifiable software techniques.
8. Develop and deploy formal-method software assurance tools and processes.

In the following sections, we will examine 11 specific types of attacks against embedded systems which could be mitigated using the above list of recommendations:

1. Stuxnet, the first true cyber-weapon, created by nation-states to attack and destroy physical equipment half a world away. This was a sophisticated cyber attack against SCADA controllers driving nuclear weapon plutonium refinement. Examining how this attack worked is extremely helpful in thinking how other embedded devices might be attacked and how to prevent it. Stuxnet destroyed the misconception of the myriad people who had been saying "we are air-gapped and so not susceptible like everyone else is."
2. Flame, Gauss, and Dudu are all derivatives of Stuxnet possibly made by different nation-states, and while not as focused on embedded devices per se, they are worth understanding in terms of their scale and sophistication. Again, knowing how these malwares work is helpful in preventing new embedded devices from being susceptible to the next attack.
3. Routers are the backbone of any network including the Internet itself. A router is a fairly simple embedded device, yet the entire network is lost if a router is compromised. It was thought that routers from major market-leading vendors were safe but once again, this overconfidence has proven to be unwarranted.
4. Aviation has embedded devices both on-board the plane and as part of the ground-based control infrastructure, any part of which if compromised could represent catastrophe. Some very simple hacks have been discovered in spite of high degrees of concern and diligence.
5. Automotive not only has a multitude of embedded devices, but it also represents one of the most complex system of systems in some vehicles reaching 100 million lines of source code. Given that Steve Mcquire reports [13] that there are at least 15 bugs per KLOC in deployed systems and approximately 10% of those can be turned into cyber vulnerabilities, these hugely complex vehicles could have 150,000 potential vulnerabilities.
6. Medical devices are embedded systems frequently with a life-critical mission. Some are embedded themselves into the human body. Security of these devices is critical for life safety of course but also to prevent the panic that would

inevitably ensue should cyber hacks occur. Well, they have occurred and quite publicly.

7. ATM Jackpotting is a hack demonstrated at a Black Hat conference. ATMs, like a lot of today's devices and machines, are running fairly standard computers and operating systems internally. ATM machines, like these other devices, allow updates over a network. Software unfortunately has flaws and as security vulnerabilities are found in the ATM software, they must be updated or they risk being hacked in increasingly creative ways like treating an ATM like a jackpot machine that spews out money like a slot machine that hit the jackpot.
8. Military is full of embedded systems from weapons systems to major platforms such as plans, tanks, ships, and submarines. Attack on these systems has not only loss of life implications of national security ones as well.
9. Infrastructure includes electric grid, water supply, communications and transportation to name just a few. All of these types of infrastructures depend heavily on embedded systems to operate. Yet many of these infrastructures are quite old and made up of legacy (and highly insecure) embedded systems. Attack on any of these infrastructures could cripple the country and thus are national security issues themselves.
10. Point-of-sale (POS) systems are included because when it comes to purely making money off of hacking, this is a favorite target of cyber attackers as shown in the Target, Home Depot, TJX, BJs and many others. These embedded devices are one of the most prevalent consumer-facing embedded devices trusted to handle highly sensitive personally identifiable data beginning with credit card numbers and they are highly vulnerable to attack.
11. Social engineering & password guessing. Password guessing must be included because it remains a common way attackers gain control and it is so easy to prevent and yet remains a prevalent weakness of embedded devices just as it is of large computer systems as well.

2.11.1 Stuxnet

Stuxnet was a 500-kilobyte computer worm that infected the software of at least 14 industrial sites in the country it was targeted at, including a uranium-enrichment plant. While a computer virus relies on an unwitting victim to install it, a worm spreads on its own, often over a computer network [14].

This worm was an unprecedentedly masterful and malicious piece of code that attacked in three phases. First, it targeted Microsoft Windows machines and networks, repeatedly replicating itself as it infected system after system. From there it sought out Siemens Step7 software, which is also Windows-based system used to program ICSs that operate equipment, such as centrifuges. Finally, it compromised the PLCs that are directly connected to and in control of the centrifuge motors. The worm's authors could thus spy on the industrial systems and even cause the fast-spinning centrifuges to tear themselves apart, unbeknownst to the human operators at the plant. The Stuxnet victim never confirmed that the attack destroyed some of its centrifuges but satellite images show 2000 centrifuges being rolled out of the facility into the trash heap.

Stuxnet could spread stealthily between computers running Windows—even those not connected to the Internet. If a worker stuck a USB thumb drive into an infected machine, Stuxnet could jump onto it, then spread onto the next machine that read that USB drive. Because someone could unsuspectingly infect a machine this way, letting the worm proliferate over local area networks, experts feared that the malware had perhaps gone wild across the world. Basically it did but Stuxnet was highly selective in terms of location, host, and SCADA controller it had its sights set on so its spread was harmless except to one facility it was designed to target.

For several years the authors of Stuxnet were not officially acknowledged, but the size and sophistication of the worm led experts to believe that it could have been created only with the sponsorship of a nation-state, leading candidates being the United States and one of its allies [15]. Then in Jun. 2012, Obama announced that the United States had been involved. Since the discovery of Stuxnet many computer-security engineers have been fighting off other weaponized viruses based on Stuxnet design principles, such as Duqu, Flame, and Gauss, an onslaught that shows no signs of abating. The callout outlines the 12 steps that the Stuxnet malware went through. The most detailed analysis of how Stuxnet worked was done by Ralph Langer [16].

Put the following in a callout:

1. Stuxnet was initially launched into the wild as early as Jun. 2009, and its creator updated and refined it over time, releasing three different versions. An indication of how determined they were to remain anonymous was that one of the virus's driver files used a valid signed certificate stolen from RealTek Semiconductor, a hardware maker in Taiwan, in order to fool systems into thinking the malware was a trusted program from RealTek.
2. Several layers of masking obscured the zero-day exploit inside, requiring work to reach it, and the malware was huge—500 k bytes, as opposed to the usual 10–15 k. Generally malware this large contained a space-hogging image file, such as a fake online banking page that popped up on infected computers to trick users into revealing their banking login credentials. But there was no image in Stuxnet, and no extraneous fat either. The code appeared to be a dense and efficient orchestra of data and commands.
3. Normally, Windows functions are loaded as needed from a DLL file stored on the hard drive. Doing the same with malicious files, however, would be a giveaway to antivirus software. Instead, Stuxnet stored its decrypted malicious DLL file only in memory as a kind of virtual file with a specially crafted name. It then reprogrammed the Windows API—the interface between the operating system and the programs that run on top of it—so that every time a program tried to load a function from a library with that specially crafted name, it would pull it from memory instead of the hard drive. Stuxnet was essentially creating

an entirely new breed of ghost file that would not be stored on the hard drive at all, and hence would be almost impossible to find.

4. Each time Stuxnet infected a system, it "phoned home" to one of two domains—http://www.mypremierfutbol.com and http://www.todaysfutbol.com hosted on servers in Malaysia and Denmark—to report information about the infected machines. This included the machine's internal and external IP addresses, the computer name, its operating system and version, and whether Siemens Simatic WinCC Step7 software, also known simply as Step7, was installed on the machine. This approach let the attackers update Stuxnet on infected machines with new functionality or even install more malicious files on systems.
5. Out of the initial 38,000 infections, about 22,000 were in the targeted country. The next most infected country was a distant second, with about 6700 infections, and the third-most had about 3700 infections. The United States had fewer than 400.
6. All told, Stuxnet was found to have at least four zero-day attacks in it. That had never been seen before. One was a Windows print spooler vulnerability that allowed the worm to spread across networks using a shared printer. Another attacked vulnerabilities in a Windows keyboard driver and a Task Scheduler process to perform a privilege escalation to root. Finally, it was exploiting a static password Siemens had hard-coded into its Step7 software, which it used to gain access to and infect a server hosting a database used with the Step7 programming system.
7. Stuxnet was not spreading via the Internet like every other worm before it had. It was targeting systems that were not normally connected to the Internet at all (air-gapped) and instead was spreading via USB thumb drives physically inserted into new machines.
8. The payload Stuxnet carried was three main parts and 15 components all wrapped together in layers of encryption which would only be decrypted and extracted when the conditions on the newly infected machine were just right. When it found its target—a Windows machine running Siemens Step7—it would install a new DLL that impersonated a legitimate one Step7 used.
9. Step7 was a Windows-based programming environment for the Siemens PLC used in many industrial control automation applications. The malicious DLL would intercept commands going from Step7 to the PLC and replace them with its own commands. Another portion of Stuxnet disabled alarms as a result of the malicious commands. Finally, it intercepted status messages sent from the PLC to the Step7 machine stripping out any signs of the malicious commands. This meant workers monitoring the PLC from the Step7 machine would see only legitimate commands and have no clue the PLC was being told to do very different things from what they thought Step7 had told it to do. At this moment, the first ever worm designed to do physical sabotage was born.
10. Stuxnet, it turns out, was a precision weapon sabotaging a specific facility. It carried a very specific dossier with details of the target configuration at the

facility it sought. Any system not matching would be unharmed. This made it clear the attackers were a government with detailed inside knowledge of its target.

11. In the final stage of Stuxnet's attack, it searched for the unique part number for a Profibus and then for one of two frequency converters that control motors. The malware would sit quietly on the system doing reconnaissance for about 2 weeks, then launch its attack swiftly and quietly, increasing the frequency of the converters to 1410 Hz for 15 min, before restoring them to a normal frequency of 1064 Hz. The frequency would remain at this level for 27 days, before Stuxnet would kick in again and drop the frequency down to 2 Hz for 50 min. The drives would remain untouched for another 27 days, before Stuxnet would attack again with the same sequence. The extreme range of frequencies suggested Stuxnet was trying to destroy whatever was on the other end of the converters. Satellite images show that approximately 2000 centrifuges—one-fifth of their total—were removed from the targeted country's nuclear facility suggesting they were indeed destroyed.
12. Recently it was discovered there were even older versions of Stuxnet that went undiscovered for many years. The first targeted gas valves in nuclear reactors and the second targeted the reactors' cores.

The 2010 Stuxnet worm used at least three separate zero-day exploits—an unprecedented feat—to damage industrial controllers and disrupt the uranium enrichment facility.

The zero-day vulnerabilities included the following from the list of CVEs [9]:

- CVE-2010-2568—Executes arbitrary code when user opens a folder with maliciously crafted.LNK or .PIF file.
- CVE-2010-2729—Executes arbitrary code when attacker sends a specially crafted remote procedure call message.
- CVE-2010-2772—Allows local users to access a back-end database and gain privileges in Siemens Simatic WinCC and PCS 7 SCADA system.

The attack also exploited already patched vulnerabilities, suggesting that the attackers knew the systems likely would not have been updated.

2.11.2 Flame, Gauss, and Duqu

Flame, Gauss, and Duqu are cousins of Stuxnet but did not focus on embedded systems. Still it is useful to understand what serious attackers are interested in doing and because the distinction between embedded systems and nonembedded systems will lesson over time.

Flame

Flame also known as Flamer and Skywiper, is computer malware discovered in 2012 that attacks computers running the Windows operating system. The program is being used for targeted cyber espionage in Middle Eastern countries. The internal code has few similarities with other malware, but exploits two of the same security vulnerabilities used previously by Stuxnet to infect systems.

Its discovery was announced by Kaspersky Lab and CrySyS Lab of the Budapest University of Technology and Economics, which stated that Flame "is certainly the most sophisticated malware we encountered during our practice; arguably, it is the most complex malware ever found" [17].

Flame can spread to other systems over a local network or via USB stick. It can record audio, screenshots, keyboard activity, and network traffic. The program also records Skype conversations and can turn infected computers into Bluetooth beacons, which attempt to download contact information from nearby Bluetooth-enabled devices. This data, along with locally stored documents, is sent on to one of several C&C servers that are scattered around the world. The program then awaits further instructions from these servers.

According to estimates by Kaspersky in May 2012, Flame had initially infected approximately 1000 machines, with victims including governmental organizations, educational institutions, and private individuals [18]. At that time, 65% of the infections happened in the Middle East and North Africa. Flame has also been reported in Europe and North America. Flame supports a "kill" command, which wipes all traces of the malware from the computer. The initial infections of Flame stopped operating after its public exposure, and the "kill" command was sent [19].

Gauss

Kaspersky Lab researchers also discovered a complex cyber-espionage toolkit called Gauss, which is a nation-state sponsored malware attack closely related to Flame and Stuxnet, but blends nation-state cyber-surveillance with an online banking Trojan [20]. It can steal access credentials for various online banking systems and payment methods and various kinds of data from infected Windows machines such as specifics of network interfaces, computer's drives, and even information about BIOS. It can steal browser history, social network, and instant messaging info and passwords, and searches for and intercepts cookies from PayPal, Citibank, MasterCard, American Express, Visa, eBay, Gmail, Hotmail, Yahoo, Facebook, Amazon, and some other Middle Eastern banks. Additionally Gauss includes an unknown, encrypted payload, which is activated on certain specific system configurations.

Gauss is a nation state sponsored banking Trojan, which carries a warhead of unknown designation according to Kaspersky. The payload is run by infected USB sticks and is designed to surgically target a certain system (or systems), which has a specific program installed. One can only speculate on the purpose of this mysterious payload. The malware copies itself onto any clean USB inserted into an infected personal computer, then collects data if inserted into another machine, before uploading the stolen data when reinserted into a Gauss-infected machine.

The main Gauss module is only about 200 k, which is one-third the size of the main Flame module, but it has the ability to load other plugins, which altogether count for about 2 MB of code. Like Flame and Duqu, Gauss is programmed with a built in time-to-live (TTL). When Gauss infects an USB memory stick, it sets a certain flag to "30." This TTL flag is decremented every time the payload is executed from the stick. Once it reaches 0, the data stealing payload cleans itself from the USB stick. This probably means it was built with an air-gapped network in mind.

There were seven domains being used to gather data, but the five C&C servers went offline before Kaspersky could investigate them. Kaspersky says they do not know if the people behind Duqu switched to Gauss at that time but they are quite sure they are related: Gauss is related to Flame, Flame is related to Stuxnet, and Stuxnet is related to Duqu. Hence, Gauss is related to Duqu. On the other hand, it is hard to believe that a nation state would rely on these banking Trojan techniques to finance a cyber-war or cyber-espionage operation.

So far, Gauss has infected more than 2500 systems in 25 countries with the majority—1660 infected machines—being located in a Middle Eastern country. Gauss started operating around Aug.–Sep. 2011 and it is almost certainly still in operation. Deep analysis of Stuxnet, Duqu, and Flame, leads these experts to believe with a high degree of certainty that Gauss comes from the same "factory" or "factories." All these attack toolkits represent the high end of nation-state sponsored cyber-espionage and cyberwar operations, pretty much defining the meaning of "sophisticated malware" [21].

Duqu

Duqu looks for information that could be useful in attacking ICSs [22]. Its purpose seems not to be destructive but rather to gather information. However, based on the modular structure of Duqu, a special payload could be used to attack any type of computer system by any means and thus cyber-physical attacks based on Duqu might be possible. However, use on personal computer systems has been found to delete all recent information entered on that system, and, in some cases, total deletion of the computer's hard drive has occurred. One of Duqu's actions is to steal digital certificates (and corresponding private keys, as used in public-key cryptography) from attacked computers apparently to help future viruses appear as secure software. Duqu uses a 54 × 54 pixel jpeg file and encrypted dummy files as containers to smuggle data to its C&C center. Security experts are still analyzing the code to determine what information the communications contain. Initial research indicates that the original malware sample automatically removes itself after 36 days, which limits its detection.

Duqu has been called "nearly identical to Stuxnet, but with a completely different purpose." Thus, Symantec believes that Duqu was created by the same authors as Stuxnet, or that the authors had access to the source code of Stuxnet [23]. The worm, like Stuxnet, has a valid, but abused digital signature, and collects information to prepare for future attacks. Duqu's kernel driver, JMINET7.SYS, is so similar to Stuxnet's MRXCLS.SYS that some systems thought it was Stuxnet.

2.11.3 Routers

Routers are attractive to hackers because they operate outside the perimeter of firewalls, antivirus, behavioral detection software, and other security tools that organizations use to safeguard data traffic. Until now, they were considered vulnerable only to sustained denial-of-service attacks using barrages of millions of packets of data. It has been "common knowledge" that they were not very vulnerable to outright takeover. However, a recent very dangerous attack against routers called SYNful has proven this commonly held belief is wrong. SYNful is a reference to SYN, the signal a router sends when it starts to communicate with another router, a process which the implant exploited.

If one seizes control of the router, that attacker owns the data of all the companies and government organizations that sit behind that router. Takeover of the router is considered the ultimate spying tool, the ultimate corporate espionage tool, the ultimate cybercrime tool. Router attacks have hit multiple industries and government agencies.

Cisco has confirmed the SYNful attacks but said they were not due to any vulnerability in its own software. Instead, the attackers stole valid network administration credentials from targeted organizations or managed to gain physical access to the routers. Mandiant has found at least 14 instances of the router implants [24]. Because the attacks actually replace the basic software controlling the routers, infections persist when devices are shut off and restarted. If found to be infected, basic software used to control those routers has to be re-imaged, a time-consuming task for IT engineers.

SYNful Knock is a stealthy modification of the router's firmware image that can be used to maintain persistence within a victim's network. It is customizable and modular in nature and thus can be updated once implanted. Even the presence of the backdoor can be difficult to detect as it uses nonstandard packets as a form of pseudo-authentication.

The initial infection vector does not appear to leverage a zero-day vulnerability. It is believed that the credentials are either default or discovered by the attacker in order to install the backdoor. However, the router's position in the network makes it an ideal target for re-entry or further infection. It is a significant challenge to find backdoors within one's network but finding a router implant is even more so. The impact of finding this implant on one's network is severe and most likely indicates the presence of other footholds or compromised systems. This backdoor provides ample capability for the attacker to propagate and compromise other hosts and critical data using this as a very stealthy beachhead.

The implant consists of a modified Cisco IOS image that allows the attacker to load different functional modules from the anonymity of the internet [25]. The implant also provides unrestricted access using a secret backdoor password. Each of the modules are enabled via the HTTP (not HTTPS), using specially crafted Transfer Control Protocol (TCP) packets sent to the routers interface. The packets have a nonstandard sequence and corresponding acknowledgment numbers. The modules can manifest themselves as independent executable code or hooks within the routers IOS that provide functionality similar to the backdoor password. The backdoor password provides access to the router through the console and Telnet.

This very sophisticated, powerful, and dangerous attack on such an important embedded device is emblematic of what any embedded device developer has to be prepared for.

Aviation

There have been a number of incidents in recent years that demonstrate that worrisome cyber security vulnerabilities exist in the civil aviation system, some of which are in embedded devices within the airplane or attacks on ground systems affected the embedded devices on board. Some incidents of note are:

- an attack on the Internet in 2006 that forced the US Federal Aviation Administration (FAA) to shut down some of its ATC systems in Alaska;
- a cyber-attack that led to the shutdown of the passport control systems at the departure terminals at Istanbul Atatürk and Sabiha Gökçen airports in Jul. 2013, causing many flights to be delayed; and
- an apparent cyber-attack that possibly involved malicious hacking and phishing targeted at 75 airports in the United States in 2013.

More directly related to the vulnerability of on-board embedded systems is the loss of Malaysia Airlines Flight MH370, which is fueling a discussion of whether it is possible to hack into an airplane and gain complete control of on-board systems. In the case of MH370, speculation persists that it was through the entertainment system. This is not far-fetched. The former scientific adviser to the UK's Home Office, Sally Leivesley, revealed Boeing 777 controls could be coopted with a radio signal sent from a small device [26]. A major vulnerability on many aircraft is the lack of any separation between the in-flight entertainment systems and critical control systems. This "open door" for hackers has USB ports and come with Ethernet. Recent modifications Boeing requested the FAA approve added a "network extension device" to separate the various systems from each other.

It was disclosed at a Hack In The Box security conference that it is possible to hack the navigation system within an airplane with an Android Smartphone. This is particularly alarming because just by using such a common and simple device, a hacker is able to take control of the entire on-board control system including plane navigation and cockpit systems. The researcher demonstrated that just using an exploit framework, called Simon, and an Android app, it is possible to gain remote control of the airplane's on-board control system [27].

On the ground, hackers have been able to target ground systems and have successfully grounded 1400 passengers on LOT Polish Airways. Hackers breached the airline's ground computers used to issue flight plans blocking the airline's ability to create flight plans for outbound flights from its Warsaw hub. The airline canceled 20 flights and several others were delayed. The airline's CEO warned all airlines are vulnerable.

The FAA's highly vaunted NextGen system has already been shown to have serious vulnerabilities. The cornerstone of this new system is automatic-dependent surveillance-broadcast or ADS-B, where planes will be equipped with GPS and will constantly send out radio broadcasts announcing to the world who they are and where they are. It turns out that ADS-B signals look a lot like little bits of computer code that are unencrypted and unauthenticated. For computer security experts, these are red flags. Concerned white hat hackers have learned they can spoof these signals and create fake "ghost planes" in the sky. A fake plane could cause a real pilot to swerve—or a series of ghost planes could shut down an airport.

Automotive

Vulnerabilities in car automation systems have been exposed by several groups of security researchers to demonstrate how a networked car's acceleration, braking, and other vital systems could be sabotaged. Researchers have also studied the risk of remote attacks against networked vehicles. A very highly publicized event was a hack on a Jeep, after which Chrysler announced that it was issuing a formal recall for 1.4 million vehicles that may be affected by a hackable software vulnerability in Chrysler's Uconnect dashboard computers [28]. Hacks have demonstrated getting into the controls of some popular vehicles, causing them to suddenly accelerate, turn, kill the brakes, activate the horn, control the headlights, and modify the speedometer and gas gauge readings.

A team from University of California demonstrated that computer systems embedded in automobiles can be hacked to compromise safety features. They have even found ways to remotely hack into these systems using Bluetooth and MP3 files. These and other researchers have been exploring security holes in electronic vehicle controls as automakers develop increasingly complicated in-car computers and Internet-connected entertainment systems with dozens of processors and millions of lines of code. These results have been presented to the National Academy of Sciences Committee on Electronic Vehicle Controls and Unintended Acceleration [29].

Most new cars have some kind of a computer system that controls basic functions, such as brakes and engine performance, as well as advanced features such as Bluetooth wireless and built-in connectors for cell phones, MP3 players, and other devices. All new American-made cars are federally mandated to have a controller area network bus for diagnostics, and several automakers have rolled out cellular technology such as General Motors' OnStar and Ford's Sync services.

The cellular and Bluetooth connectivity is proving to be a particularly fruitful area for hackers. They have demonstrated ability to control the car's brakes, locks, and computerized dashboard display by accessing the on-board computer using the Bluetooth wireless and OnStar and Sync's cellular networks. The hacker team (friendly in this case) was also able to access GPS data and vehicle identification numbers. They broke through the cellular network's authentication system to upload an audio file containing malware. Then they played an MP3 file containing some malicious code over the car's stereo to alter the firmware. They found a vulnerability in the way Bluetooth was implemented that allowed them to execute malicious code by using an app installed on a smart phone that had been "paired" with the car's Bluetooth system. Their work showed that attackers could search for desired models of cars, identify their locations using GPS tracking, and unlock them without laying a hand on the car. They could also sabotage such a car by disabling its brakes.

Another group decided to go after the tire pressure monitors built into modern cars and showed them to be insecure. The wireless sensors, compulsory in new automobiles in the United States since 2008, contain unique IDs, so merely eavesdropping enabled the researchers to identify and track vehicles remotely. Beyond this, they could alter and forge the readings to cause warning lights on the dashboard to turn on, or even crash the engine control unit completely. The tire pressure monitors are notable because they're wireless, allowing attacks to be made from adjacent vehicles while driving down the road. The researchers used equipment costing $1500, including radio sensors and special software, to eavesdrop on, and interfere with, two different tire pressure monitoring systems. While these attacks are more of a nuisance than any real danger—the tire sensors only send a message every 60–90 s, giving attackers little opportunity to compromise systems or cause any real damage—the point is, several teams have now demonstrated that in-car computers have been designed with ineffective security measures.

Medical

More than 2.5 million people rely upon implantable medical devices to treat conditions ranging from cardiac arrhythmias to diabetes to Parkinson's disease. Increasingly, these electronic devices, each with a processor on-board, are connected to some network and are becoming part of the IoT Inevitably, organized crime will turn its attention to the computers inside of us, whether for financial gain, for attention, or simply to cause fear.

Pace makers

Pacemakers from several manufacturers can be commanded to deliver a deadly, 830-V shock from someone on a laptop up to 50 ft away, the result of poor software programming by medical device companies. This was first reported by well-known white hat hacker Barnaby Jack[6] of security vendor IOActive, known for his analysis of other medical equipment such as insulin-delivering devices. The flaw lies with the programming of the wireless transmitters used to give instructions to pacemakers and implantable cardioverter-defibrillators (ICDs), which detect irregular heart contractions and deliver an electric shock to avert a heart attack. A successful attack using the flaw could definitely result in fatalities. Jack made a video demonstration showing how he could remotely cause a pacemaker to deliver an 830-V shock, which could be heard with a crisp audible pop.

As many as 4.6 million pacemakers and ICDs were sold between 2006 and 2011 in the United States alone. In the past, pacemakers and ICDs were reprogrammed by medical staff using a wand that had to pass within a couple of meters of a patient who

---

[6]Barnaby Jack was a New Zealand hacker, programmer and computer security expert. He was known for his presentation at the Black Hat computer security conference in 2010, during which he exploited two ATMs and made them dispense fake paper currency on the stage. Among his other most notable works were the exploitation of various medical devices, including pacemakers and insulin pumps. Jack was renowned among industry experts for his influence in the medical and financial security fields. In 2012 his testimony led the United States Food and Drug Administration to change regulations regarding wireless medical devices. At the time of his death at age 35, Jack was the Director of Embedded Device Security at IOActive.

had one of the devices installed. The wand flips a software switch that would allow it to accept new instructions. But the trend is now to go wireless. Several medical manufacturers are now selling bedside transmitters that replace the wand and have a wireless range of up to 30–50 ft. In 2006, the US Food and Drug Administration approved full radio-frequency-based implantable devices operating in the 400 MHz range.

With that wide transmitting range, remote attacks against the software become more feasible. Jack found the devices would give up their model and serial number after he wirelessly contacted one with a special command. With the serial and model numbers, Jack could then reprogram the firmware of a transmitter, which in turn allowed reprogramming of a pacemaker or ICD in a person's body.

Other problems found with the devices included the fact they often contain personal data about patients, such as their name and their doctor and access to remote servers used to develop the software. It is possible to upload specially crafted firmware to a company's servers that would infect multiple pacemakers and ICDs, spreading through their systems like a real virus. Jack painted a doomsday scenario saying, "We are potentially looking at a worm with the ability to commit mass murder." Ironically, both the implants and the wireless transmitters are capable of using AES (advance encryption standard) encryption, but it was not enabled. The devices also were built with "backdoors," or ways that programmers can get access to them without the standard authentication using a serial and model number.

There is a legitimate medical need for a backdoor since without one, you might have to perform otherwise unnecessary surgery. It is vital when designing a backdoor that extreme care be taken. In this case, it has to be embedded deep inside the ICD core. Ultimately the flaws in the pacemaker could mean an attacker could perform a fairly anonymous assassination from 50 ft away turning a simple laptop into a murder weapon.

And it doesn't take a professional highly advanced hacker like Barnaby Jack to construct these attacks. A University of Alabama group showed that a pacemaker or insulin pump attack can be done by a student with basic information technology and computer science background. The student attackers had no penetration testing skills, but successfully launched brute force and DoS attacks as well as attacks on security controls of a pacemaker. Their attacks included a DoS attack using HPING3 and using Reaver for a brute force attack against Wi-Fi Protected Setup register PIN numbers. In the students second attempt, they were able to crack the pacemaker's passphrase in 9528 s or 2 h 38 min and 48 s.

Diabetes glucose monitors and insulin pumps

Insulin pumps are used to treat patients with diabetes by infusing their bodies with insulin, which, in healthy individuals, is secreted by the pancreas. When insulin levels are too low, people suffer from excessive blood sugar levels, a condition known as hyperglycemia. When insulin levels are too high, they suffer from hypoglycemia, a condition that can result in death if left unchecked. A glucose monitor, until recently just an external device a patient or his/her doctor uses to test blood glucose levels, can now be surgically inserted under the skin to continuously monitor levels without a skin prick. Continuous monitoring that provides fine-grained control over the insulin pump is much healthier because spikes and valleys in blood glucose levels are not good for the patient.

In addition to his pacemaker attack, Barnaby Jack succeeded in taking control of both an insulin pump's radio control and vibrating alert safety mode. Jack's hacking kit included a special piece of software and a custom-built antenna that has a scan range of 300 ft and for which the operator does not need to know the serial number. This meant the latest models of insulin pumps, equipped with small radio transmitters allowing medics to adjust function, can become easy prey to this hacking invention that scans around for insulin pumps. Once the hacker's code takes control of the targeted machine, he can then disable the warning function and/or make it disperse 45 days worth of insulin all at once—a dose that will most likely kill the patient.

After these attacks were brought to light, manufacturers are working to improve the security of the medical devices by evaluating encryption and other protections that can be added to their design. Medical device manufacturers have promised to set up an industry working group to establish a set of standard security practices.

2.12 ATM "JACKPOTTING"

Barnaby Jack has also demonstrated attacks against two unpatched models from two of the world's biggest ATM makers. One exploited software that uses the internet or phone lines to remotely administer a machine made by Tranax Technologies. Once Jack was in, he was able to install a rootkit that allowed him to view administrative passwords and account PINs and to force the machine to spit out a steady stream of dollar bills, something Jack called "jackpotting."

In a second attack against a machine from Triton Systems, Jack used a key available for sale over the internet to access the model's internal components. He was then able to install his rootkit by inserting a USB drive that was preloaded with the malicious program.

Both Triton and Tranax have patched the vulnerabilities that were exploited in the demos. But in a press conference after his revelations, Jack said he was confident he could find similarly devastating flaws—including in machines made by other manufacturers as well. Jack said he wasn't aware of real-world attacks that used his exploits.

To streamline his work, Jack developed an exploit kit he calls Dillinger, named after the 1930s bank robber. It can be used to access ATMs that are connected to the internet or the telephone system, which Jack said is true of most machines. The researcher has developed a rootkit dubbed Scrooge, which is installed once Dillinger has successfully penetrated a machine.

Jack said vulnerable ATMs can be located by war-dialing large numbers of phone numbers or sending specific queries to IP addresses. Those connected to ATMs will send responses that hackers can easily recognize. Jack called on manufacturers to do a better job securing their machines. Upgrades for physical locks, executable signing at the operating system kernel level and more rigorous code reviews should all be implemented, he said.

2.13 MILITARY

Despite a 2013 Pentagon report warning of major vulnerabilities in the cyber security of weapons systems, [30] the US House of Representatives believes the military is lagging behind in closing those software holes. Meanwhile the US Senate proposes a 3-year cyber vulnerability assessment of all major weapons systems focusing on whether they can be hacked. The chief Pentagon weapons tester James Michael Gilmore found after a year of running dozens of tests and simulations on over 40 military weapons systems that almost all of them have some kind of major cybersecurity weakness [31].

Hackers have accessed designs for more than two dozen major US weapons systems according to a classified Defense Science Board report. A partial list of compromised designs includes the F-35 fifth-generation fighter jet, the V-22 Osprey, THAAD missile defense, Patriot missile defense, AIM-120 Advanced Medium-Range Air-to-Air Missile, and the Global Hawk high-altitude surveillance drone.

The benefits to an attacker using cyber exploits are potentially spectacular. Should the United States find itself in a full-scale conflict with a peer adversary, attacks would be expected to include DoS, data corruption, supply chain corruption, traitorous insiders, and kinetic and related nonkinetic attacks at all altitudes from underwater to space. US guns, missiles, and bombs may not fire, or may be directed against our own troops. Hackers could theoretically infiltrate the cryptographic intranet communications systems of an F-35 and render it essentially inoperable. Resupply, including food, water, ammunition, and fuel may not arrive when or where needed. Military Commanders may rapidly lose trust in the information and ability to control US systems and forces. Once lost, that trust is very difficult to regain.

Analysts at IOActive have shown that communications devices from Harris, Hughes, Cobham, Thuraya, JRC, and Iridium are all highly vulnerable to attack. The security flaws are numerous but the most important one—the one that's the most consistent across the systems—is back doors, special points that engineers design into the systems to allow fast access. Cobham defends the practice of providing a back door claiming, as many manufacturers do, that such a back door was a "feature" that the helps ensure ease of maintenance. They believe that you have to be physically present at the terminal to use the maintenance port. But experienced hackers dispute that and while you do need physical access to pull off certain attacks, other vulnerabilities within the swift broadband unit (SBU) can be attacked through the Wi-Fi. This author insists that back doors should never be allowed into critical systems. Another common security flaw in these communication devices used in many weapons systems is hardcoded credentials, which allows multiple users access to a system via a single login identity. Another serious no-no that no customer should allow in devices they purchase.

The most serious vulnerability found on Cobham's equipment allowed a hacker access to the system's SBU and the satellite data unit (SDU). This means any of the systems connected to these elements, such as the multifunction control display unit (MCDU), could be impacted by a successful attack. Meanwhile, the MCDU unit provides information on such vital areas as the amount of fuel left in the plane. A hacker could give the pilot a lot of bad information that could imperil the aircraft, as happened in 2001 aboard Transat Flight 236, when a mechanical error did not inform the pilots that fuel was being diverted to a leaky tank. The pilots didn't know the severity of the mechanical problem until there was a massive power failure in mid-air. This example shows the similarity between "naturally occurring" events and cyber attacks and this similarity is not lost on sophisticated attackers who try to disguise some or all of their attack as a naturally occurring event or in association with such an event such as a storm in the case of a power grid attack.

The military sea-borne systems aren't any safer. Friendly hackers have shown that they can access the SAILOR 6000 satellite communications device, also manufactured by Cobham, which is used in naval settings by countries, including the United States, participating in the Global Maritime Distress and Safety System, an international framework to allow for better communication among maritime actors. This system, which the world's maritime nations—including the United States—have implemented, is based upon a combination of satellite and terrestrial radio services and has changed international distress communications from being primarily ship-to-ship-based to primarily ship-to-shore-based according to the Department of Homeland Security.

Also, closely examined were SATCOMS used by ground forces [32]. SATCOMS are in wide use among the US and NATO militaries used by troops to communicate with units beyond the line of sight. Traffic over these systems is encrypted so there's no threat of enemies listening in on restricted calls. But IOActive showed that it is possible not only to disrupt the communication but also discover troop locations. This means that a would-be enemy could exploit design flaws in some very common pieces of military communications equipment used by soldiers on the front lines to block calls for help, or potentially reveal troop positions.

2.14 INFRASTRUCTURE

The Director of National Intelligence James Clapper has identified cyberattacks as the greatest threat to US national security [33]. Critical infrastructure—the physical and virtual assets, systems, and networks vital to national and economic security, health, and safety such as utilities, refineries, military defense systems, water treatment plants and other facilities on which we depend every day—has been shown to be vulnerable to cyberattacks by foreign governments, criminal entities, and lone actors. Due to the increasingly sophisticated, frequent, and disruptive nature of cyberattacks, such an attack on critical infrastructure could be significantly disruptive or potentially devastating. Policymakers and cybersecurity experts contend that energy is the most vulnerable industry; a large-scale attack could temporarily halt the supply of water, electricity, and gas, hinder transportation and communication, and cripple financial institutions. America's critical infrastructure has become its soft underbelly, the place where we are now most vulnerable to attack.

Over the past 25 years, hundreds of thousands of analog controls in these facilities have been replaced with digital systems. Digital controls provide facility operators and managers with remote visibility and control over every aspect of their operations, including the flows and pressures in refineries, the generation and transmission of power in the electrical grid, and the temperatures in nuclear cooling towers. In doing so, they have made industrial facilities more efficient and more productive. But the same connectivity that managers use to collect data and control devices allows cyber attackers to get into control system networks to steal sensitive information, disrupt processes, and cause damage to equipment. Hackers have taken notice. While early control system breaches were random accidental infections, ICSs today have become the object of targeted attacks by skilled and persistent adversaries.

ICS is a general term that encompasses several types of control systems used in industrial production, including SCADA systems, distributed control systems, and other smaller control system configurations such as PLCs often found in the industrial sectors and critical infrastructures. ICSs are typically used in industries such as electrical, water, oil, gas, and data. Based on data received from remote stations, automated or operator-driven supervisory commands can be pushed to remote station control devices, which are often referred to as field devices. Field devices control local operations such as opening and closing valves and breakers, collecting data from sensor systems, and monitoring the local environment for alarm conditions.

The 2010 discovery of the Stuxnet worm demonstrated the vulnerability of these systems to cyber attack.

Attackers target ICSs because they might want to disrupt critical operations, create public panic, harm a company or the nation economically, or may be stealing information from these devices.

The recently discovered ICS modules of the HAVEX trojan are one example. This malware infiltrated an indeterminate number of critical facilities by attaching itself to software updates distributed by control system manufacturers. When facilities downloaded the updates to their network, HAVEX used open communication standards to collect information from control devices and sent that information to the attackers for analysis. This type of attack represents a significant threat to confidential production data and corporate intellectual property and may also be an early indicator of an advanced targeted attack on an organization's production control systems.

Other hacks represent a direct threat to the safety of US citizens. Recently the FBI released information on Ugly Gorilla, an attacker who invaded the control systems of utilities in the United States. While the FBI suspects this was a scouting mission, Ugly Gorilla gained the cyber keys necessary for access to systems that regulate the flow of natural gas [34].

Considering that cyber attackers are numerous and persistent—for every one you see there are a hundred you don't—those developments should sound alarms among executives at companies using industrial controls and with the people responsible for protecting American citizens from attacks. To a limited extent both businesses and the US government have begun to take action; however, no one is adequately addressing the core of the issue. That issue is that every computer system we have deployed for the last 70 years is highly vulnerable to attack no matter how many patches we employ or how many layers of perimeter we lay down.

2.14.1 Electric grid

The people who carried out the Dec. 2015 first known hacker-caused power outage, used highly destructive malware to gain a foothold into multiple regional distribution power companies in an Eastern European country and delay restoration efforts once electricity had been shut off. The malware, known as BlackEnergy, allowed the attackers to gain a foothold on the power company systems, said the report, which was published by a member of the SANS ICSs team. The still-unknown attackers then used that access to open circuit breakers that cut power. After that, they likely used a wiper utility called KillDisk to thwart recovery efforts and then waged denial-of-service attacks to prevent power company personnel from receiving customer reports of outages.

The attackers demonstrated planning, coordination, and the ability to use malware and possible direct remote access to blind system dispatchers, cause undesirable state changes to the distribution electricity infrastructure, and attempt to delay the restoration by wiping SCADA servers after they caused the outage. This attack consisted of at least three components: the malware, a DoS to the phone systems, and the missing piece of evidence of the final cause of the impact. Analysis indicates that the missing component was direct interaction from the adversary and not the work of malware. Or in other words, the attack was enabled via malware but consisted of at least three distinct efforts.

2.15 POINT-OF-SALE SYSTEMS

POS systems are embedded devices at checkout points in retail stores. Since this is where consumer's credit cards are processed, this has become a favorite target for hackers. The most famous POS attack was against Target during the Christmas holiday shopping season in 2013.

Target disclosed that the initial intrusion into its systems, on Nov. 15, 2013, was traced back to network credentials that were stolen from a third-party HVAC contractor. This was the same HVAC as Trader Joe's, Whole Foods and BJ's Wholesale Club locations in Pennsylvania, Maryland, Ohio, Virginia, and West Virginia. These large retailers, with very large buildings, commonly have an HVAC team that routinely monitors energy consumption and temperatures in stores to save on costs (particularly at night) and to alert store managers if temperatures in the stores fluctuate outside of an acceptable range that could prevent customers from shopping at the store. To support this temperature monitoring, the HVAC vendors need to be able to remote into the system to do maintenance (updates, patches, etc.) or to troubleshoot glitches and connectivity issues with the software.

Between Nov. 15 and Nov. 28 (Thanksgiving, the day before Black Friday—the biggest shopping day of the year), the attackers succeeded in uploading their card-stealing malicious software to a small number of cash registers within Target stores. Two days later, the intruders had pushed their malware to a majority of Target's POS devices, and were actively collecting card records from live customer transactions. Target acknowledged that the breach exposed approximately 40 million debit and credit card accounts between Nov. 27 and Dec. 15, 2013.

The issues here are that the HVAC contractor was seemingly an easy target for the hackers. This is not too surprising given that most small businesses are not able to afford the highest standards of network security nor do they have sufficient IT expertise to even know what they do not know. But more distressing is the fact that Target (and many retailers) do not, and are not required even by published industry standards to, maintain separate networks for payment and nonpayment operations. Those industry standards do require merchants to incorporate two-factor authentication for remote network access originating from outside the network by personnel and all third parties—including vendor access for support or maintenance. Whether the HVAC contractor was required to use two-factor authentication for access to Target's system has not been disclosed.

Callout box with the title: The Target POS attack by the numbers [35].

40 million—The number of credit and debit card thieves stole from Target between Nov. 27 and Dec. 15, 2013.
70 million—The number of records stolen that included the name, address, email address, and phone number of Target shoppers.
46—The percentage drop in profits at Target in the fourth quarter of 2013, compared with the year before.
200 million—Estimated dollar cost to credit unions and community banks for reissuing 21.8 million cards—about half of the total stolen in the Target breach.
100 million—The number of dollars Target says it will spend upgrading their payment terminals to support Chip-and-PIN-enabled cards.
18.00–35.70—The median price range (in dollars) per card stolen from Target and resold on the black market.
1 million–3 million—The estimated number of cards stolen from Target that were successfully sold on the black market and used for fraud before issuing banks got around to canceling the rest.
53.7 million—The income that hackers likely generated from the sale of 2 million cards stolen from Target and sold at the mid-range price of $26.85.
55 million—The number of dollars outgoing CEO Gregg Steinhafel got in executive compensation and other benefits on his departure as Target's chief executive.

2.16 SOCIAL ENGINEERING

Social engineering is when one person tricks another into sharing confidential information, for example, by posing as someone authorized to have access to that information. Social engineering can take many other forms. Indeed, any one-to-one communication medium can be used to perform social engineering attacks. Here we will briefly explore three forms of social engineering: spoofing email, phishing, and password guessing all of which can be a step toward an attack against embedded systems.

2.16.1 Spoofing email

An example of spoofing email might be that an attacker changes the name in the From field to be the name of the network administrator and sends the message to the CEO's assistant. The message says the IT department is having trouble with the CEO's account and to please help them out by supplying the CEO's confidential password. Seeing the correct name of the network administrator and not knowing how to look deeper at the inner workings of the email message, the admin is likely to comply and provide the password. With the CEO's credentials messages can be sent to many people in the company that they would believe that might allow the attacker to get at critical embedded systems within the company's operations.

2.16.2 Phishing

Phishing is closely related to spoofing email. Phishing is used to install and execute CryptoLocker, which is a very malicious form of ransomware that encrypts files on the user's machine and across their entire connected network and demands a ransom to provide the key to decrypt those files. One way it enters is via a spoofed email message that pretends to originate from a source the user trusts and has a relationship with. The message has an attached Word document with a name and a message in the email body that makes the user want to open the document. Once the document is opened, it starts running some embedded Word macros. As those macros begin to execute CryptoLocker installs itself into the user's Documents and Settings folder, using a randomly generated name, and adds itself to the list of programs in the registry that Windows loads automatically every time one logs on. It produces a lengthy list of random-looking server names in the domains .biz, .co.uk, .com, .info, .net, .org, and .ru. It tries to make a web connection to each of these server names in turn, trying one each second until it finds one that responds. Once it has found a server that it can reach, it uploads a small file that is essentially the "CryptoLocker ID." The server then generates a public-private key pair unique to that ID, and sends the public key part back to the source computer. The private key part is tucked away impossible for the victim to access. The malware on the computer uses this public key to encrypt all the files it can find that match an extensive list of extensions, covering file types such as images, documents, and spreadsheets. At this point, the bad guys won and the victim has lost. CryptoLocker then pops up a "pay page," giving the victim a limited time, typically 72 hours, to buy back the private key for their data, typically for 2 bitcoins or $300. Since the ransomers are using modern cryptography, once they have encrypted the victim's disk, there are no back doors or shortcuts to help the victim out. What the ransomers have encrypted can only be decrypted using their private key. Therefore, if they don't provide that private key, the data is as good as gone. There are many situations where this could directly or indirectly affect embedded devices that are within the same network as the original victimized machine. Also other types of phishing can more directly target embedded devices such as those used widely in critical infrastructure.

2.16.3 Password guessing

It may not sound like an attack category but by far the easiest attack against embedded systems is still password guessing. Many devices like routers, PLCs, and even automotive diagnostic ports have factory-defined passwords that are supposed to be changed by the owner at the time of installation but frequently are not. Thus, passwords like "12345," "password," and "admin" are prevalent on millions of devices. As these devices increasingly are put on the Internet for legitimate purposes, those passwords leave a wide open pathway from the outside world into whatever the embedded device is attached to.

Many people use very simple easy to remember passwords such as "123456" or their pet's name. These are extremely easy to guess. This type of behavior will affect consumer IoT devices going forward and with the enormous number of these devices projected into the future, weak passwords could create great problems if a coordinated attack were launched against a lot of these devices simultaneously.

Weak passwords keep getting weaker. That's because ordinary desktop computers can now test over a hundred million passwords per second using password cracking tools that run on a general purpose CPU and over 350 billion passwords per second using GPU-based password cracking tools. A user-selected eight-character password with numbers, mixed case, and symbols, reaches an estimated 30-bit strength, according to NIST; 2○30 is only 1 billion permutations and would take an average of 16 min to crack. If a hacker can access the encrypted passwords (Unix used to place them in /etc/password) then they can bypass any schemes that limit the number of failed attempts before a lockout. All they have to do is run a high-speed password cracker and encrypt that and compare the result to the items in the encrypted password file.

Until two-factor authentication becomes ubiquitous and most device's sole protection is passwords, the best plan of action is to use a password generator (such as SecureSafe) that stores the password in a reliable digital vault that does require two-factor authentication. That vault should have a very strong password that is not stored anywhere. The recommendation for generating solid passwords that have to be remembered is to use a passphrase approach such as a poem you know well where the first letter of each of 12–16 words is remembered and where a few letters are turned into numbers and special characters. For example, You could turn an "l" into a 1, an "o" into a 0, an "a" into a & and separate two phrases with an "=" or ";".

Now that we have looked at what types of embedded devices have had high profile attacks against them and how those attacks have worked, in the next section we will put into context how bad things could get with certain types of attacks against key embedded devices.

2.17 HOW BAD COULD THIS BE?

The network connectivity that the United States has used to tremendous advantage economically and militarily over the past 20 years has made the country more vulnerable than ever to cyber attacks. At the same time, our adversaries are far more capable of conducting such attacks.

Symantec reports blocking over 5.5 billion attacks with its software in 2011 alone finding that the average breach exposed 1.1 million identities and nearly 5000 new vulnerabilities were identified in the calendar year. Over 400 million unique variants of malware attempted to take advantage of those vulnerabilities, up to 40% from 2010. Attack toolkits are easy to find and available in web forums or on the underground black-market and cost only $40–$4000 to procure. Use of these widely available tools allows almost anyone to exploit any known and uncorrected vulnerability.

At the same time, the cyber world has moved from exploitation and disruption to destruction.

The impact of a destructive cyber attack on the civilian population would be frightening as it could mean no electricity, money, communications, TV, radio, or fuel (electrically pumped). In a short time, food and medicine distribution systems would be ineffective; transportation would fail or become so chaotic as to be useless. Law enforcement, medical staff, and emergency personnel capabilities could be expected to be barely functional in the short term and dysfunctional over sustained periods. If the attack's effects were reversible, damage could be limited to an impact equivalent to a power outage lasting a few days. If an attack's effects cause physical damage to control systems, pumps, engines, generators, controllers, etc., the unavailability of parts and manufacturing capacity could mean months to years are required to rebuild and reestablish basic infrastructure operation.

Let's look at a few real instances of attacks and extrapolate that to the growth of the connected embedded devices projected over the next decade.

2.17.1 Heartbleed

Heartbleed is a security bug disclosed in Apr. 2014 in the OpenSSL cryptography library, which is a widely used implementation of the transport layer security (TLS) protocol. SSL (secure sockets layer) and TLS are frequently used interchangeably but TLS is the newer version of the protocol. The "s" in https is from SSL and is what displays the padlock symbol in one's browser and protects transmission of confidential or private data such as social security or credit card numbers. Heartbleed has been shown to have affected 17% of the servers on the Internet and the clients that connect to them. Heartbleed may be exploited regardless of whether the party using a vulnerable OpenSSL instance for TLS is a server or a client. It results from improper input validation (due to a missing bounds check) in the implementation of the TLS heartbeat extension [36] designed to keep sessions in browsers alive and thus the bug's name derives from "heartbeat." The vulnerability is classified as a buffer over-read, a situation where more data can be read than should be allowed, meaning data can be exfiltrated.

The Heartbeat Extension keeps the session alive by having a computer at one end of a connection to send a "Heartbeat Request" message, consisting of a payload, typically a text string, along with the payload's length as a 16-bit integer. The receiving computer then must send exactly the same payload back to the sender. The affected versions of OpenSSL allocate a memory buffer for the message to be returned based on the length field in the requesting message, without regard to the actual size of that message's payload. Because of this failure to do proper bounds checking, the message returned consists of the payload, possibly followed by whatever else happened to be in memory after the allocated buffer.

Heartbleed is exploited by sending a malformed heartbeat request with a small payload and large length field to the vulnerable party (usually a server) in order to elicit the victim's response, permitting attackers to read up to 64 kilobytes of the victim's memory that was likely to have been used previously by OpenSSL. Where a Heartbeat Request might ask a party to "send back the four-letter word 'bird'," resulting in a response of "bird," a "Heartbleed Request" (a malicious heartbeat request) of "send back the 500-letter word 'bird'" would cause the victim to return "bird" followed by whatever 496 characters the victim happened to have in active memory. Attackers in this way could receive sensitive data, compromising the confidentiality of the victim's communications. Although an attacker has some control over the disclosed memory block's size, it has no control over its location, and therefore cannot choose what content is revealed.

The problem can be fixed very simply by ignoring Heartbeat Request messages that ask for more data than their payload needs.

Heartbleed is registered in the CVEs system as CVE-2014-0160. A fixed version of OpenSSL was released on Apr. 7, 2014, on the same day Heartbleed was publicly disclosed.

At the time of disclosure, some 66% of the Internet's secure web servers certified by trusted authorities were believed to be vulnerable to the attack, allowing theft of the servers' private keys and users' session cookies and passwords. The Electronic Frontier Foundation, Ars Technica, and Bruce Schneier [37] all deemed the Heartbleed bug "catastrophic." Forbes cybersecurity columnist Joseph Steinberg wrote, "Some might argue that [Heartbleed] is the worst vulnerability found (at least in terms of its potential impact) since commercial traffic began to flow on the Internet" [38].

Although the bug received more attention due to the threat it represents for servers, TLS clients using affected OpenSSL instances are also vulnerable. Relevant to the embedded device world, a "reverse Heartbleed" has malicious servers exploiting Heartbleed to read data from a vulnerable client's memory. This shows that Heartbleed is not just a server-side vulnerability but also a client-side vulnerability because the server, or whomever you connect to, is as able to ask you for a heartbeat back as you are to ask them. Those clients are in many cases embedded systems so contrary to many people's initial assumptions, Heartbleed is not just an issue for Web servers and other purely computer systems. OpenSSL is very popular in client software and somewhat popular in networked appliances which have most inertia in getting updates.

To wit, Cisco Systems has identified 75 of its products as vulnerable, including IP phone systems and telepresence (video conferencing) systems [39]. Games including Steam, Minecraft, Wargaming.net, League of Legends, GOG.com, Origin, Sony Online Entertainment, Humble Bundle, and Path of Exile were affected and subsequently fixed as well [40].

2.17.2 Shellshock

Shellshock, also known as Bashdoor, is a family of security bugs in the widely used Unix Bash shell, the first of which was disclosed on Sep. 24, 2014. Many Internet-facing services, such as some web server deployments, use Bash to process certain requests, allowing an attacker to cause vulnerable versions of Bash to execute arbitrary commands. This can allow an attacker to gain unauthorized access to a computer system [41].

In Unix-based operating systems, and in other operating systems that Bash supports, each running program has its own list of name/value pairs called environment variables. When one program starts another program, it provides an initial list of environment variables for the new program. Separately from these, Bash also maintains an internal list of functions, which are named scripts that can be executed from within the program. Since Bash operates both as a command interpreter and as a command, it is possible to execute Bash from within itself. When this happens, the original instance can export environment variables and function definitions into the new instance. Function definitions are exported by encoding them within the environment variable list as variables whose values begin with parentheses ("()") followed by a function definition. The new instance of Bash, upon starting, scans its environment variable list for values in this format and converts them back into internal functions. It performs this conversion by creating a fragment of code from the value and executing it, thereby creating the function "on-the-fly," but affected versions do not verify that the fragment is a valid function definition. Therefore, given the opportunity to execute Bash with a chosen value in its environment variable list, an attacker can execute arbitrary commands or exploit other bugs that may exist in Bash's command interpreter. Increasingly, embedded systems run Unix variants and some run the bash shell making this dangerous bug relevant to embedded systems.

The discoverer of Shellshock contacted Bash's maintainer reporting their discovery of the original bug called "Bashdoor." Together the discoverer and maintainer developed a patch. The bug was assigned the CVE identifier CVE-2014-6271. Once Bash updates with the fix were ready for distribution, it was announced to the public. Within days of the publication of this, intense scrutiny of the underlying design flaws discovered a variety of related vulnerabilities which the developers addressed with a series of further patches.

Attackers exploited Shellshock within hours of the initial disclosure by creating botnets of compromised computers to perform distributed denial-of-service attacks and vulnerability scanning. Security companies recorded millions of attacks and probes related to the bug in the days following the disclosure.

Shellshock could potentially compromise millions of unpatched servers and other systems. Accordingly, it has been compared to the Heartbleed bug in its severity [42].

2.17.3 Blackouts

One of the most critical applications of embedded systems has to be across the broad expanse of the US power grid. However this system is quite old and vulnerable to attack. In Oct. 2012, US defense secretary Leon Panetta warned that the United States was vulnerable to a "cyber Pearl Harbor" that could derail trains, poison water supplies, and cripple power grids. The next month, Chevron confirmed the speculation by becoming the first US Corporation to admit that Stuxnet had spread across its machines.

A doomsday blackout scenario is not far-fetched. Meanwhile, the asset owners across the power grid know that their SCADA controllers are vulnerable but will not move without regulations forcing them too. Their profit margins are thin and no one wants to incur the costs of cyber-motivated upgrades that would cost them dearly while their competitors wait and don't incur those costs.

Let's examine the potential impact of a massive blackout by examining the (noncyber-attack instigated) Aug. 2003 Northeast blackout. On that Aug. 14, an electrical disruption caused a loss of electrical service to consumers in New York, Michigan, Ohio, Pennsylvania, New Jersey, Connecticut, Vermont, Massachusetts, and substantial parts of the Canadian province of Ontario. In total, about 50 million US residents lost electrical power. Many also lost water service, and still more found it difficult or impossible to get gasoline for automobiles and trucks. The total impact on US workers, consumers, and taxpayers was a loss of approximately $6.4 billion.

If we extrapolate to a full US blackout that lasts for a week the numbers are staggering. Instead of affecting 50M people it would affect all 319M people in the United States. Assume worst case that it hits on a Monday morning which, if it was intentional, is what would happen, and further assume it lasts a full 7 days before we can recover. That makes the economic impact jump to $185.8B in 2003 dollars. Adjusting to 2015 (assumes 27.6% cumulative rate of inflation) increases that number to $237B in 2015 dollars. Attempting to take a guess at how much the increased dependence on computers and the Internet has on every aspect of the economy, we will assume $3\times$ what it was in 2003 and the final damage estimate is at least $711B. That represents 4% of 2014's GDP. To put that in perspective, the worst quarterly drop of the 2009 Great Recession was about 4%, so a blackout of that magnitude would send the country into a tailspin of unprecedented proportions leading perhaps to something much worse than the Great Recession.

What may be most frightening is that the United States could suffer a coast-to-coast blackout if saboteurs knocked out just nine of the country's 55,000 electric-transmission substations on a scorching summer day, according to a study by the Federal Energy Regulatory Commission, which concluded that coordinated attacks in each of the nation's three separate electric systems could cause the entire power network to collapse [43]. Frighteningly, the FERC analysis indicates that knocking out nine of those key substations could plunge the country into darkness for weeks, if not months. We know that nation-state adversaries have probed the electric grid and have demonstrated the ability to attack various parts of the system making many believe they could shut down parts or all of it if they were prepared for the consequences of doing so.

About once every 4 days, part of the power grid is struck by a cyber or physical attack. Although the repeated security breaches have never resulted in the type of cascading outage that swept across the Northeast in 2003, they have sharpened concerns about vulnerabilities in the electric system. A widespread outage lasting even a few days could disable devices ranging from ATMs to cellphones to traffic lights, and could threaten lives if heating, air conditioning, and health care systems exhaust their backup power supplies.

3 SECURITY & COMPUTER ARCHITECTURE

In this section, we will discuss a new class of inherently secure processor ideally suited for embedded devices that attempts to thwart a broad class of attacks that are typically used in the worst types of cyber incursions. The processor described evolved out of a DARPA program initiated in 2010 as a reaction to the Stuxnet attack against the SCADA controllers and the centrifuges they controlled, which resulted in an estimated 2000 of these plutonium enrichment centrifuges being irreparably destroyed.

3.1 PROCESSOR ARCHITECTURES AND SECURITY FLAWS

The Von Neumann architecture, also known as the Princeton architecture, is a computer architecture based on that described in 1945 by the mathematician and physicist John Von Neumann. He described an architecture for an electronic digital computer with parts consisting of a processing unit containing an arithmetic logic unit (ALU) and processor registers, a control unit containing an instruction register and program counter (PC), a memory to store both data and instructions, external mass storage, and input and output mechanisms. The meaning has evolved to be any stored-program computer in which an instruction fetch and a data operation cannot occur at the same time because they share a common bus.

The design of a Von Neumann architecture is simpler than the more modern Harvard architecture which is also a stored-program system but has one dedicated set of address and data buses for reading data from and writing data to memory, and another set of address and data buses for fetching instructions. A stored-program digital computer is one that keeps its program instructions, as well as its data, in read-write, random-access memory.

A stored-program design also allows for self-modifying code. One early motivation for such a facility was the need for a program to increment or otherwise modify the address portion of instructions, which had to be done manually in early designs. This became less important when index registers and indirect addressing became usual features of machine architecture. Another use was to embed frequently used data in the instruction stream using immediate addressing. Self-modifying code has largely fallen out of favor, since it is usually hard to understand and debug, as well as being inefficient under modern processor pipelining and caching schemes.

On a large scale, the ability to treat instructions as data is what makes assemblers, compilers, linkers, loaders, and other automated programming tools possible. One can "write programs which write programs." On a smaller scale, repetitive I/O-intensive operations such as the BITBLT image manipulation primitive or pixel & vertex shaders in modern 3D graphics were considered inefficient to run without custom hardware. These operations could be accelerated on general purpose processors with "on the fly compilation" ("just-in-time compilation") technology, e.g., code-generating programs—one form of self-modifying code that has remained popular.

There are drawbacks to the Von Neumann design especially when it comes to security, which was not even conceived as a problem until the 1980s. Program modifications can be quite harmful, either by accident or design. Since the processor just executes the word the PC points to, there is effectively no distinction between instructions and data. This is precisely the design flaw that attackers use to perform code injection attacks and it leads to the theme of the inherently secure processor: the processor cooperates in security.

3.2 SOLVING THE PROCESSOR ARCHITECTURE PROBLEM

Since vulnerabilities are inevitable using today's—and the foreseeable future's—programming methods, and they make our systems unsafe by default, how can we win? What if the underlying processor didn't allow the vulnerabilities to become an attack vector? What if the processor was inherently secure even if the software running on it has vulnerabilities? Inherent security means safe by default even when programmers make mistakes.

This is achievable. It is possible to change the architecture of our processors, use transistors for a security purpose, and build a new computing foundation that is inherently secure without giving up backwards compatibility to existing processor instruction sets, operating systems, and programming languages.

We can add metadata tags to every word in memory, a way to process those tags in parallel with instruction processing, and we will define a language of micro-policies that enforce security policies based on information in the metadata tags. A good way to think about what these security/safety micro-policies do for our current inherently insecure processors is to add an interlock. When you build a nuclear power plant, you include (several levels) of safety interlocks to prevent the worst things from happening. Interlocks here include temperature sensors to shut things down if they get too hot, and shunt valves to dump reactants. Machines have physical interlock buttons to shut down dangerous cutters if they are opened (including lowly paper shredders). Radiation machines have physical interlocks to prevent removal of the shield when radiation beam is set too high. Even highways have physical guard rails and barriers to prevent drivers from driving off ramps and bridges into oncoming traffic even if the driver falls asleep. So, our goal is to provide a set of interlocks in the processor that detect bad things that shouldn't happen and prevent the most egregious things from happening. Let's look in some detail now at the three key ideas that create these security interlocks: programmable metadata tags, a processor interlocks for policy enforcement (PIPE), and a set of micro-policies.

3.3 FULLY EXPLOITING METADATA TAGS

The first architectural modification to apply to the inherently secure processor is programmable metadata tags. Here is where we spend the most silicon (transistors) to support security. Here the metadata tag is a word of memory attached to and permanently associated with a "normal" word of memory. Call that normal word the payload. A payload might be an instruction for the processor, values to be processed, pointers to variables stored in memory. A tag attached to a payload word of memory is not a new concept. Lisp machines[7] used it for identifying types [44]. A machine made by a long-lost computer company called Burroughs had a tag bit in each word to identify the word as a control word or numeric word. This was partially a security mechanism to stop programs from being able to corrupt control words on the stack. Later it was realized that the 1-bit control word/numeric distinction was a powerful idea and this was extended to three bits outside of the 48 bit native word into a more general tag used for things like "this is the top of the stack," "this is a return address," "this is code," "this is data," "this is the index value for a loop," and other similar identifications [45]. Tagged architectures have been the subject of several research papers [46,47].

Recently some major processor manufacturers are recognizing the importance of using tagged memory to protect memory. Intel is introducing new MPX (memory protection extensions) instructions as a set of extensions to the x86 instruction set architecture

With compiler, runtime library, and operating system support, these instructions are designed to check pointer references whose normal compile-time intentions are maliciously exploited at runtime due to buffer overflows. With MPX there are new bounds registers, and new instruction set extensions that operate on these registers. The result is $\times 86$ support for memory protection that, once compilers are modified to use them, will protect against buffer overflows. Also recently, Oracle/SUN announced the Sparc M7 which uses the highest four bits of a pointer to keep track of 24 (or 16) different regions of memory (buffers) to protect them from reads or writes going out of bounds on them. This is a step in the right direction but only the simplest program creates only 16 distinct buffers so this does not sufficiently address the buffer overflow vulnerability. While both

[7]Lisp machines were general-purpose computers designed to efficiently run the Lisp programming language in the late 1980s.

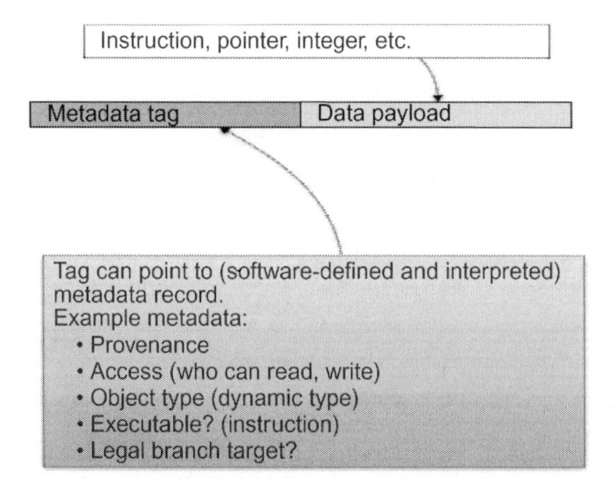

FIG. 2

Metadata tags are added to every word in memory including program counter (PC), instructions, registers, and main memory. Tags describe critical attributes about the Data payload used by the micro-policies. Metadata tags are highly protected.

these developments are encouraging because it means the major vendors are cognizant of the huge problem, they have a long way to go in addressing all of the other already identified serious Common Vulnerabilities and Exploits.

For a real solution, we need to employ a comprehensive approach to metadata tagging as shown in Fig. 2. These tags can point to software-defined and interpreted metadata records. This metadata can express provenance, access control (who can read or write the payload word), executability (does the payload contain an instruction), and legal branch target (does the payload contain address that it is legal for the program to branch to and continue execution). Very important is the fact that these metadata tags are permanently bonded to the payload word, are uninterpreted by the hardware, can be a pointer to an arbitrary data structure, and are not accessible from the application. The hardware component that will process these tags is called a PIPE.

3.4 PROCESSOR INTERLOCKS FOR POLICY ENFORCEMENT

To process these tags in parallel with the standard ALU—the heart of the processing pipeline of any standard processor—we create a PIPE, the second architectural modification to the processor, and add it into the instruction processing pipeline such that it operates in parallel with the ALU (see Fig. 3). Note that in the diagram, the PC, the instruction-store, the register file, and all the general-purpose memory have the extra word of metadata associated with them shown in green. As the ALU executes an instruction, the PIPE operates in parallel checking and updating appropriate tags. The hardware makes sure that payloads are sent to the ALU while tags are sent to the PIPE so that parallel processing of tags happens at the same speed as instruction processing.

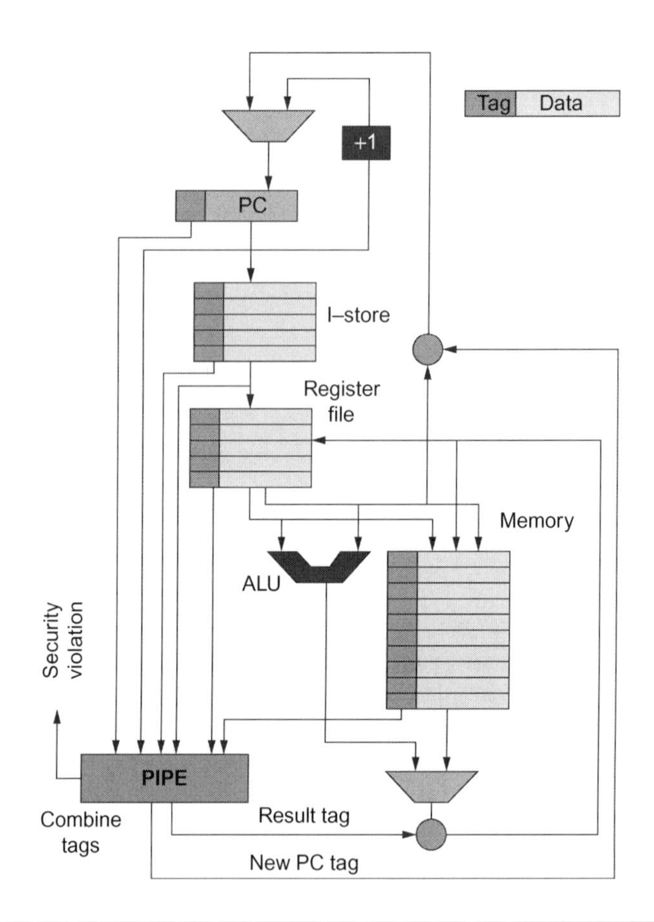

FIG. 3

The standard processor pipeline continues to operate with no instructions added or subtracted. But every component of a computation has a metadata tag which are the inputs to the PIPE which acts as an interlock on the processor preventing it from violating any of the active policies.

3.5 MICRO-POLICIES

The "secret sauce" to make the metadata maximally flexible and powerful is a set of micro-policies (abbreviated as μ-policies hereon in). These policies define what operations are allowed and specify how metadata is updated as a result of an instruction being processed. Examples of policies include:

Access control	Fine-grained control over who has what kind of access to a piece of data
Type safety	Making sure types declared in the program are manipulated as those types not just as raw data

Instruction permission	Protects special sections of code from being executed by user-level applications; part of the ISP self-protection mechanism
Memory safety	Protecting buffers in memory from overread or overwrite
Control-flow integrity	Guaranteeing that only jumps to program-defined locations are made at run-time
Taint tracking/information flow control	Tracking the influences of values through a computation to prevent untrusted values from influencing critical decisions and to limit the flow of sensitive data (e.g., guarantee encryption if data leaves the system)

As the PIPE processes tags associated with an instruction, it takes the relevant metadata and sends it to the installed set of software-defined policies where the metadata is checked against those policies and, if the result is allowed, determines the result tags. If the result is not allowed by policy, that instruction results in a security violation and the instruction is voided.

3.5.1 μ-Policies enforce security

The programming model for μ-policies involves abstracting the hardware. At the hardware level, as we have shown, there are metadata bits attached to each word. There is a hardware PIPE whose job it is to resolve this metadata.

The programmer should not have to worry about limits such as the number of bits of metadata or the complexity of the rule logic inside the PIPE. The metadata tag will be viewed as a pointer that can point to a data structure of arbitrary size.

A policy therefore is a collection of rules that take the form

$$(\text{opcode}, \text{PC}_{\text{tag}}, \text{INST}_{\text{tag}}, \text{OP1}_{\text{tag}}, \text{OP2}_{\text{tag}}, \text{MR}_{\text{tag}}) \Rightarrow (\text{allow?}, \text{PC}_{\text{tag}}, \text{Result}_{\text{tag}})$$

which we call a transfer function. The abstract function of the PIPE is diagramed here

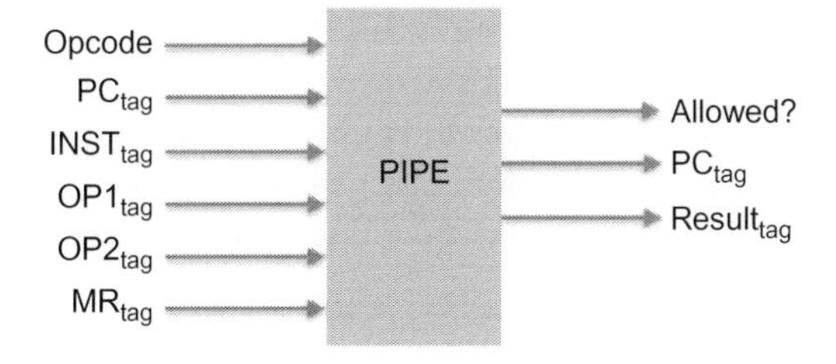

3.5.2 Memory safety μ-policy

The memory safety policy is designed to enforce spatial and temporal safety. Since in the rankings of the top vulnerability categories of 2014, according to Verizon, memory misuse accounted for 70% of all exploits, this is the first and probably most important one to consider.

This policy will work for both heap-allocated data as well as compiler stack-based allocations. We want this policy to protect against both spatial safety violations (e.g., accessing an array out of its bounds) and temporal safety violations (e.g., referencing through a pointer after the region has been freed). Such violations are a common source of serious security vulnerabilities such as heap-based buffer overflows, confidential data leaks, and exploitable use-after-free, and double-free bugs.

Many μ-policies will require assistance to perform their functions. That assistance is called a monitor service. For memory safety, the monitor services for heap memory are the allocation and freeing routines that are part of the set of operating system services. The allocation and freeing monitor services are parameterized by two functions, malloc() and free(), that are assumed to satisfy certain high-level properties: the malloc monitor service first searches the list of block descriptors for a free block of at least the required size, cuts off the excess if needed, generates a fresh color, initializes the new memory block with each word's metadata having that color, and returns the address of the start of the block which is a pointer to the newly allocated memory (or buffer). The free monitor service reads the pointer color, deallocates the corresponding block, tags its cells with a special freed tag F, and updates the block descriptors. The F tags prevent any remaining pointers to the deallocated block from being used to access it after deallocation. If a later allocation reuses the same memory, it will be tagged with a different (larger) color, so these dangling pointers will still be unusable.

The method we will use for this μ-policy, shown in Fig. 4, is to give each pointer a unique "color" and then to color each memory slot in the allocated buffer with the same color. When the memory is later freed all slots are re-colored to the uncolored value (i.e., 0). A "color" is just a numeric value equal to 2number-of-bits where number-of-bits is the size of the metadata tag available for this purpose (typically 32 bits which provides over 4 billion unique values).

The transfer function for this μ-policy is

$$(\mathrm{LOAD}, \mathrm{PC}_{\mathrm{tag}}, -, \mathrm{R1}_{\mathrm{tag}}, -, \mathrm{MR}_{\mathrm{tag}}) \Rightarrow (\mathrm{MR}_{\mathrm{tag}} == \mathrm{R1}_{\mathrm{tag}})\&\&(\mathrm{PC}_{\mathrm{tag}} == \mathrm{INST}_{\mathrm{tag}})), -, -)$$

which says that the LOAD instruction being executed will be allowed only if the tag on the memory reference (MR, the address in memory that is about to be written) is the same (represented as) as the tag on the pointer (R1). We additionally require that the PC tag matches the color of the block to which the PC points. This ensures that the PC cannot be used to leak information about inaccessible frames by loading instructions from them.

All other slots in the formula are do-not-cares for this instruction. We do allow adding and subtracting integers from pointers. The result of such pointer arithmetic is a pointer with the same color. The new pointer is not necessarily in bounds, but the rules for LOAD and STORE opcodes will prevent invalid accesses. (Computing an out-of-bounds pointer is not a violation per se, reading or writing through it is and will be handled by the rules for LOAD and STORE.)

This simple μ-policy, in conjunction with support from the operating system malloc() and free(), and from the compiler as it creates stack frames as functions are

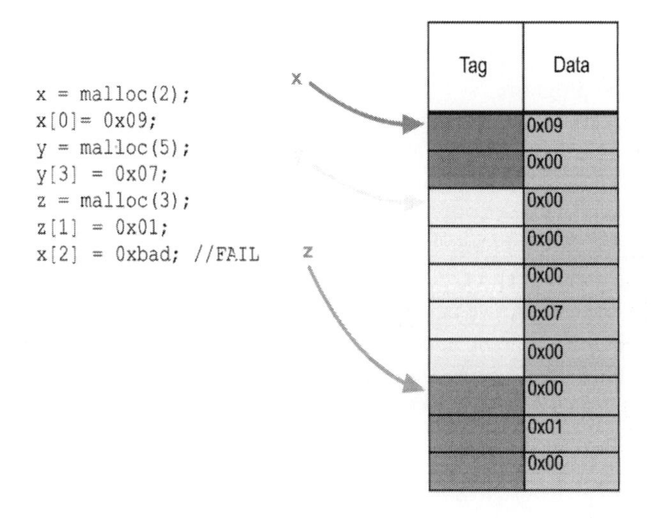

FIG. 4

C language calls to malloc() and subsequent references to the buffers so created are shown on the left. On the right are shown the metatags and the payload data each metadata tag describes. Each tag is assigned a unique color as is the pointer returned from malloc(). This information enables the memory safety policy to determine if a memory reference is legal and should be allowed to complete.

called, provides very powerful memory safety, which prevents buffer overflows when both reading and writing into memory.

3.5.3 Control flow integrity μ-policy

The memory safety policy just discussed leads people to apply the policy to make code nonwriteable. However, that didn't stop attackers. They figured out that they could use your own code against you using return-oriented programming (ROP) attacks. These are a relatively new type of attack that has grown in popularity with attackers and is very dangerous. ROP is a computer security exploit technique that allows an attacker to execute code even when memory protection is in place. In this technique, an attacker gains control of the call stack to hijack program control flow and then executes carefully chosen machine instruction sequences, called "gadgets." Each gadget typically ends in a return instruction and is located in a subroutine within the existing program and/or shared library code. Chaining these gadgets together enables an attacker to perform any computation just with these return sequences. To understand the policy to prevent ROP attacks, it is important to understand that these attacks work by returning into the middle of sequence of code—returns (i.e., control-flow-branches) that did not exist in the original program. Current processors will blindly run the code for the attacker.

The control flow integrity (CFI) policy dynamically enforces that all indirect control flows (computed jumps) adhere to a fixed control flow graph (CFG) most commonly obtained from a compiler. This policy is another interlock that prevents control-flow-hijacking attacks by locking down control transfers to only those intended by the program.

The sample code we will use to illustrate this is shown in Fig. 5. While in function foo(), function bar() is called. The address at which that call occurs is t1. The address where bar() begins is t2. And the address in bar() where it is about to return to its caller is t3. Somewhere else in the program assume bar() is also called from address t42 as well as a few other places. Thus the full list of legal address locations bar() can be called from includes t1, t42, and a few others as shown at the bottom of the figure. As you will see, CFI uses tags to distinguish the memory locations containing instructions and the sources and targets of indirect jumps, while using the PC tag to track execution history (the sources of indirect jumps).

On a call to a the function bar() from the address t1 the μ-policy that governs what happens is described by the following transfer function

$$(\text{CALL}, \text{none}, \text{t1}, \text{R1}_{\text{tag}}, -, -) \Rightarrow (\text{true}, \text{t1}, -)$$

which declares that the tag from the call instruction (which is t1) must be copied to the tag for the PC as a result of the CALL instruction (that is, the PC which is now at the first instruction for bar() which is t2). The second half of CFI enforcement comes from the following transfer function

$$(/\text{CALL}, \text{t1}, \text{t2}, -, -, -) \Rightarrow (\text{t1} \, \text{in} \, \text{t2}, \text{none}, -)$$

which says whenever the processor is not executing a CALL instruction and the PC is tagged (in this case t1 because we are in a previous CALL that began at t1), check that the tag on the PC is in the list of "legal caller tags" (t2 is in that list) on the current instruction (which must be the target of a call). The PIPE will also untag the PC as shown by the none in the PC tag spot on the right-hand side of the transfer function. These two transfer functions direct the PIPE to strictly enforce the CFG and only

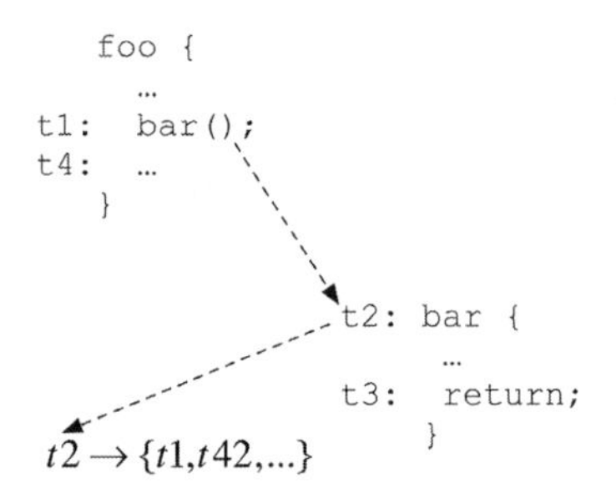
foo {
...
t1: bar();
t4: ...
}
t2: bar {
...
t3: return;
}
t2 → {t1,t42,...}

FIG. 5

The compiler knows the legal call graph and when the Control Flow Integrity policy is active, the control flow is tracked through the metadata tag on the PC. Here as bar() is called, the list of legal addresses where bar() can be called from is tracked such that if a return is corrupted by an attacker, the PIPE knows the processor is trying to return to an illegal address and halts that instruction.

allow control transfers to those locations specified in the program thus creating another interlock that thwarts ROP attacks.

3.5.4 Taint tracking μ-policy

Here we will show how a taint tracking policy can be used quite generally to "taint" data such that certain rules can be enforced throughout the system. Taint tracking can enforce rules like "if the data is not public it may not leave the system unencrypted" and "at this critical decision juncture, if the data about to be used is untrusted do not proceed." Taint tracking can also be used for a safety policy that might be used in a medical device like "if you are about to pump insulin and more than N milligrams have been pumped in 24 h do not pump any, alert the user, and shut down the processor." The policy, as applied to the ADD instruction, would be represented by the following transfer function

$$(\mathrm{ADD}, \mathrm{PC}_{\mathrm{tag}}, \mathrm{INST}_{\mathrm{tag}}, \mathrm{OP1}_{\mathrm{tag}}, \mathrm{OP2}_{\mathrm{tag}}, -) \Rightarrow (\mathrm{true}, \mathrm{PC}, \mathrm{union}(\mathrm{PC}, \mathrm{INST}, \mathrm{OP1}, \mathrm{OP2}))$$

where the tag on the Result is formed from the union of the tag on all the operands including the PC, INST, OP1, and OP2.

3.5.5 Composite policies

It is important that for any given metadata tag there not be just one policy that can be enforced. An arbitrary number of policies may be required and this is easily supported by our model by using the tag as a pointer to a tuple of μ-policies. There is no hardware limit on the number of μ-policies supported in this fashion.

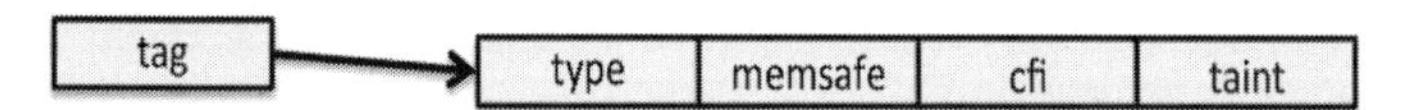

To propagate tags efficiently, the processor is augmented with a rule cache that operates in parallel with instruction execution. On a rule cache miss, control is transferred to a trusted miss handler which, given the tags of the instruction's arguments, decides whether the current operation should be allowed and, if so, computes appropriate tags for its results. It then adds this set of argument and result tags to the rule cache so that when the same situation is encountered in the future, the rule can be applied without slowing down the processor. In performance tests on accurate machine simulators, the overall performance overhead for security enforcement is less than 10%.

A partial list of the types of safety and security policies that can be implemented using the mechanisms just described is listed in the following table.

• Type safety	• Mandatory access control
• Memory safety	• Classification levels
• Control-flow integrity	• Lightweight compartmentalization
• Stack safety	• Software fault isolation
• Unforgettable resource identifiers	• Sandboxing
• Abstract types	• Access control
• Immutability	• Capabilities

- Linearity
- Software architecture enforcement
- Units
- Signing
- Sealing
- Endorsement
- Taint
- Confidentiality
- Integrity
- Bignums
- Provenance
- Full/Empty bits
- Concurrency: race detection
- Debugging
- Data tracing
- Introspection
- Audit
- Reference monitors
- Garbage collection

Returning to the analogy of metadata+PIPE+μ-policies being interlocks, one of their advantages is that policies are small (few 100–1000 lines of code), but they are protecting millions of lines of code. So, while a programmer has no chance of having millions of lines of code be bug free, she does have a chance of making the interlocks be bug free. Just to be certain, since these bits of software are small and highly important, they are amenable to and worth verifying formally.

3.6 SELF-PROTECTION

For an inherently secure processor to maintain that designation, it is vital that it strongly protects itself against malfeasance of any sort. In this case, that means protecting the metadata, the PIPE, the mechanisms of μ-policies, having a secure boot, and providing protection from physical tampering.

3.6.1 Metadata protection

In our architecture, data and metadata do not mix. Metadata is not addressable; not by the user-level application, not by the operating system. In the hardware, data paths for data and metadata do not cross. No user-accessible instructions read or write metadata. Metadata is only transformed through the PIPE.

Separation of data and metadata is important for maintaining strong protection of the PIPE and policy mechanisms. Metadata rule misses could be processed on a separate processor. We could store the metadata structures that the metadata tags point to in separate memory. Then this separate metadata processor would only need to access metadata memory and would never access standard data (and vice versa for the standard processing structures). But to avoid a second processor we want to process a miss on the same processor albeit in an isolated subsystem. In this metadata processing subsystem metadata tags become "data," that is, they are just pointers into metadata memory space.

The mechanism to accomplish this on the RISC-V uses special PIPE control and status registers (CSRs) for rule inputs and outputs. On a PIPE miss trap the PIPE tag inputs are stored in the PIPE CSRs. Now these tags have just become data to the metadata processing subsystem. Here the PIPE CSRs are read and processed. If a result is allowed the tag result is written to the PIPE CSRs which triggers a rule insertion for this PIPE transfer function (same inputs mapped to same outputs) making this a simple lookup next time this rule hits the PIPE.

3.6.2 PIPE protection

We can use the PIPE itself both to implement the symbolic μ-policy and, at the same time, to enforce the restrictions above (which we call monitor self-protection). To achieve this, we use the special Monitor tag to mark all of the monitor's code and data (remember, malloc() and free() are monitors), allowing the miss handler to detect when untrusted code is trying to tamper with it.

Every instruction causes a rule cache lookup, which results in a fault if no corresponding rule is present (i.e., a miss). Since the machine has no special "privileged mode," this applies even to monitor code. To ensure that monitor code can do its job, we set up cache ground rules (one for each opcode) saying that the machine can step whenever the PC and INST tags in the input vector of a rule are tagged Monitor; in this case, the next PC and any result of the instruction are also tagged Monitor. Monitor code never changes or overrides these rules.

3.6.3 The Dover processor

Draper Laboratory in Cambridge Massachusetts is an independent not-for-profit research and development laboratory that is developing an inherently secure processor for embedded systems called Dover (short for do-over). In early 2016, the processor achieved its first stage of demonstrability by showing it was immune to the Heartbleed attack and could automatically protect memory from a buffer overflow. The processor will be developed as either an FPGA or an ASIC depending on the requirements of the embedded system it is intended for. As embedded devices adopt an inherently secure processor more and more broadly, infrastructure, the IoT, transportation systems, weapons systems and all other embedded systems will become much more difficult for attackers to overwhelm.

REFERENCES

[1] Shon Harris, Fernando Maymi (Eds.), CISSP All-in-One Exam Guide, Seventh Edition, McGraw-Hill Education, 2015. ISBN 0071849262.

[2] Netcraft April 2014 Web Server Survey.

[3] http://www.wsj.com/articles/SB123914805204099085.

[4] 2014 Verizon Data Breach Report.

[5] http://www.thedailybeast.com/articles/2015/03/18/china-reveals-its-cyber-war-secrets.html.

[6] Verisign Domain Name Industry Brief.

[7] http://www.cnn.com/2014/01/20/us/money-target-breach/.

[8] https://en.wikipedia.org/wiki/Operation_Aurora.

[9] http://cve.mitre.org.

[10] Verizon Incidents Report for 2014.

[11] http://www.wired.com/2015/07/hackers-can-disable-sniper-rifleor-change-target/.

[12] http://www.wired.com/2015/08/hackers-can-seize-control-of-electric-skateboards-and-toss-riders-boosted-revo/.
[13] S. Mcguire, Writing Solid Code, 2nd. ed., Greyden Press, Dayton, OH, ISBN 978-1570740558, 2013.
[14] http://spectrum.ieee.org/telecom/security/the-real-story-of-stuxnet.
[15] http://www.businessinsider.com/stuxnet-was-far-more-dangerous-than-previous-thought-2013-11.
[16] http://www.langner.com/en/wp-content/uploads/2013/11/To-kill-a-centrifuge.pdf.
[17] https://www.crysys.hu/skywiper/skywiper.pdf.
[18] http://www.kaspersky.com/flame.
[19] https://en.wikipedia.org/wiki/Flame_%28malware%29.
[20] http://www.kaspersky.com/gauss.
[21] http://www.computerworld.com/article/2597456/security0/gauss-malware–nation-state-cyber-espionage-banking-trojan-related-to-flame–stuxnet.html.
[22] https://en.wikipedia.org/wiki/Duqu.
[23] http://www.symantec.com/connect/blogs/duqu-20-reemergence-aggressive-cyberespionage-threat.
[24] https://www.fireeye.com/blog/threat-research/2015/09/synful_knock_-_acis.html.
[25] http://www.cisco.com/web/about/security/intelligence/ERP_SYNfulKnock.html.
[26] http://www.scientificamerican.com/article/how-do-you-hide-a-boeing-777/.
[27] http://www.darkreading.com/vulnerabilities-and-threats/airplane-takeover-demonstrated-via-android-app/d/d-id/1109503?.
[28] http://www.computerworld.com/article/2952186/mobile-security/chrysler-recalls-14m-vehicles-after-jeep-hack.html.
[29] http://www.eweek.com/c/a/Security/GMs-OnStar-Ford-Sync-MP3-Bluetooth-Possible-Attack-Vectors-for-Cars-420601#sthash.gX4SHD8k.dpuf.
[30] http://www.dodig.mil/PUBS/managementChallenges/pdfs/FY2014_DoDIG_SummaryManagement_Challenges.pdf.
[31] http://www.dote.osd.mil/pub/reports/FY2014/pdf/other/2014DOTEAnnualReport.pdf.
[32] http://www.ioactive.com/pdfs/IOActive_SATCOM_Security_WhitePaper.pdf.
[33] http://www.washingtontimes.com/news/2015/feb/26/james-clapper-intel-chief-cyber-ranks-highest-worl/?page=all.
[34] http://www.bloomberg.com/news/articles/2014-06-13/uglygorilla-hack-of-u-s-utility-exposes-cyberwar-threat.
[35] http://krebsonsecurity.com/2015/09/inside-target-corp-days-after-2013-breach/.
[36] Cyberoam Security Advisory—Heartbleed Vulnerability in OpenSSL, April 11, 2014.
[37] Schneier on Security: Heartbleed, Schneier on Security, April 11, 2014.
[38] J. Steinberg, Massive Internet Security Vulnerability—Here's What You Need To Do, Forbes, Jersey City, New Jersey, 2014.
[39] OpenSSL Heartbeat Extension Vulnerability in Multiple Cisco Products, Cisco Systems, 2014.
[40] P. Younger, PC Game Services Affected by Heartbleed and Actions You Need to Take, IncGamers, 5 Kings Court, Falkirk, FK1 1PG, 2014.
[41] L. Seltzer, Shellshock Makes Heartbleed Look Insignificant, ZDNet, 2014. CBS Interactive. http://www.zdnet.com/article/shellshock-makes-heartbleed-look-insignificant/
[42] C. Cerrudo, Why the Shellshock Bug Is Worse than Heartbleed, MIT Technology Review, MIT, Cambridge, MA, 2014.

[43] http://www.wsj.com/articles/SB10001424052702304020104579433670284061220.
[44] M. Ash, Friday Q&A 2012–07-27: Let's Build Tagged Pointers.
[45] E.A. Feustel, On the advantages of tagged architecture (PDF), IEEE Trans. Comput. (1973) 644–656.
[46] E.A. Feustel, The Rice Research Computer—a tagged architecture, in: Proceedings of the 1972 Spring Joint Computer Conference, American Federation of Information Processing Societies (AFIPS), 1972, pp. 369–377.
[47] H. Shrobe, T. Knight, A. DeHon. TIARA: Trustmanagement, intrusion-tolerance, accountability, and reconstitution architecture, Technical Report MIT-CSAIL-TR-2007-028, MIT CSAIL, 2007.

CHAPTER

7 Reliable electrical systems for micro aerial vehicles and insect-scale robots: Challenges and progress

X. Zhang
Washington University, St. Louis, MO, United States

1 INTRODUCTION

The dream to fly has sent men on an ardent pursuit to build ever more sophisticated aerial vehicles. Over the years, steady progress has been made to miniaturize such systems and has led to prolific development and maturation of drones. Recent research advancement endeavors to drive the miniaturization further to the micro-scale, and promising prototypes are being developed to explore this exciting yet untrodden frontier. In this chapter, we present the progress our research team has made on building an insect-scale aerial vehicle and zoom into the critical reliability issues associated with this system. Since many such systems are envisioned to operate in harsh environment hostile or harmful to human beings and have stringent design requirements and operation constraints, reliability becomes a first-class design consideration for these microrobotic systems.

We will first give a brief background on micro aerial vehicles (MAV), the opportunities they offer, and the challenges they present, followed by an introductory overview on the insect-scale MAV prototype called RobeBee, which is currently under our development to demonstrate autonomous flight. Next, we will dive into the detailed implementation of its electronic control system and elaborate the reliability concerns and considerations in our design. Improved performance and reliability have been confirmed in both simulation and experiment after applying codesign strategy between supply regulation and clock generation. Finally, we will conclude with our vision on the future of MAV from a reliability perspective.

Rugged Embedded Systems. http://dx.doi.org/10.1016/B978-0-12-802459-1.00007-5

2 BACKGROUND OF MICRO AERIAL VEHICLE

2.1 WHAT IS MAV?

Unmanned aerial vehicle technology has indeed taken off and its tremendous commercial success and wide adoption in many fields has also fueled increasing recent interest in MAV, which loosely refer to air craft with size less than 15 cm in length, width, or height and weigh less than 100 g. These systems are envisioned for applications including reconnaissance, hazardous environment exploration, and search-and-rescue, and therefore may require various morphologies that can be broken into a number of classes such as fixed wing, flapping wing, or rotary wing.

The history of MAV dates back to 1997, when the United States Defense Advanced Research Projects Agency (DARPA) announced its "micro air vehicle" program [1]. The technology advances that propelled this bold marching step are the maturation of microsensors, the rapid evolution of microelectromechanical system, also known as MEMS, as well as the continued exponential improvement of computing technology. In 2005, DARPA again pushed the limits of aerial robotics by announcing its "nano air vehicle" program with tighter requirements of 10 g or less and within 7.5 cm dimension. These programs have led to successful MAV prototypes including Microbat [2], Nano Hummingbird [3], and inspired a number of recent commercially available flapping-wing toy ornithopters and RC helicopters on the scale of MAVs.

Taking the pursuit of miniaturization to the next level, researchers are now working on "pico" air vehicles that have a maximum takeoff mass of 500 mg or less and maximum dimension of 5 cm or less [4]. As this size and weight range falls into the scale of most flying insects, it is no surprise that many successful MAV systems developed at this scale are modeled after insects. The Harvard RoboBee project, which is the focus of this chapter, is one example prototype of a "pico" air vehicle.

2.2 WHY MAV?

Why are we interested in the extreme art of building insect-scale MAVs? Rodney Brooks, a renowned roboticist, has put it quite eloquently in his forward-looking paper in 1989, titled "Fast, Cheap, and Out of Control: A Robot Invasion of the Solar System" [5], where he laid out the benefits of employing small robots for space exploration. Fast forward 25 years, rapid development in information technology and advanced manufacturing has pushed the applications of robots well beyond the space exploration missions, yet many of the same benefits Brooks argued in his seminal work remain:

Fast development and deployment time: it used to be that "big" complex systems take years of planning to take shape and the turnaround time to discover any critical problems in the system design can be prohibitive. Similarly, even after the system has been designed, developed, and debugged, its deployment can be equally time-consuming because of its complexity and the intricacies involved to interface it with other complex systems and humans. Therefore, only huge organizations, such as the military and government agencies, could afford the resources and man power required for such missions over an extended period of time, which severely limited the adoption of the traditional large-scale robotic technology. As we miniaturize robots to the microscale by leveraging established design methodology from the IC industry and development practices from software engineering, the time it takes from conception to implementation of the robotic system can be drastically reduced, resulting in faster design iterations and ultimately superior system performance and reliability.

Cheap prototyping and manufacturing cost: if you watched the latest DARPA Robotics Challenge like I did, you are probably blown away with awe too by the bipedal humanoid robot called Atlas that stands 6 foot tall and weighs 330 pounds. You might be asking the same question—"Why can't I get that?!" The answer probably has most to do with its whopping price tag of $500,000. It is obvious that while robotic technology has been making steady progress in the military and industrial setting, the cost of a large-scale system makes it hard for the same technology to spill over to our everyday life. However, things are rapidly changing in the MAV arena. Moore's law and its counterpart in sensors have driven high-performance embedded processors and high-resolution cameras as inexpensive commodities; 3D printers, laser cutters, and computer numerical control (CNC) machines have enabled desktop fabrication; and the open source and open standard movements have built a vibrant and resourceful community that is ripe for innovating. Once again, miniaturization has torn down the cost barrier of traditional robots and is poised to release the technology's full potential.

Collective power of the swarm: one tiny aerial robot that can perform simple tasks may not seem much at first glance, but if you put large number of these autonomous agents together in a coordinated way, they may be more effective and capable than a few large complex individual robots. For example, consider a search-and-rescue scenario. A rescue worker could release a box of 1000 micro aerial robots, each weighing less than 1 g, at the site of a natural disaster to search for heat, sound, or exhaled carbon dioxide signature of survivors. Many of them may fail, but if only a few of these robots succeed, they would have accomplished their mission with significant cost and deployment advantages over current generation of $100,000 rescue robots. This indeed highlights another interesting aspect of MAV reliability that Brooks emphasized—"...that large numbers of robots can change the trade-off between reliability of individual components and overall mission success."

In addition to the three key advantages offered by miniature robots in general as listed above, an insect-scale MAV can serve as a scientific platform for studying flight mechanics and flight control. Shaped and tuned by nature over millions of years of evolution, insects exhibit unmatched agility and prowess in their flying ability. There are many mysteries yet to be revealed regarding the aerodynamics, the sensorimotor coordination, and the control strategy involved in flight at such small scale. The kind of MAVs we are interested in provides the much-needed alternative methods and tools to investigate these open scientific questions in a controlled, repeatable, and efficient manner.

2.3 MAV CHALLENGES

Miniaturization of autonomous aerial vehicles presents unique technological challenges, especially as we shrink down to the "pico" aerial vehicle scale [4].

First, fluid mechanics changes as a function of characteristic length and velocity. Bird-sized MAVs with Reynolds number (*Re*) larger than 10,000 that is envisioned in the original DARPA program exist in a regime of turbulent flow and steady lift to drag ratios greater than 10, while "nano air vehicle" ($1000 < Re < 10,000$) may start to be affected by the impact of boundary layer separation. For insect-scale MAVs ($Re < 3000$), unsteady mechanism beyond constant velocity such as wing flapping can be employed to generate lift, because the flow is almost entirely laminar. However, the energetic cost increases with smaller characteristic length, which penalize pico air vehicles significantly on its flight time. Whereas a larger-scale aircraft may stay aloft based on passive stability for hours or even days, flight time for MAV is expected to be on the order of minutes.

Second, the scaling trend for device manufacturing can be another hurdle. Feature size and characteristic size scale in tandem in aerial vehicle fabrication and assembly. The former refers to the smallest dimension of the mechanical components of the system, such as gear teeth pitch, constituent material thickness, and flexure length, and the latter describes the overall size of the vehicle and thus often relates to wingspan and chord length. As we approach manufacturing of MAVs at the insect scale, the standard machining and assembly tools and "off-the-shelf" components no longer apply, and novel methods have to be developed from scratch. For example, high-resolution CNC mills with positioning accuracy down to 1 μm require rare end mills below 100 μm that are hard to come by. Furthermore, area-dependent forces, such as friction, electrostatic, and van der Waals force become dominant as feature size shrinks, degrading the loss in more traditional bearing joints with respect to power transmission. And the same trend is true with transducers. Friction losses and current density limits worsen the effectiveness of electromagnetic motors with diminishing feature size. Although at first glance, it may seem that MEMS surface micromachining techniques could be the solution to achieve micron-order feature sizes, such methods are often hindered by the time-consuming serial process steps, limited three dimensional capabilities, and high prototyping cost using specialized MEMS foundries. Therefore insect-scale MAVs require innovative solutions to device fabrication and assembly.

Finally, flight control experiences distinctive transitions from larger-scale aerial vehicles to microscale ones. The former can often take advantages of passive stability mechanisms such as positive wing dihedral and enjoy larger mass and power capacity to accommodate various sensors and processors, whereas the latter must work under significantly reduced payload capacity and wrestle for the stringent share of size, weight, and power budget. Therefore, the control challenge in pico aerial vehicles is centered on flight stabilization using constrained sensing and computation capabilities, instead of higher-level control problems such as autonomous navigation and multiple vehicle coordination.

The design and development of MAVs present exciting application prospects, as well as challenging engineering and scientific questions that pique our research interest. In the reminder of this chapter, we will dive into the RoboBee system as a case study of an insect-scale MAV to discuss various design strategies and considerations in building a miniature autonomous flying robot.

3 OVERVIEW OF RoboBee

3.1 MOTIVATED BY A GLOBAL CRISIS

Our intrepid pursuit to build an insect-scale aerial robot is driven with a mission to address a global crisis—the decline of pollinator populations in nature. Bees and other animal pollinators are responsible for the pollination of 60–80% of the world's flowering plants and 35% of crop production, and hence the continuing decline could cost the global economy more than 200 billion dollars and spell devastating effect on human nutrition that is beyond the measure of monetary values. To these days, the exact cause of the decline remains to be determined, but its disastrous consequence calls for awareness and actions.

While the near-term remedy to this problem will certainly have to come out of the toolbox of entomologists, ecologists, and environmental scientists, as engineers and technologists, we would like to approach it with a longer-term out-of-the-box proposal by looking at the feasibility of building artificial robotic pollinators at the insect scale. We consider it a moonshot idea whose merits lie mostly in the creativity it spurs and the discovery it enables rather than its immediate practicality. It is against this backdrop that we, a team of investigators from Harvard University's John A. Paulson School of Engineering and Applied Sciences and Department of Organismic and Evolutionary Biology and Northeastern University's Department of Biology, conceived the project of RoboBee.

3.2 INSPIRED BY NATURE

It is no coincidence that we look to nature for inspiration to build tiny autonomous flying apparatus. As we mentioned earlier in the chapter, larger-scale man-made aerial vehicles can take advantage of passive stability that is associated with a large Reynolds number, whereas smaller-sized MAVs commonly employ unsteady mechanisms such as wing flapping to sustain flight. Insects are among the most agile flying creatures on Earth, and we probably have all experienced this first hand at failed attempts to swat an evasively maneuvering fly or mosquito.

Since no existing vehicles have been demonstrated before to achieve comparable maneuverability at the insect scale, we modeled the form and functionality of RoboBee directly after the morphology of *Diptera* (flies), because Dipteran flight has been well-studied and documented in the past [6].

3.3 BASIC DESIGN AND IMPLEMENTATION

3.3.1 Fabrication

RoboBee requires mechanical components with feature sizes between micrometers and centimeters that fall between the gap of conventional machining and assembly methods and MEMS fabrication. To tackle this problem, the mechanical experts on our team developed a design and manufacturing methodology called "smart composite microstructure" (SCM), which stacks different material layers together with adhesives and applies laser-micromachining and lamination to bring them into desired shape. We are able to employ this monolithic planar process to manufacture all the electromechanical elements of the robotic fly, including flight muscles, thorax, skeleton, and wings. High stiffness-to-weight-ratio carbon fiber-reinforced composites are used for the structural elements, while polyimide film flexure hinges are used for articulation to emulate low-friction revolute joints.

3.3.2 Actuation

Now that we are able to reliably and efficiently fabricate the body of the flying robot with the SCM method, the next step is to determine the actuation scheme. Due to the lack of passive stability mechanism and the unfavorable scaling of transducers at small scale, flapping-wing flight can be quite costly energetically. After surveying [7] a variety of actuation technologies, we have identified piezoelectric ceramics technology as the most promising candidate for delivering oscillatory power to the robot, instead of rotary electromagnetic motors that are prevalent in larger classes of flying vehicles.

RoboBee's flight muscles consist of voltage-driven piezoelectric bimorphs that generate bidirectional forces. A four-bar linkage acts as a lever arm to amplify the small displacement of the piezoelectric flight muscle. The complex flapping motion of the flies has been simplified in our implementation of the robot by using passive compliant flexures to regulate the pitch rotation. This is sufficient to mimic the wing kinematics in insect flight and generate downward propulsive force over a full stroke cycle.

Efficient actuation of the wings is achieved by exciting the piezoelectric bimorphs with sinusoidal waveforms near the resonant frequency of the coupled muscle-thorax-wing system. The thrust is modulated through amplitude modulation of the sinusoidal waveform, and a dual-actuator design is selected to independently drive each individual wing. It is worth mentioning that although the frequency of the drive waveform is modestly around 100 Hz, its amplitude can be as high as 300 V, which calls for novel power electronics architecture.

3.3.3 Maneuver

Equipped with the independently controlled dual-actuator wings, RoboBee is able to enjoy three-rotational degrees of freedom in its aerial maneuvers. Roll torque is generated by flapping one wing with larger stroke amplitude than the other, inducing differential thrust force; Pitch torque is generated by moving the mean stroke angle of both wings forward or backward to offset the thrust vector away from the center of mass, mimicking the method observed in *Drosophila* (fruit fly); Yaw torque is generated by cyclically modulating stroke velocity in a "split-cycle" scheme to induce an imbalanced drag force per stroke cycle.

Modulation of the thrust force and three body torques (roll, pitch, and, yaw) allows the robot to be controllable in unconstrained flight, and based on the desired wing movements, we can derive the corresponding drive signal waveforms to be generated by the power electronics. The next question is how to achieve stable flight.

3.3.4 Sensing and control

Since the dynamics of our insect-scale vehicle are fast and unstable, an active control strategy is implemented. As an initial start point, the sensing, and controller computation is performed off-board. The state of the robotic fly is sensed by an external array of motion-capture cameras. Retroreflective tracking markers are put on the RoboBee body to estimate position and orientation from the images. Desktop computers running Matlab are used to process the position and orientation information and control the bench-top signal generator and power amplifier to generate the power and control signals. A wire tether consisting of four bundled, 51-gauge copper wire is connected to the robot to send the power and control signals. Successful demonstration of unconstrained stable hovering and basic controlled flight maneuvers has been performed using this off-board configuration.

RoboBee's tethered flight with off-board sensing and control gives us the confidence to take the next leap towards an autonomous flight with fully on-board components. However, it is by no means a straightforward migration from off-board components to on-board because the bulk size and power consumption of the off-board solution simply could not fit onto the bee-sized robot. We have to rethink and redesign the entire sensing and control scheme.

Once again the inspiration comes from nature. Among all the sensory modalities natural insects employ in flight control, visual stimuli are essential for maintaining stability and avoiding obstacles. In unidirectional flight, insects use optical flow to navigate. It is the visual motion of objects, surfaces, and edges in a scene caused by the relative motion between the observer and the scene. To achieve light-weight on-board vision sensing, we select an optical flow sensor, designed, and fabricated by Centeye, as the "eyes" for RoboBee. It is used to provide the position information to the flight controller [8]. In addition to the vision sensor, an inertial measurement unit (IMU) chip is integrated to supplement the optical flow data with accelerometer and gyro scope readings. Experiments have shown that IMU enabled upright stability using the RoboBee platform [9]. With the sensors in place, we turn our attention now to the miniaturization of the central microprocessor and the piezoelectric driver system, which will be the main focus of this chapter. The goal is to pack the same computation performance and power conversion function of the bench-top equipment into a highly-integrated chip module.

3.4 VISION FOR FUTURE

We are at the final stage of assembling the on-board sensing and control system to demonstrate autonomous flight in RoboBee. It represents the culmination of years of engineering and development efforts on this research project that led to one of the world's first functioning prototypes of an insect-scale aerial vehicle with fully on-board sensing and control.

Obviously, many future improvements of the current system are cut out for us. For example, advances in small, high-energy-density power sources could significantly increase the duration of the flight time; low power reliable bee-to-bee and bee-to-hive communication module could greatly facilitate coordination at the colony-level; and implementation of higher-level control will enable the robot to perform more sophisticated tasks and deliver more useful functionalities in general.

With this crucial proof-of-concept step, we are now closer to realizing the promising potentials of miniature aerial vehicles, such as for applications in reconnaissance, hazardous environment exploration, search-and-rescue, and assisted agriculture as our initial purpose in conceiving the RoboBee project. The ultimate goal is to use swarms of these small, agile, and potentially disposable robots for these applications to achieve better coverage of the mission and enhanced robustness to robot failure, as compared to larger, more complex individual robots.

Finally, it is worth pointing out that the endeavor to build insect-scale robots stretches well beyond the accomplishment of applicability. The greater impact of this line of work is to provide a set of new tools and methodologies for open scientific questions on flight mechanics and control strategies at the insect scale and the motivating technological context and challenge to spark ideas and drive creativity.

4 BrainSoC

As described in the previous section, building the RoboBee system is truly a multidisciplinary undertaking, in this section we will zoom into its electronic subsystem and in particular dive into the custom designed system-on-chip (SoC) acting as the "brain" for RoboBee, which we nicknamed BrainSoC.

4.1 RoboBee's ELECTRONIC SUBSYSTEM

The ultimate goal of autonomous flight requires converting the external bench-top test equipment into customized electronic components that the robot can carry within its tight payload budget. Towards this end, we designed an energy-efficient BrainSoC to process sensor data and send wing flapping control signals to a power electronics unit (PEU) that generate 200–300 V sinusoids for driving a pair of piezoelectric actuators to flap each individual wing [10]. Fig. 1 is a cartoon illustration of the connections between the BrainSoC and the power electronics.

To understand the basic operation of the power electronics, we first have to revisit the mechanism of exciting the flapping wings with piezoelectric actuators (Fig. 2). Electrically, these layered actuators can be modeled as capacitors. To generate sufficient wing flapping amplitude, these actuators need to be driven by a sinusoidal waveform of 200–300 V at the mechanical resonant frequency of the robot, approximately at 100 Hz. In one of our schemes, we fix the bias across the top and bottom layers of the actuator and use the PEU to drive the middle node of the actuator with a sinusoidal signal.

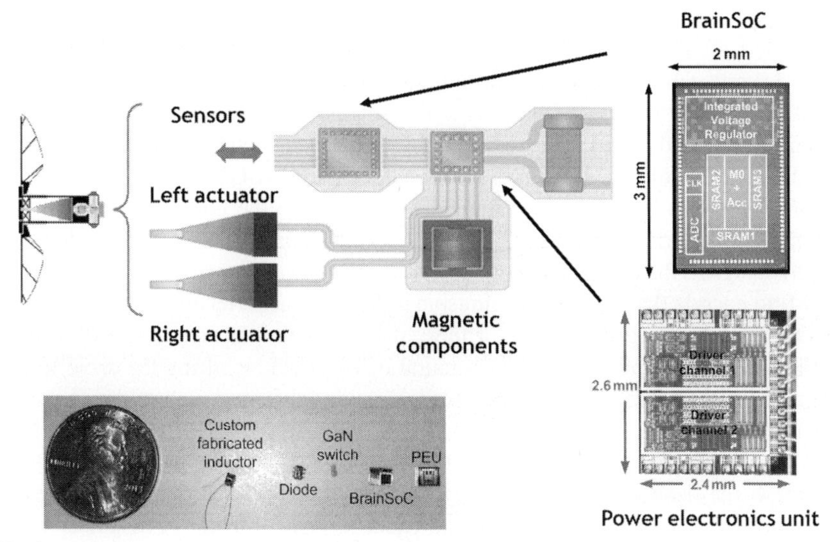

FIG. 1

Illustration of different functional components connected through the substrate flexible PCB, including the BrainSoC chip and the high voltage IC chip for the power electronics unit (PEU), in the RoboBee electrical system. (inset) The relative foot print of the components compared to a U.S. five-cent coin.

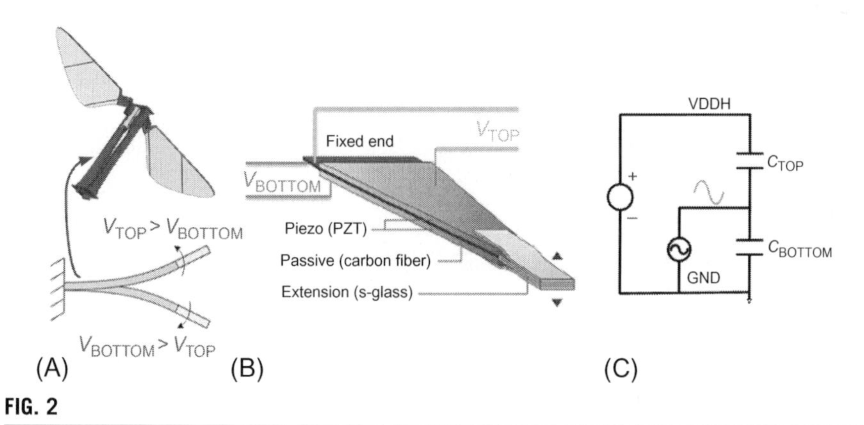

FIG. 2

(A) The voltage-driven piezoelectric bimorph that generates bidirectional wing-flapping forces in RoboBee; (B) the layered structure in the piezoelectric bimorph; and (C) the equivalent electrical circuit model for the piezoelectric actuator.

To provide the required drive signal in an efficient manner, a two-stage design is adopted for the PEU. The first stage is a tapped-inductor boost converter built with discrete components, and it outputs the high voltage bias VDDH in the 200–300 V range for the second stage. The second stage is implemented as a high voltage integrated circuit chip using double diffused metal oxide semiconductor (DMOS) transistors in a 0.8 μm 300 V BCD (Bipolar-CMOS-DMOS) process. It consists of two linear driver channels, connecting to the middle node of the left and right actuators, respectively. To generate the desired sinusoidal waveform for the drive signal, the controller of the linear drivers uses pulse frequency modulation (PFM) to encode the slope of the sinusoidal drive signals. Recall from the previous section that changing the shapes of these drive signals with respect to each other could result in roll, pitch, and, yaw rotation of the robot—shifting the drive signals up causes the robot to pitch forward, while applying different amplitudes to the left and right actuator results in roll; skewing the upward and downward slew of the drive signals result in yaw rotation (Fig. 3). As indicated by the block diagram in Fig. 4, the desired drive signal is generated by a feedback loop that relies on an embedded controller to calculate the exact sequence of the PFM pulses based on the current signal level and the high-level rotation command. More technical details on the design consideration and performance of the power electronic unit can be found in our recent paper [11].

We approach the RoboBee electronic system design with a multichip strategy, because the power electronics require specialized DMOS process that can tolerate extremely high breakdown voltages, while the real-time computational demand for autonomous flight control calls for faster digital logic circuits using more advanced CMOS process at smaller technology node. At the same time, since the multichip system has to fit within stringent weight, size, and power budget, we would like to minimize the use of external discrete components and to directly power off battery.

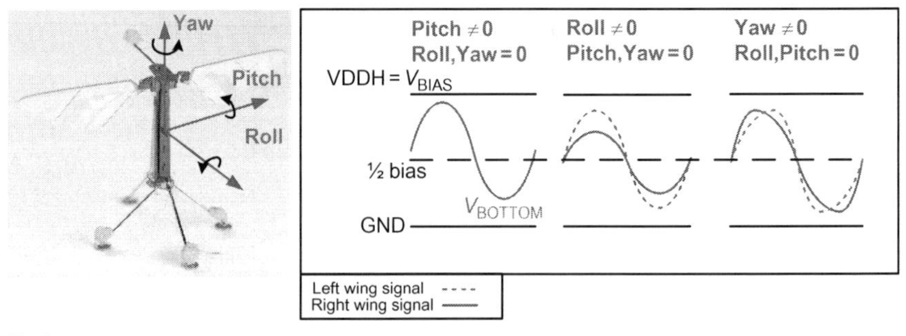

FIG. 3

Drive signals for the piezoelectric actuator to generate three degree-of-freedom maneuver—roll, pitch, and yaw.

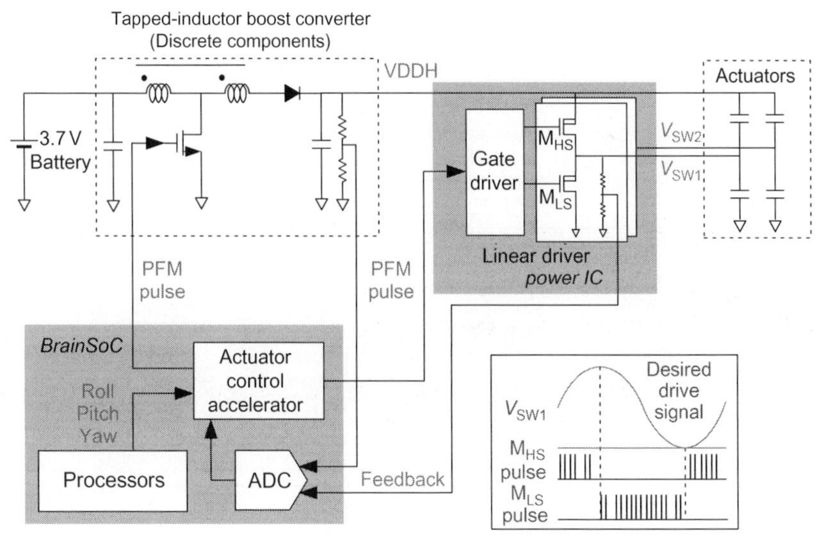

FIG. 4

Block diagram of the two-stage PEU used for the RoboBee. (inset) The drive signal generated by the PEU for the piezoelectric actuator and the corresponding pulse modulation signals.

All these design constraints lead to the decision of a custom-designed SoC that embeds a multitude of functionalities into a single chip including integrated voltage regulator (IVR), analog to digital converters (ADC), multiple clock generators, and a computational core with heterogeneous architecture.

4.2 BrainSoC ARCHITECTURE

As the second chip in the multichip electronic system design for RoboBee, the BrainSoC (Fig. 5) integrates peripheral support circuits to obviate external components other than a battery.

First, RoboBee uses a 3.7 V lithium-ion battery as its power source, while the central control digital logics on the BrainSoC requires a digital supply voltage below 0.9 V. Due to the limited weight budget allocated for the SoC, neither external regulator module nor discrete components such as capacitors and inductors can be afforded, which leads to the integration of a two stage 4:1 on-chip switched-capacitor regulator as part of the SoC. The switched-capacitor topology is selected for the IVR because it does not need external inductors unlike the buck topology and it delivers much better conversion efficiency compared to a linear regulator, which suffers intrinsic low efficiency when the output voltage is only a fraction of the input voltage. The IVR steps the 3.7 V battery voltage down to 1.8 and 0.9 V, resulting in three distinct voltage domains.

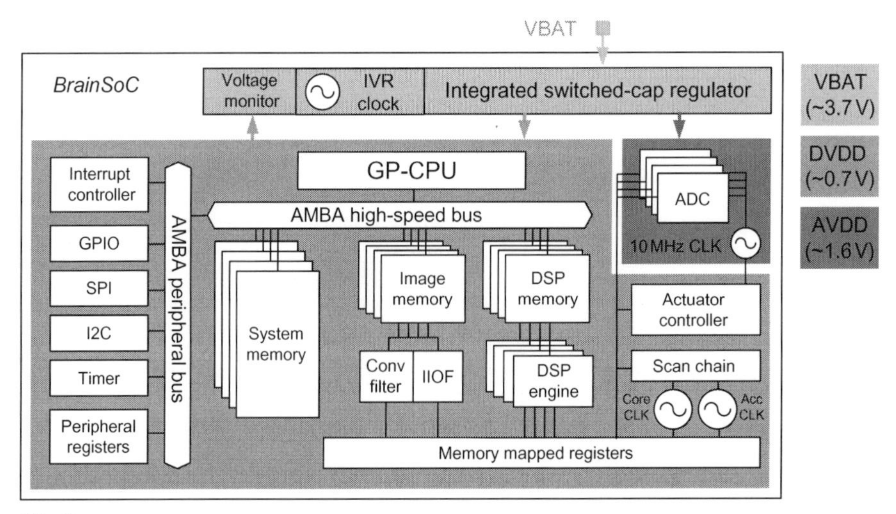

FIG. 5

Block diagram of the BrainSoC used for the RoboBee.

Second, in addition to voltage conversion and regulation, clock generation is yet another function that must be entirely integrated for this microrobotic application. Our design budget does not allow room for adding external crystal oscillator component, and therefore all the clock signals have to be generated on chip. There are several different types of clock generators in the BrainSoC to fulfill different needs for the system: a high-precision 10 MHz relaxation oscillator sets the wing-flapping frequency to best match its mechanical resonance frequency; a 16-phase supply-invariant differential ring oscillator is designed for the switching control logics in the IVR that burns extremely low power and is directly supplied by the battery; multiple digitally controlled oscillators (DCO) generate the clock signals for the general-purpose microprocessor, memory, accelerators, and peripherals.

Next, the BrainSoC is equipped with a diverse set of peripheral circuits to interface with a multitude of sensor chips. For sensors with analog I/O channels, an embedded ADC is integrated that comprises of four 8-bit successive approximation ADCs channels multiplexed between 16 analog inputs; and for sensors with digital serial communication interfaces, we incorporate standard serial protocol controllers such as I2C, SPI, and GPIO.

Finally, the computational core inside BrainSoC employs a heterogeneous architecture. It consists of a 32-bit ARM Cortex-M0 microcontroller that handles general computing needs and acts as the master of the AMBA high-speed bus connecting to various memories. In order to meet the real-time performance demands of autonomous flight in an energy efficient way, we have to supplement the general-purpose computing system with various hardware accelerators. Since in our initial exploratory experiment, vision sensor and IMU are the essential sensing mechanism for RoboBee, accelerators for image processing and rotation control are built into the hardware with dedicated memory and coordinated by M0 via memory-mapped register.

5 SUPPLY RESILIENCE IN MICROROBOTIC SoC

Among the numerous challenges surrounding the SoC design for microrobotic applications, reliability, and performance of the system in the presence of supply noise is one critical problem to be addressed. Similar to many integrated computing systems, a microrobotic SoC employs synchronous digital logics in its central control unit and thus is susceptible to disturbance on the supply voltage.

However, the crucial weight and form factor constraints set the microrobotic SoC apart from conventional systems. Given the extremely stringent weight budget, extra external components must be avoided at all cost, which leads to the integration of on-chip DC-DC converter and the absence of external frequency reference. With such IVRs powered directly off a discharging battery, the microrobotic SoC experiences supply-noise characteristics different from conventional digital systems, where existing supply-noise mitigation techniques cannot be easily applied.

In this section, we describe how an adaptive-frequency clocking scheme is used in our BrainSoC design to exploit the synergy between IVR and clock generation. The resulting supply-noise resilience and performance improvement has been demonstrated by a prototype SoC developed prior to the BrainSoC [12]. Our proposed adaptive clocking scheme not only delivers better reliability and performance, but also extends the error-free operation to a wider battery voltage range, which is beneficial to a microrobotic system.

BACKGROUND ON SUPPLY NOISE

Digital computing systems based on synchronous logic circuits typically employ a fixed frequency clock. To guarantee correct operation, final outputs from the datapath must arrive at the next flip-flop stage before the next clock edge by some time margin known as the "setup time". Since the datapath delay is a function of the supply voltage, it is susceptible to noise on the supply line.

Supply noise is the result of nonideal power delivery system and load current fluctuation under varying computation workload. It can come from the parasitic resistance, inductance, and capacitance in the power delivery network, and manifests itself as static IR-drop, which is the static voltage drop due to power grid resistance, as well as dynamic $L\,di/dt$ drop, which is the transient voltage fluctuation caused by the inductance and capacitance in response to load current changes. Also, for systems with integrated switching regulators, the intrinsic voltage ripple of the regulator contributes additional noise to the supply. The existence of supply noise can modulate the datapath delay, which may lead to setup time margin violation and eventually computation errors. In order to ensure sufficient delay margins under all operating conditions, the most straightforward approach is to lower the clock frequency and provide a "guardband" to tolerate even the worst supply-noise scenario.

SUPPLY RESILIENCE IN CONVENTIONAL COMPUTING SYSTEMS

Conservative design strategy such as timing guardband is the most commonly applied to combat supply- noise in conventional computing systems. It may incur hefty performance loss and thus is highly undesirable. Instead, a number of alternative techniques have been proposed to mitigate supply-noise with less performance penalty. The active management of timing guardband [13] in a prototype IBM POWER7 processor is an example of adaptive clocking: a digital phase-locked loop (DPLL) adjusts the processor core's clock frequency based on the timing guardband sensed by a critical path timing monitor [14]. Since resonant noise caused by the LC tank between the package inductance and the die capacitance has been identified as the dominant component of supply noise in high-performance microprocessors [15], many studies have focused on this particular type of supply noise by proposing adaptive phase-shifting PLL [16] and clock drivers [17]. Following the duality between the clock frequency and the supply voltage in synchronous digital systems, the other approach to optimize performance in the presence of supply noise is adaptively adjusting the voltage level delivered by the power supply at different desirable operating frequency. Despite their different implementations, both adaptive clocking and adaptive supply are along a similar vein of technical route that applies closed feedback loop to adjust frequency and/or supply based on monitored timing margin of the system, and therefore are subject to the bandwidth limitation of the feedback loop.

In addition to the above-mentioned systems and techniques, there exist other classes of logic implementations such as asynchronous logics and self-timed logics [18] that do not rely on a global clock for their operations. Unlike synchronous logics, these systems are intrinsically delay-insensitive and thus immune from the negative impact of supply noise. However, these logic implementations lack the full support of standard libraries, IPs, and EDA tools and thus are difficult to incorporate into the digital design flow of a sophisticated SoC.

5.1 UNIQUE RELIABILITY CHALLENGE FOR RoboBee

In many ways, the microrobotic SoC, such as the BrainSoC, suffers similar setup time violations due to supply noise as other types of conventional computing systems, but its weight and form factor constraints present unique design challenges that require different supply-noise mitigation techniques from those employed in a typical microprocessor.

In the BrainSoC, the supply voltage generated by the fully integrated switched-capacitor voltage regulator (SC-IVR), which is directly powered off a battery, can have different noise characteristics from the resonant-noise-dominated supply experienced by the microprocessor. In fact, for the IVR-enabled microrobotic SoC, the worst supply-noise is often caused by sudden load current steps in a very small time scale instead of LC resonance. Therefore, the mitigation techniques developed earlier for slow-changing or periodic supply-noise in microprocessors [13,16,17] are not applicable to the fast-changing supply-noise induced by the load current in a microrobotic SoC.

On the other hand, unlike the microprocessor, RoboBee is a self-sustained autonomous system with no need to synchronize with other systems for communication or I/O transactions, therefore the timing jitter/phase noise requirement of its clock signal can be relaxed from the specifications in typical high-speed I/O interfaces that demand PLL-generated clean clock. Moreover, the lack of external frequency source as the reference signal renders the implementation of a PLL impractical. We therefore conclude that a free-running oscillator is a better candidate as the clock generator for the BrainSoC.

The above discussion explains that the microrobotic SoC differs significantly from conventional digital systems in its integration of a battery-connected IVR and its ability to operate with a free-running clock. The former suggests distinctive supply-noise characteristics dominated by fast load current changes and slow battery discharge, while the latter provides opportunity for an adaptive clocking scheme.

In the context of the RoboBee's BrainSoC, the goal is thus to optimize the flight time with respect to the total energy available in its battery and the associated battery discharge profile. However, using the RoboBee system as a platform, we are able to investigate effective supply-noise mitigation techniques that can be applied to more general microrobotic systems with minimal performance penalty. Along this vein, we explore the relative merits of different operational modes offered by the supply regulation mechanisms and the clock generation schemes. Fig. 6 illustrates two possible modes of supply regulation with respect to a typical lithium-ion battery discharge profile. The first one (Fig.6A) is closed-loop operation at a fixed voltage. In this case, with the help of feedback control loop, the SC-IVR can provide a constant supply voltage that is resilient to input battery (V_{BAT}) and output load (I_{LOAD}) conditions. One advantage associated with the fixed voltage operation is that it provides a relatively constant operating frequency. However, for a target output voltage level (V_{REF}), the SC-IVR's operating range is limited to $V_{BAT} > 4V_{REF}$. In contrast, open-loop operation with variable unregulated voltage (Fig. 6B) exhibits an entirely different set of attributes; with no feedback control, the SC-IVR's output voltage is roughly 1/4th the input battery voltage, but varies with both the discharge profile and load fluctuations. While open-loop SC-IVR mode allows the system to operate over a wider range, down to the minimum voltage limit of the digital load, performance and energy efficiency depend on the clocking strategy used. The choices are between two clocking schemes: fixed-frequency and adaptive-frequency clocking. Out of the four total combinations, we compare the following three: (1) fixed regulated voltage, fixed frequency; (2) fixed regulated voltage, adaptive frequency; and (3) variable

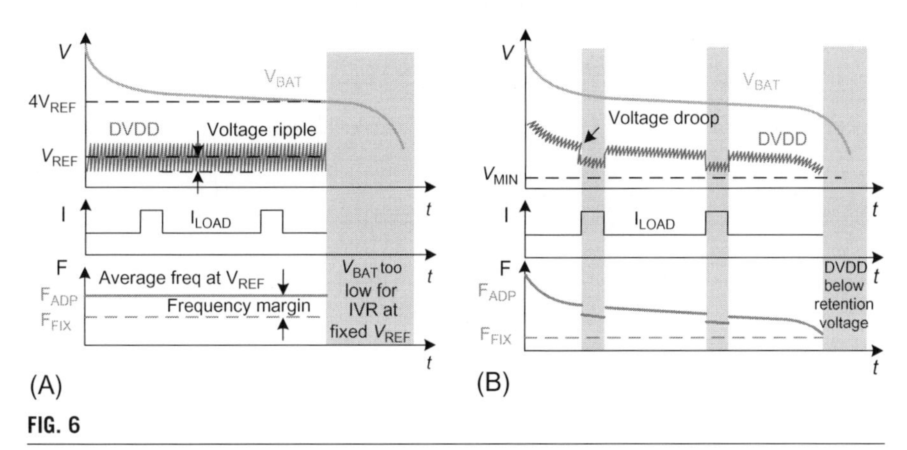

FIG. 6

Illustration of two SC-IVR modes of operations versus typical battery discharge profile. (A) Closed-loop regulation and (B) open-loop operation.

unregulated voltage, adaptive frequency. Intuitively speaking, fixed-frequency (F_{FIX}) clocking requires extra timing margins to account for nonnegligible worst-case voltage ripple, which is an intrinsic artifact of the SC-IVR's feedback loop; while, alternatively, an adaptive-frequency (F_{ADP}) clocking scheme that allows the clock period to track the changes in the supply voltage could offer higher average frequency. Adaptive-frequency clocking also works well for open-loop SC-IVR mode, because it maximizes performance with respect to battery and load conditions.

5.2 TIMING SLACK ANALYSIS ON ADAPTIVE CLOCKING

To drill deeper on the above intuitive explanation of the supply-noise impact on timing margin and the potential beneficial compensation effect when the clock period can match the datapath delay by tracking supply voltages, we now employ a more rigorous analysis of timing slack for broad-band supply-noise with fast transients.

Fig. 7 shows one stage of a pipeline circuit that is clocked by a free-running DCO. The clock signal (CLK) is generated by a clock edge propagating through the delay cells of the DCO, and is then buffered to trigger the flip-flops at the input and output of the datapath. The buffered clock signals are labeled as CP1 and CP2. Using similar definition proposed by previous work [4,9], the timing slack can be calculated as:

$$\text{slack} = t_{clk} + t_{cp2} - t_{cp1} - t_d \tag{1}$$

Here, $t=0$ is the time the first clock edge is launched, and it takes t_{cp1} to travel through the clock buffers and reach the first flip-flop as CP1. In the meantime, the first clock edge propagates through the delay cells and, at $t=t_{clk}$, it completes the round trip in the DCO and the second clock edge is launched at CLK and takes t_{cp2} to reach the second flip-flop as CP2. Instead of resorting to small signals, we simply represent the supply voltage as a function of time with $v(t)$. Without loss of generality, let us assume each circuit block (X) has a unique function $f_x(v(t))$ that measures the rate of propagation delay accumulation as a function of the supply voltage, such that the propagation delay of the circuit t_x can be expressed as:

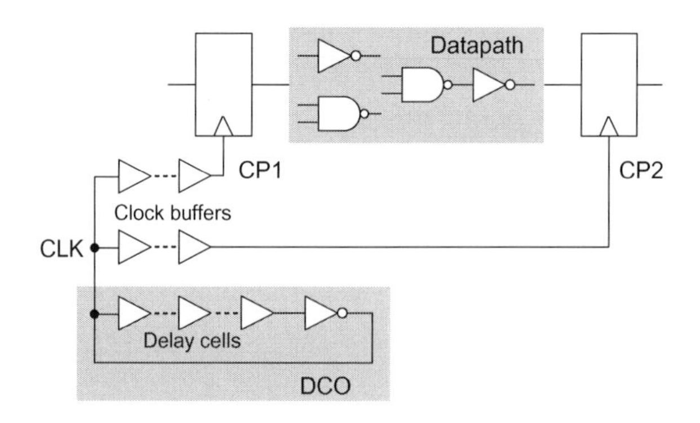

FIG. 7

Simplified diagram of a pipeline circuits.

$$t_x = \int_0^{t_x} f_x(v(t))dx \tag{2}$$

In this way, we can re-write the delay parameters in Eq. (1) as the following:

$$\begin{aligned} t_{\mathrm{clk}} &= \int_0^{t_{\mathrm{clk}}} f_{\mathrm{clk}}(v(t))dx \\ t_{\mathrm{cp2}} &= \int_{t_{\mathrm{clk}}}^{t_{\mathrm{clk}}+t_{\mathrm{cp2}}} f_{\mathrm{cp2}}(v(t))dx \\ t_{\mathrm{cp1}} &= \int_0^{t_{\mathrm{cp1}}} f_{\mathrm{cp1}}(v(t))dx \\ t_{\mathrm{d}} &= \int_{t_{\mathrm{cp1}}}^{t_{\mathrm{cp1}}+t_{\mathrm{d}}} f_{\mathrm{d}}(v(t))dx \end{aligned} \tag{3}$$

The important finding derived from Eq. (3) is that to ensure constant positive timing slack under all supply-noise conditions is to match all the delay accumulation functions ($f_{\mathrm{clk}}, f_{\mathrm{cp1}}, f_{\mathrm{cp2}}, f_{\mathrm{d}}$), rather than simply the DCO and the datapath. Intuitively, this is because the delay is accumulated first at the DCO and then at the clock driver for CP2, whereas it is first at the clock driver and then at the datapath for CP1. If there is any mismatch between the clock driver and the DCO or the datapath, the impact of the supply noise cannot be fully compensated. Therefore, our design uses the fanout-of-4 delay tracking DCO as a reasonably good approximation to track the typical delay in both the datapaths and the clock drivers, rather than a precise implementation to perfectly match either the datapaths or the clock drivers.

5.3 SYSTEM IMPLEMENTATION

The prototype microrobotic SoC designed as the precursor of BrainSoC is not a full-fledged implementation of all the functionalities for the RoboBee, but it captures the most essential components in such a SoC for our investigation of the interaction between supply noise and clocking scheme. Shown in Fig. 8, the prototype SoC contains a fully integrated two-stage 4:1 SC-IVR, a 32-bit ARM Cortex-M0 general-purpose processor, two identical 64 KB memories, and a programmable DCO that generates the voltage-tracking adaptive-frequency clock. To gather measurement data for performance evaluation, the chip also includes numerous blocks for testing and debug purposes: a built-in self-test (BIST) block allows thorough testing of the two memory blocks; a scan chain configures the digital blocks; a voltage monitor block probes internal voltages to record fast transients on the supply line; and a current-load generator enables different testing scenarios for load

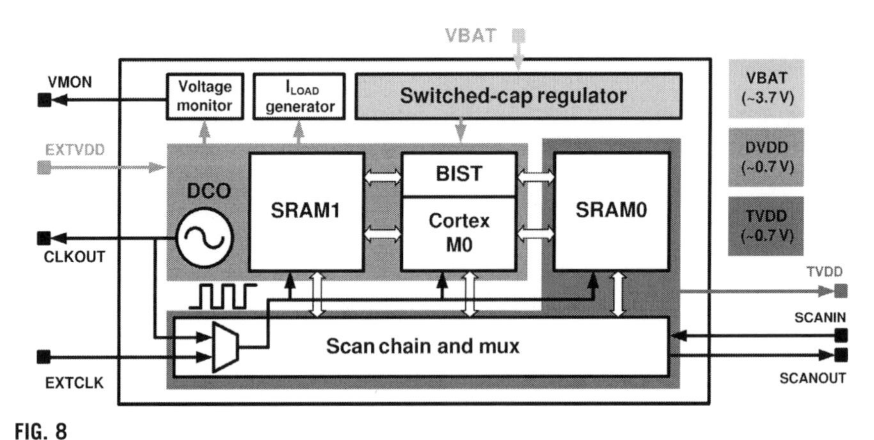

FIG. 8

Block diagram of the fully integrated prototype system-on-chip.

current-induced supply-noise. Lastly, the prototype chip provides direct interfaces for external power and clock sources to set up different operating modes.

5.3.1 Switched-capacitor integrated voltage regulator

The SC-IVR converts the battery voltage (V_{BAT}=3.7 V) down to the digital supply (DVDD=0.7 V). It consists of a cascade of two 2:1 switched-capacitor converters that are respectively optimized for high input voltage tolerance and fast load response and individually tuned to maximize conversion efficiency, and each converter stage employs a 16-phase topology to reduce voltage ripple. A low-boundary feedback control loop can regulate DVDD to a desired voltage level. A thorough discussion of the SC-IVR and its implementation details can be found in [19].

While this SC-IVR can achieve high conversion efficiency, 70% at its optimal operating point, it is subject to the inherent limitations of any switched-capacitor based DC-DC converter; efficiency varies with respect to input and output voltages and the load current level. As an example, Fig. 9 plots the SC-IVR's output voltages at different battery voltage levels for both open- and closed-loop operation, and the peak efficiency is labeled at each point. These efficiency numbers and output voltage levels are derived by sweeping the output voltage to find the peak conversion efficiency and its corresponding output voltage level at each fixed input battery voltage, and hence each point presents the best possible efficiency for each input and output voltage combinations. The data in Fig. 9 clearly shows that open-loop operation consistently offers 2–3% higher conversion efficiency and 16–30 mV higher output voltage levels than closed-loop operation, resulting in both higher efficiency and higher performance for the SoC.

Fig. 10 plots the transient behavior of the IVR with respect to load current steps between 3 and 50 mA for both open- and closed-loop operation. According to the transient waveform of the output voltage, the closed-loop operation quickly responds to avoid the steep voltage droop otherwise seen in the open-loop case; however, it

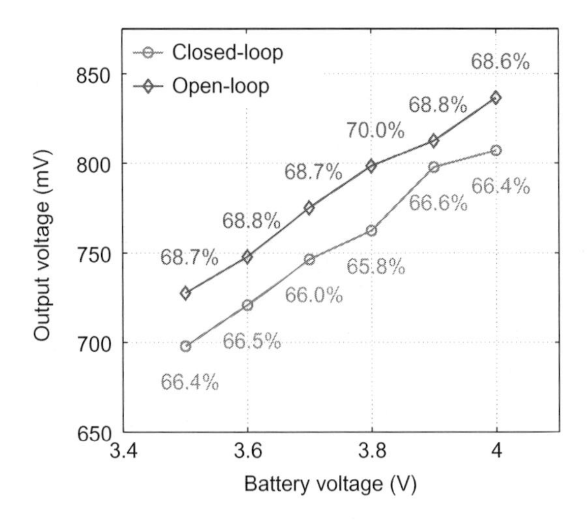

FIG. 9

Output voltages at peak conversion efficiency versus battery voltage for open- and closed-loop operation.

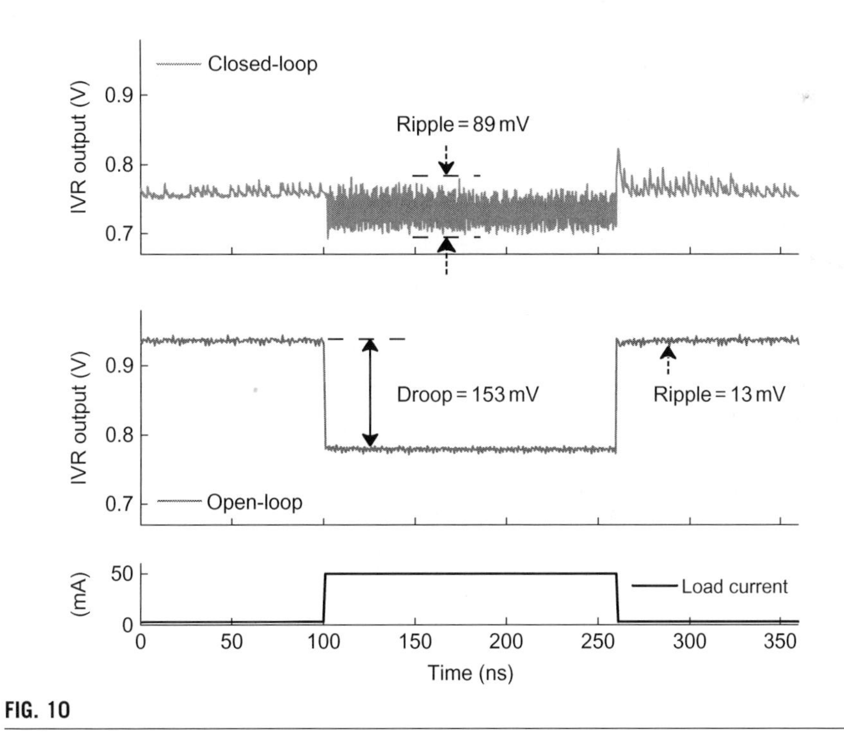

FIG. 10

Transient response of SC-IVR output voltage to load current steps.

exhibits larger steady-state voltage ripple, especially for higher load currents, due to the control loop implementation and feedback delay.

The voltage ripple in closed-loop operation is caused by the delay of the feedback control that prevents the regulator from instantaneous reaction when its output voltage drops below the reference voltage and thus can result in both undershoot and overshoot in the output voltage. Even with sophisticated feedback strategy and fast transistors in advanced technology node, it is difficult to reduce the feedback delay beyond 1 ns, and this fundamentally determines the magnitude of the steady-state closed-loop ripple. On the other hand, the voltage ripple in open-loop operation is due to charge sharing between the flying capacitor and the decoupling capacitor at the output node. It can be reduced by phase interleaving, and therefore does not experience the same fundamental limit as the closed-loop ripple. This is why in our 16-phase interleaved implementation of the IVR, the open-loop ripple magnitude is significantly smaller than the closed-loop ripple.

5.3.2 Digitally controlled oscillator

Our proposed adaptive-frequency clocking scheme needs a clock generator whose frequency tracks closely with changes in supply voltage. This allows the operating frequency of the digital load circuitry to appropriately scale with voltage fluctuations, providing intrinsic resilience to supply noise. There are numerous examples of critical-path-tracking circuits for local timing generation [14,20]. Instead, we use a programmable DCO to generate the system clock. The DCO contains a ring of programmable delay cells comprising of transmission and NAND gates that approximate a typical fanout-of-4 inverter delay.

As shown in Fig. 11A, a 7-bit digital code, $D = D_6,\ldots,D_1D_0$, sets the DCO frequency by selecting the number of delay cells in the oscillator loop. While our implementation uses 7 bits of control code, measurement results show that the lower 4 bits are sufficient for the normal operating range of the digital system. Fig. 11B plots DCO frequency versus supply voltage (DVDD) across a range of the digital control codes. Over the measured voltage range (0.6–1 V), frequency scales roughly linearly with supply voltage, but slightly flattens out for voltages below 750 mV. The uneven frequency spacing with respect to the control code D results from the delay cell's asymmetric design and should be improved for future implementation.

5.3.3 Cortex-M0 and memory

Both the Cortex-M0 microprocessor and the memories used in the prototype SoC are IP blocks provided by our collaborators. To facilitate our testing strategy for different operation modes, the Cortex-M0 and one of the 64 KB memories (SRAM1) are intentionally designed to share the voltage domain (DVDD) with the DCO, while the other memory (SRAM0) sits in another separate voltage domain (TVDD) with the rest of the test peripheral circuits. Except for being in a different voltage domain, SRAM0 shares the same physical design as SRAM1 as both are using the same hard memory IP.

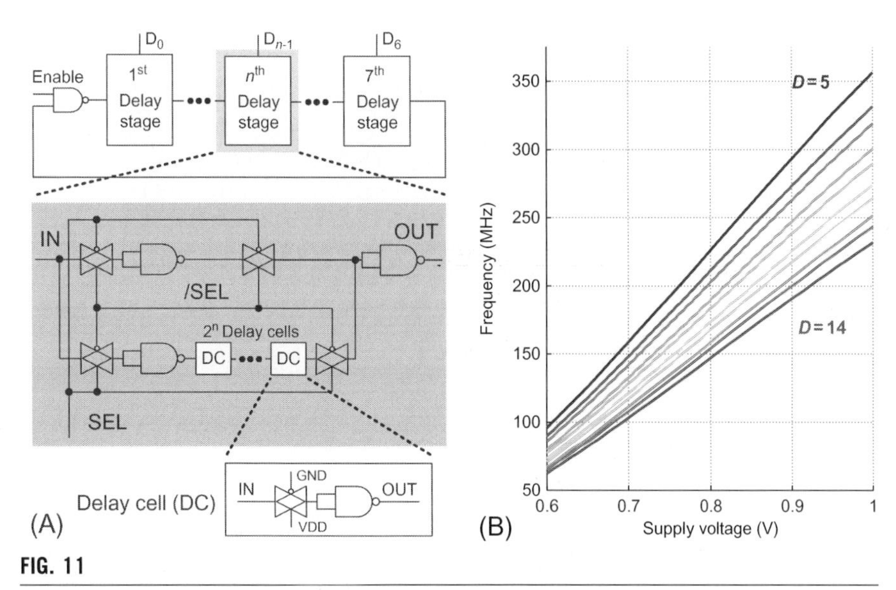

FIG. 11

Digitally controlled oscillator (DCO) schematic and measured characteristics. (A) DCO schematic and (B) DCO frequency versus DVDD at digital control code D.

5.3.4 Built-in self-test module

A BIST module performs at-speed test for both SRAMs. To make sure full test coverage of all the memory cells, the BIST performs a modified MARCH-C routine that writes to and reads different data patterns from each memory address in the SRAMs successively. At the conclusion of the MARCH-C routine, the BIST module would raise a pass/fail flag if any read or write error has been detected. Due to the limited storage for testing vectors, only the address and data of the last failure are recorded for posttest analysis.

The test peripheral circuits operated off a TVDD include a voltage monitor block that captures fast nanosecond-scale transient changes of the supply line and a programmable load current generator that is made up of differently weighted switched current sources. Although external supply and clock interfaces such as TVDD, EXTVDD, and EXTCLK are included, they are for testing purposes only, as the SoC is capable of operating directly off the battery without any external supply or clock reference.

5.4 PROTOTYPE SoC EXPERIMENTAL EVALUATION

To demonstrate improved resilience and performance of our proposed adaptive clocking across a wide range of supply voltage, measurement results were obtained from the prototype SoC chip (Fig. 12) fabricated in TSMC's 40 nm CMOS technology. We use the maximum error-free operating frequency of the memory performing BIST as a proxy metric, because it is often the on-chip SRAM sharing the same

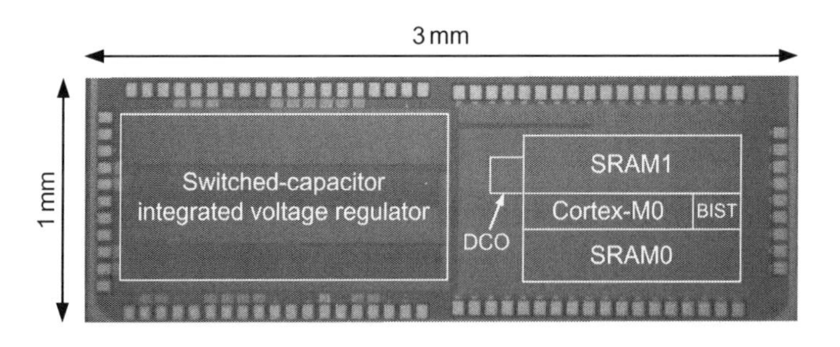

FIG. 12

Die photo of the fully integrated prototype system-on-chip.

voltage domain with the digital logic that limits the system performance at lower supply levels. Also, the retention voltage of the SRAM cells typically determines the minimum operating voltage of the system [21].

First, we characterize the voltage versus frequency relationship of the SRAMs using external sources in order to determine the efficacy of using the DCO for adaptive-frequency clocking. Then, we compare the fixed- and variable-frequency clocking schemes with a regulated voltage generated by the SC-IVR in closed-loop operation. Lastly, we present the advantages of combining adaptive clocking with a variable voltage provided by operating the SC-IVR in open loop.

5.4.1 Frequency versus voltage characterization

The on-chip SRAMs were characterized at static supply voltage levels provided externally via EXTVDD, in order to determine the SRAM's voltage to frequency relationship under quiet supply conditions. Using an external clock (EXTCLK) at different fixed frequencies, we obtained the Shmoo plot in Fig. 13A. It shows (1) the minimum retention voltage of SRAM cell is between 0.6 and 0.65 V; (2) the maximum SRAM frequency scales roughly linear with supply voltage and ranges from 68 at 0.65 to 256 MHz at 1.0 V; and (3) the maximum SRAM frequency closely correlates with DCO frequency plot for control code D = 10. This correlation suggests the FO4-delay-based DCO tracks the critical path delay in the SRAM across a wide supply range and should enable error-free memory BIST for control word D above 10. We experimentally verified this by turning on the internal DCO to provide the system clock instead of using EXCLK and sweeping the same voltage range via EXTVDD. The resulting Shmoo plot in Fig. 13B further demonstrates the DCO's ability to track SRAM delay at different static supply voltage levels. Considering process and temperature variations, the same control word D = 10 may not apply to chips from all process corners and over all temperature ranges. However, since the process and temperature conditions are relatively static, and our results show that a fixed control word can cover the range of fast-changing supply variation, it is possible to determine this fixed control word during the calibration phase before the normal operation of the robotic system starts.

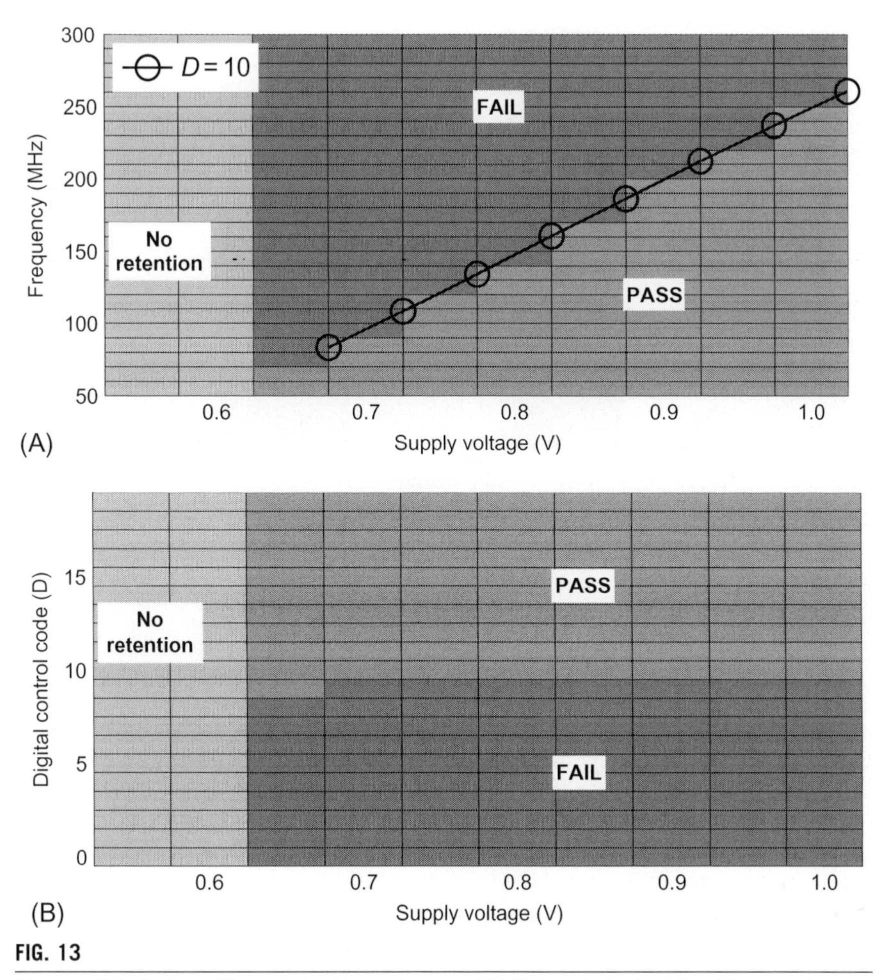

FIG. 13

Shmoo plots for two different clocking schemes. (A) External clock at fixed frequencies and (B) internal adaptive clock generated by the DCO at different control code D.

5.4.2 Fixed versus adaptive clocking with regulated voltage

Having verified the DCO, we now compare fixed- and adaptive-frequency clocking schemes for a system that operates off a regulated voltage with the SC-IVR operating in closed loop. We also emulate noisy operating conditions using the on-chip I_{LOAD} generator that switches between 0 and 15 mA at 1 MHz. Measurements made via the on-die voltage monitoring circuit showed approximately $\pm$70 mV worst-case ripple about a mean voltage of 0.714 V.

For the conventional fixed-frequency clocking scheme, the maximum operating frequency ought to depend on the worst- case voltage droop, measured to be 0.647 V. Using the measured relationship in Fig. 13A, the maximum frequency cannot exceed

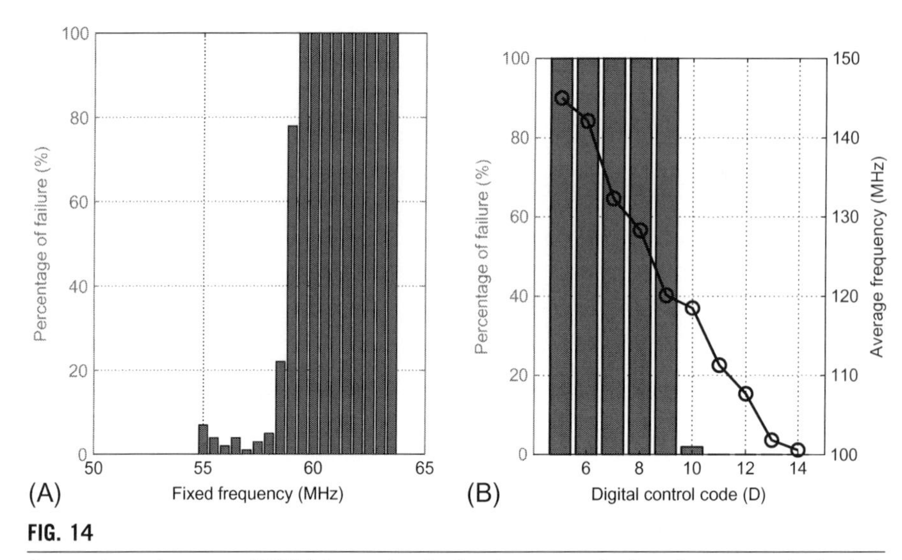

FIG. 14

Comparison of memory BIST failure rates for fixed- versus adaptive- frequency clocks, but under the same IVR closed-loop regulation. (A) Fixed-frequency externally and (B) adaptive-frequency from DCO.

68 MHz. To measure the actual maximum error-free frequency, we performed 100 independent BIST runs using the EXTCLK, set to a fixed frequency and recorded the failure rate. Fig. 14A summarizes the measured failure rates across different externally driven operating frequencies. These results show that the maximum error-free frequency is below 55 MHz for the fixed-frequency clocking scheme, which is even lower than the anticipated 68 MHz, perhaps attributable to the additional noise injection.

Using the same IVR configuration and test conditions, Fig. 14B plots the failure rates versus different digital control codes of the DCO. At D = 10, there were intermittent failures, attributable to the additional noise not present in the prior experimental results of Fig. 13B. The adaptive-frequency clocking scheme delivers consistent and reliable operation at D = 11. Based on the average DCO frequency measured during the tests and also plotted in Fig. 14B, corresponds to an average frequency of 111 MHz, which is 2 × the fixed-frequency clocking scenario. Here we use the average frequency instead of the worst frequency as a measure of system performance, because the operating frequency, which tracks the noisy supply, changes at much faster time scale than the task performance requirement of the robot, which means that the aggregated throughput measured by the average frequency is the more meaningful metric and occasional drop to lower frequency does not have destructive effect on the system.

We attribute this large frequency difference between the two clocking scenarios to a couple of factors. Fixed-frequency clocking requires sufficiently large guardbands to guarantee operation under the worst-possible voltage droop condition. In contrast, adaptive-frequency clocking allows both the clock period and load circuit delays to fluctuate together as long as both vary with voltage in a similar manner. Hence, the guardband must only cover voltage-tracking deviations between the DCO and load circuit delay paths across the operating voltage range of interest, and can be built into the DCO. Another factor that penalizes the performance of the fixed-frequency clocking comes from the additional noise on the EXTCLK signal for crossing the TVDD to DVDD boundary.

5.4.3 Adaptive-frequency clocking with unregulated voltage

We now turn our attention to how adaptive-frequency clocking performs with an unregulated voltage generated by the SC-IVR operating in open loop. We used the same test setup with and noise injection via the on-chip generator. The failure rates and the average frequencies are captured in Fig. 15. Compared to the measured results in Fig. 14B, average frequencies are much higher, because DVDD settles to higher values (≈0.8 V) when the SC-IVR operates in open loop. Despite the high susceptibility to fluctuations on DVDD to load current steps as seen in Figs. 10 and 15A shows zero errors occurred even for D = 10. The higher DVDD voltage provides more cushion to avoid intermittent retention failure.

In order to illustrate the extended operating range offered by running the SC-IVR in open loop, Fig. 15B plots the average DCO frequency and average DVDD voltage for error-free operation versus battery voltage. These measurements were again made with 0–15 mA current load steps. As expected, the open-loop SC-IVR's average output voltage scales proportional to the battery voltage. Moreover, the system

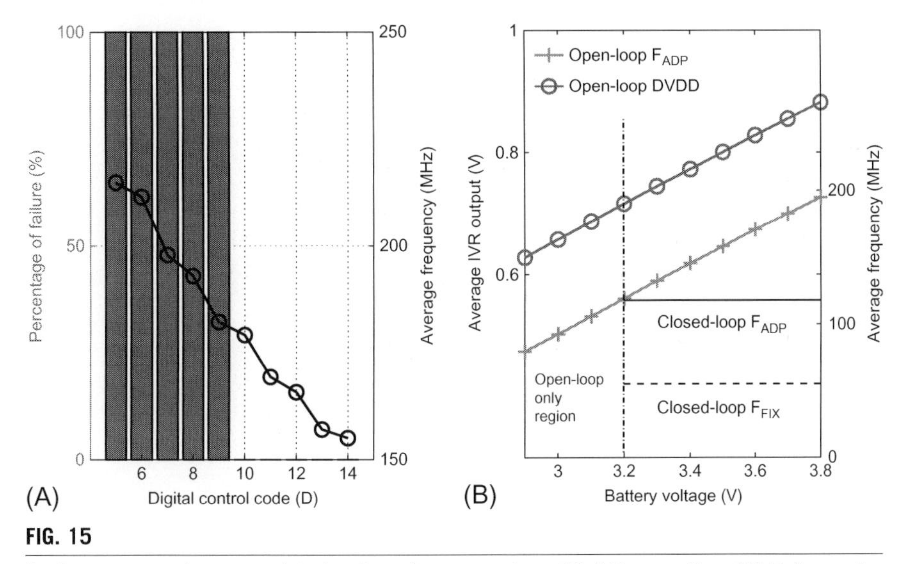

FIG. 15

Performance under unregulated voltage from open-loop SC-IVR operation. (A) Failure rate and average frequency versus DCO settings and (B) measured average clock frequency and output voltage.

can operate error-free even for battery voltages below 3 V, which approaches the 2.5–2.7 V lower discharge limit of Li-ion batteries. In comparison, assuming a target SC-IVR regulated voltage of 0.7 V, the system would only operate down to a battery voltage of 3.2 V and at a lower frequency across the battery discharge profile even with the adaptive-frequency clocking scheme. A fixed-frequency clocking scheme would lead to even lower performance.

Finally, we look at the transient waveform of the supply voltage captured by the internal voltage monitor when the SC-IVR operates in open loop with the same periodic load step condition used in Fig. 15A. Fig. 16 shows both CLKOUT, which is a divide-by-2 signal of the internal system clock, and the supply voltage (DVDD). The DCO's control code is set to 10 as suggested by Fig. 15A. In the zoom-in window of

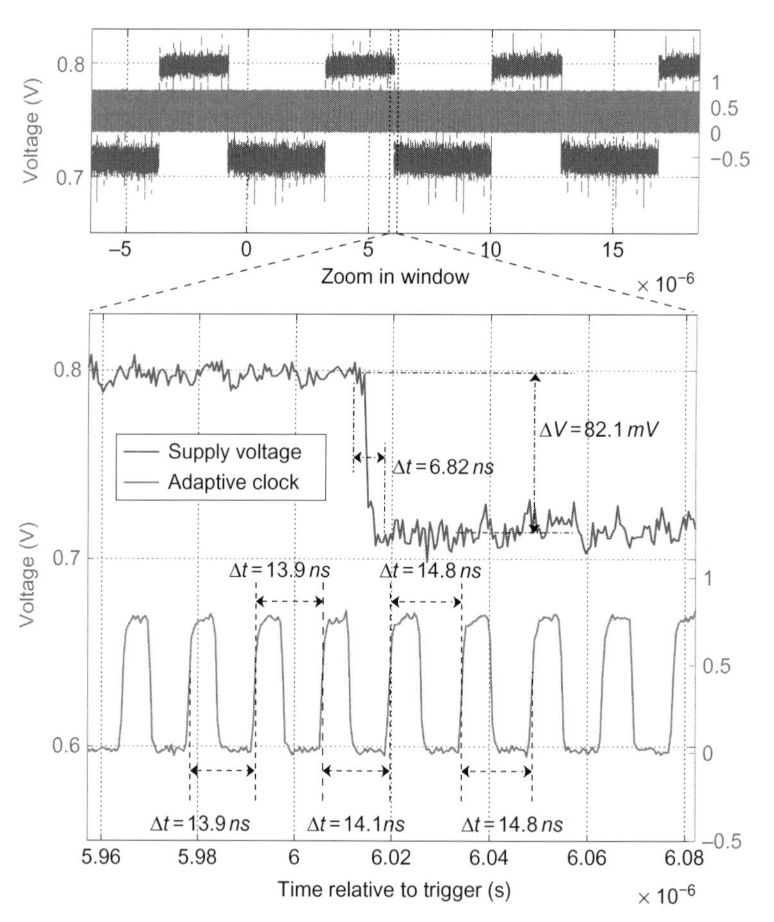

FIG. 16

Waveforms of the supply voltage (DVDD) and a divide-by-2 clock signal (CLKOUT) during IVR open-loop operation with load current (I_{LOAD}) periodically switching between 5 and 30 mA.

the waveforms, it clearly illustrates that the DCO frequency can respond promptly to a 82.1 mV supply droop within 6.82 ns, as the load current steps from 5 to 30 mA, so that no memory access error is recorded by the BIST module, suggesting superior supply-noise resilience of the SoC system due to the adaptive clocking scheme. Moreover, the waveform of the supply voltage demonstrates that supply noise in a typical microrobotic SoC with IVR is characterized more by the voltage droop and ripple in response to the load current steps and the IVR switching, instead of resonant noise cause by the LC tank in the power delivery path.

In addition to validating the resilience and the performance advantages of adaptive-frequency clocking, our experimental results also reveal the synergistic properties between the clocking scheme and the IVR design in a battery-powered microrobotic SoC. The supply-noise resilience provided by an adaptive clock alleviates design constraints imposed by voltage ripple and voltage droop. Therefore, the IVR can trade-off its transient response for better efficiency or smaller area when codesigned with adaptive-frequency clocking.

6 CONCLUSION AND FUTURE WORK

In this chapter, we have presented the design context and motivation of customized SoC for microrobotic application and have done an in-depth investigation of supply resilience in battery-powered microrobotic system using the prototype chip, which is the precursor of BrainSoC, with both analytical and experimental evaluations. The conclusion we are able to arrive at is that our proposed adaptive-frequency clocking scheme offers major advantages when combined with an IVR in a battery-powered microrobotic SoC. Our thorough evaluation has revealed the subtle but crucial trade-offs and merits associated with different configurations of the IVR and the clock—for regulated voltage operation via closed-loop IVR, adaptive-frequency clocking enables 2 × performance improvement, compared to conventional fixed-frequency clocking scheme; combining adaptive-frequency clocking with an unregulated voltage via open-loop IVR extends the operating range across a wider portion of the battery's discharge profile.

Our study has mainly focused on the effect of battery discharge and voltage regulator operation on supply noise with the running of the digital processor. However, another important aspect of supply disturbance in the context of RoboBee is the interference from the piezoelectric actuator drive, which consumes significant portion of the system power by drawing large energy packet in periodic bursts. We plan to further investigate this effect in the mounted flight experiment with the fully on-board electronic system on the RoboBee.

In addition to supply resilience and electronics robustness, there are other aspects of a robot design that warrant careful considerations in order to improve the reliability of the overall system. And the system trade-off does not stop at the individual robot level. As we have hinted at the beginning of the chapter, one facet of reliability that is unique to microrobots stems from their capability to cooperate in swarms.

Unlike conventional large-scale robot, a microrobotic swarm can tolerate multiple failures from individual robots, yet still accomplish the mission at hand. Study [22] has found that degradation of hardware reliability can be tolerated to some extent without significant affect on the swarm performance. Moreover, harsh environmental factors, such as extreme temperature and severe electromagnetic field may present yet more restricted design specifications and additional reliability considerations.

Many challenges remain to be solved before we can unleash the full power of autonomous MAV, but one message is clear—it is immensely valuable to build microrobotic platforms such as RobeBee to provide a development environment and a research testbed for all the open scientific and engineering questions that beg to be answered. I hope you have had as much fun in learning about our journey towards an autonomous insect-scale flying robot as we had in building it.

REFERENCES

[1] M.M. James, S.F. Michael, Micro air vehicles—toward a new dimension in flight, US DAPPA/TTO Report, 1997.

[2] M.T. Keennon, J.M. Grasmeyer, Development of the black widow and microbat MAVs and a vision of the future of MAV design, in: AIAA International Air and Space Symposium and Exposition: The Next 100 Years, 2003.

[3] http://www.avinc.com/nano/.

[4] R.J. Wood, B. Finio, M. Karpelson, K. Ma, N.O. Perez-Arancibia, P.S. Sreetharan, H. Tanaka, J.P. Whitney, Progress on "pico" air vehicles, Int. J. Robot. Res. 31 (11) (2012) 1292–1302.

[5] R.A. Brooks, A.M. Flynn, Fast cheap and out of control: A robot invasion of the solar system, J. Brit. Interplanetary Soc. 42 (1989) 478–485.

[6] K. Ma, P. Chirarattanon, S. Fuller, R.J. Wood, Controlled flight of a biologically inspired, insect-scale robot, Science 340 (2013) 603–607.

[7] M. Karpelson, G.-Y. Wei, R.J. Wood, A review of actuation and power electronics options for flapping-wing robotic insects, in: IEEE International Conference on Robotics and Automation, Pasadena, CA, May, 2008.

[8] P.-E. Duhamel, J. Porter, B.M. Finio, G. Barrows, D. Brooks, G.-Y. Wei, R.J. Wood, Hardware in the loop for optical flow sensing in a robotic bee, in: IEEE/RSJ International Conference on Intelligent Robots and Systems, San Francisco, CA, September, 2011.

[9] Z.E. Teoh, S.B. Fuller, P. Chirarattananon, N.O. Perez-Arancibia, J.D. Greenberg, R. J. Wood, A hovering flapping-wing microrobot with altitude control and passive upright stability, in: IEEE/RSJ International Conference on Intelligent Robots and Systems, Vilamoura, Portugal, September, 2012.

[10] X. Zhang, M. Lok, T. Tong, S. Chaput, S.K. Lee, B. Reagen, H. Lee, D. Brooks, G.-Y. Wei, A Multi-Chip System Optimized for Insect-Scale Flapping-Wing Robots, in: IEEE Symposium on VLSI Circuits (VLSIC) Digest of Technical Papers, 2015, pp. 152–153.

[11] M. Lok, X. Zhang, E.F. Helbling, R. Wood, D. Brooks, G.-Y. Wei, A power electronics unit to drive piezoelectric actuators for flying microrobots, in: Custom Integrated Circuits Conference (CICC), 2015 IEEE, 2015, pp. 1–4.

[12] X. Zhang, T. Tong, D. Brooks, G. Wei, Evaluating adaptive clocking for supply-noice resilience in battery-powered aerial microrobotic system-on-chip, IEEE Trans. Circuits Syst. I Regul. Pap. 61 (8) (2014) 2309–2317.
[13] C.R. Lefurgy, A.J. Drake, M.S. Floyd, M.S. Allen-Ware, B. Brock, J.A. Tierno, J.B. Carter, Active management of timing guardband to save energy in power7, in: Proc. 44th Annual IEEE/ACM, Int. Symp. Microarchitecture, New York, USA, ser.MICRO-44'11, 2011, pp. 1–11.
[14] A. Drake, R. Senger, H. Deogun, G. Carpenter, S. Ghiasi, T. Nguyen, N. James, M. Floyd, V. Pokala, A distributed critical-path timing monitor for a 65 nm high-performance microprocessor, in: Proc. ISSCC, 2007, pp. 398–399.
[15] Y. Kim, L.K. John, S. Pant, S. Manne, M. Schulte, W.L. Bircher, M.S.S. Govindan, Audit: stress testing the automatic way, in: Proc. MICRO, December, 2012, pp. 212–223.
[16] D. Jiao, B. Kim, C.H. Kim, Design, modeling, test of a programmable adaptive phase-shifting PLL for enhancing clock data compensation, IEEE J. Solid-State Circuits 47 (10) (2012) 2505–2516.
[17] D. Jiao, J. Gu, C.H. Kim, Circuit design and modeling techniques for enhancing the clock-data compensation effect under resonant supply noise, IEEE J. Solid-State Circuits 45 (10) (2010) 2130–2141.
[18] M.E. Dean, STRiP: A Self-Timed RISC Processor, Ph.D. Dissertation, Stanford University, Stanford, 1992.
[19] T. Tong, X. Zhang, D. Brooks, G. Wei, A fully integrated battery-connected switched-capacitor 4:1 voltage regulator with 70% peak efficiency using bottom-plate charge recycling, in: Custom Integrated Circuits Conference (CICC), September, 2013.
[20] M. Nomura, Y. Ikenaga, K. Takeda, Y. Nakazawa, Y. Aimoto, Y. Hagihara, Delay and power monitoring schemes for minimizing power consumption by means of supply and threshold voltage control in active and standby modes, IEEE J. Solid-State Circuits 41 (4) (2006) 805–814.
[21] S. Jain, S. Khare, S. Yada, V. Ambili, P. Salihundam, S. Ramani, S. Muthukumar, M. Srinivasan, A. Kumar, S.K. Gb, R. Ramanarayanan, V. Erraguntla, J. Howard, S. Vangal, S. Dighe, G. Ruhl, P. Aseron, H. Wilson, N. Borkar, V. De, S. Borkar, A 280 mV-to-1.2 V wideoperating-range IA-32 processor in 32 nm CMOS, in: Proc. ISSCC, 2012, pp. 66–68.
[22] N. Hoff, R.J. Wood, R. Nagpal, Effect of sensor and actuator quality on robot swarm algorithm performance, in: IEEE/RSJ International Conference on Intelligent Robots and Systems, San Francisco, CA, September, 2011.

CHAPTER

Rugged autonomous vehicles

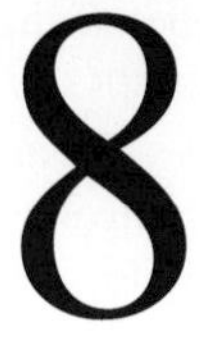

R. Zalman
Infineon Technologies, Neubiberg, Germany

1 AUTOMOTIVE EMBEDDED SYSTEM OVERVIEW

In the current automotive development domain, around 70–90% of the innovation is based on embedded systems. As a result, in conjunction with its complex mechanical capabilities, the modern automobile is becoming probably the most complex consumer product. Also, depending on many factors like market area, high or low-end, etc. the embedded systems account from 20% up to 40% from a modern car cost. The modern vehicles are developing more and more into "computers on wheels" as more and more functions and processing power are added. However, this massive computation add-on comes with a cost: the automotive embedded systems have specific domain "harsh" constraints deriving mainly from the environment, and dependability strong technical requirements. Other factors like cost, development cycles, maintainability, etc. are adding yet more specific constraints to these systems.

1.1 ELECTRONICS IN MODERN VEHICLES

Electronics controls more and more functionality in a modern vehicle. The electronic components are assembled in modules with specific requirements for electrical, mechanical, and chemical environments which are later mounted in the car. These modules are called electronic control units (ECUs). From the 1970s when electronics started penetrating the automobile industry (at that time with only a radio and some electric power generation control, e.g., alternator control), the amount of functionality taken over by electronics increased permanently.

The massive penetration of electronics in the automotive industry has mainly the following access points:

- Replacement of an existing mechanical system (e.g., the engine management control);
- Merging of mechanical functions with electronic ones in mechatronic systems (e.g., antiblocking brakes, ABS);
- New functions entirely possible only by electronics (e.g., navigation systems).

Rugged Embedded Systems. http://dx.doi.org/10.1016/B978-0-12-802459-1.00008-7

Examples of this continuous growth of electronics into the vehicle functions can be the electronic engine control in the late 1970s, the antilock braking (ABS) in the 1980s, airbags in the 1990s, and infotainment and GPS in the 2000s.

The amount of growth in the electronic domain is best seen in the segment growth comparison inside of the automotive domain: according to Ref. [1], in Germany, the motor vehicle to electronics to semiconductor ratio was 10:13:15 in 2013 and will continue in the same dynamic trend through the mid-2020s. This shows the more and more important role of electronics and semiconductors in the vehicle eco-system and strong increase of electronic based functions in modern cars.

The automotive electronics are different from consumer electronics mainly as a result of specific constraints like:

- Functioning under harsh environmental conditions like temperature range, mechanical stress (e.g., vibration), humidity, etc.
- Strict requirements for electromagnetic compatibility;
- Stringent availability and reliability requirements which are key factors for customer perception;
- Stringent safety requirements enforced by national and international regulations;
- Specific lifecycles (e.g., comparatively long product cycles).

The lifecycle of a typical new vehicle, from start to the end of life is shown in Fig. 1. As these relatively long life cycles are different from the typical ones in electronic domain, additional constraints are derived for the automotive embedded systems.

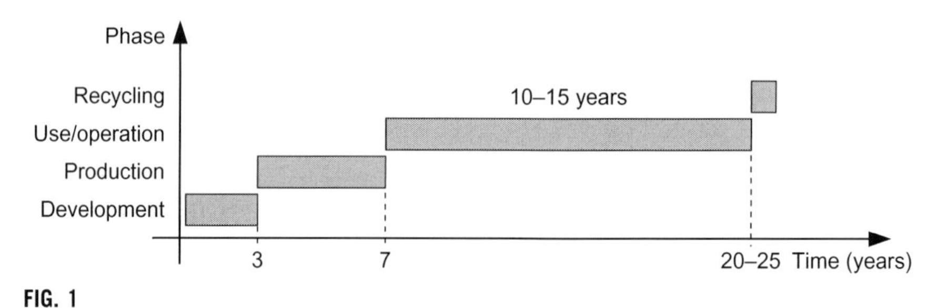

FIG. 1

Typical automobile lifecycle.

These constraints emerge especially during the long use/operation phase, where the need for spare parts makes the production of the needed ECUs challenging due to hardware availability. Hence, for the electronic component manufacturers, special product lines are needed for automotive customers together with dedicated business models allowing these timelines. These limitations are due to the need of producing the same components or "pin compatible" devices with similar (or better) datasheet parameters, which can be used during the entire lifecycle by spare parts producers. It is worth noticing that similar lifecycle requirements are propagating also to other domains of the ECU development—an example being the SW maintenance requirements, which are typically around 15 years.

In the same way, automotive electronics are different from other similar domains like rail and aerospace (which are also very different from customer area) due mainly to two important factors:

- Comparatively high series: yearly automobile production sums up to more than 60 million units (representing around 75% from total motor vehicles produced as not considering light commercial vehicles, heavy trucks, buses, coaches, etc.);
- Strong cost pressure—as the electronics account for an important part of overall vehicle costs (between 20% and 40% mainly depending on segment).

1.2 TYPE OF AUTOMOTIVE ECUS

The electronic functions are built in electronic modules, ECUs, which are interconnected with different specialized communication channels (buses). Each ECU is providing a function or group of functions while in the same time using and providing information from and to the vehicle network. Depending on the particular implementation, each ECU may have a tailored functionality which is available only to specific vehicle instances. This approach is needed for platform development (ECUs which are used in more vehicle models) or to enable optional features of the same vehicle. This approach is generating specific requirements for variant management with implications in the hardware/software architecture, the implementation and in the verification/validation phases of the development.

Some simple examples of electronically controlled functions in an automobile, grouped by their area of application:

- Powertrain:
  - Engine ignition (e.g., spark generation, timing) and fuel injection
  - Gearbox control
  - Emission control
  - Starter/alternator
- Body, chassis, and convenience:
  - Lights, adaptive light control
  - Electronic suspension control
  - Wipers, defrosters, etc.
  - Heating and climate control
  - Seat and door electric control
  - Remote keyless entry, etc.
- Safety:
  - Braking, antilock brakes system (ABS)
  - Steering, power steering
  - Suspension and transmission control
  - Airbags, passive restraint
  - Tire pressure monitor system

    - Adaptive cruise control
    - Lane departure warning
- Infotainment:
    - Navigation
    - Real-time traffic information
    - E-call
    - Multimedia features (audio/video capabilities)
    - Smartphone integration
    - Voice-based, haptic and human machine interface systems and touch
    - Wide and short range communication (e.g., GPS, Bluetooth)

1.3 TYPICAL ECU ARCHITECTURE

The majority of vehicle ECUs, especially in the domain of powertrain, chassis, and body, have a control function. Monitoring functionality is also part of control ECUs or can be integrated in standalone modules (e.g., in safety-related ECUs). For the control type of ECUs, in general, the control (of the controlled element) is based on a physical variable like temperature, voltage, pressure, rotational speed, torque, etc.

In Fig. 2, a typical overview of an ECU embedded in an abstract representation of an automotive control loop is shown.

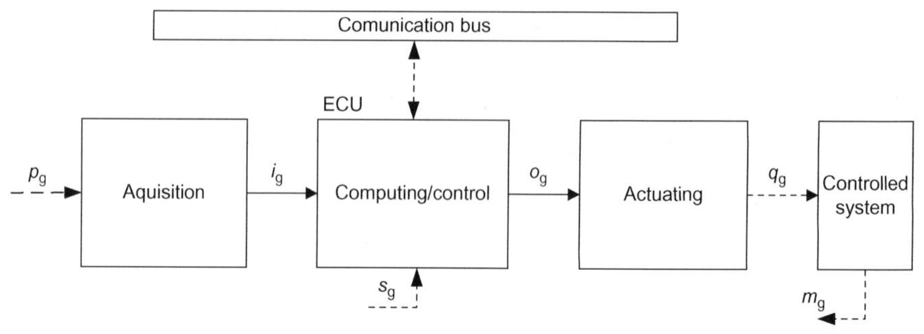

FIG. 2

A typical automotive ECU control loop.

The ECU is implementing its intended functionality by controlling a physical system (the controlled system) which can be a combustion motor, a gearbox, a light control module, etc. This control is achieved by varying a certain physical input value of the controlled system (e.g., for a combustion engine, the throttle angle being proportional with the air intake or the time a fuel injector is open being proportional with the amount of fuel injected in the cylinder). In Fig. 2 this generic value is noted with "q_g" and represents the input value[1] for the controlled system. This value

[1]The discussed values are generic variables; they can be a vector of values, e.g., $q_g = \{q_1, q_2, \ldots, q_i\}$.

however is not directly produced by the ECU but rather the result of a specialized actuation stage. These actuators are electronic or mechatronic devices which convert an electrical control signal at their input, "o_g," (which is also the output of the ECU) into the physical value needed for controlling the system. So, at the actuator stage, the outputs are in a known clear relation with the inputs:

$$q_g = f_{\text{actuator}}(o_g)$$

The control loop is based on a measurable physical value from the controlled system, "m_g" (this measurable value shall be in a defined relation with the physical property which is the subject of the control loop).

This measurable physical value is converted in a signal manageable by the ECU (e.g., electrical value or logical digital signal) by an acquisition stage. The measurable value is subject to external influences and disturbances from the environment so the input of the acquisition stage will be an approximation of the measurable value from the controlled system. By neglecting influences due to disturbances from the environment in the control loop we can write:

$$p_g \approx m_g$$

This acquisition stage consists mainly of sensors capable of executing the needed conversion from physical values to ECU input signals "i_g." This conversion is also subject to approximation and disturbances (e.g., electrical noise) and it consists in general of an analog to digital conversion. The rate and precision of this conversion need to be adapted depending on the characteristics of the controlled system.

Finally, the control loop done in the ECU is a defined relation between its input values and its output values. Another important input parameter for the control loop and the ECU is the setpoint(s) definition. The control loop is trying to maintain the measured values from the system in a defined relation with these setpoints. As an example, the setpoint given by the position of the acceleration pedal to the engine control ECU is maintaining the rotational speed (revolutions per minute, RPM) of the engine constant even in the context of varying other parameters (e.g., torque). So:

$$o_g = f_{\text{ECU}}(i_g)$$

One important property of the control loop is the type of control, which can be "closed loop" or "open loop" control. In a closed loop type of control, the desired output of the controlled system (in our case, the generic measurable "m_g") is continuously sampled and the purpose of the whole loop is to maintain it in a close relation with the required setpoint. So the control loop maintains (or tries to maintain) this relationship during execution of the control algorithm. By contrast, in an open loop system, the output variables are only depending on the input ones without taking into account the feedback signals from the controlled system.

Examples of a close loop ECU control relation (algorithm) is the PI algorithm in which the output is depending in a proportional way together with an integral one with the difference between the measured values and the setpoint values.

$$o_g(t) = k_P(i_g(t) - s_g(t)) + \int k_I(i_g(t) - s_g(t))dt$$

In other words, the output signal will be proportional with the difference between the setpoint and the measured values and also proportional with the "history" of these differences.

Inside an ECU, the architecture is very similar and is built around a microcontroller. Without showing the power supply and secondary circuitry, a typical ECU architecture seen as open/close loop controller is described in Fig. 3.

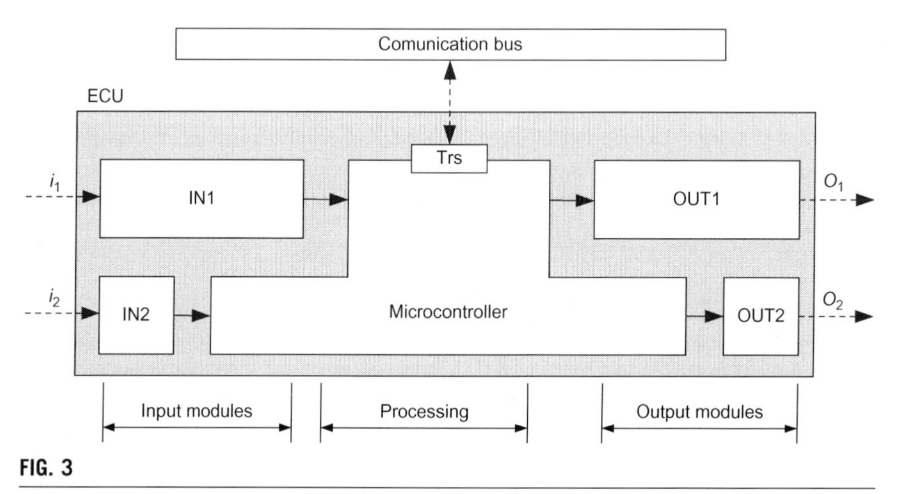

FIG. 3

ECU internal architecture abstraction.

The central unit of such an ECU is an automotive microcontroller which usually includes a high degree of peripheral integration. In Fig. 3, two different channels of I/O with similar functions but different from the microcontroller coverage are shown in an abstract view.

The first channel ($i_1 - o_1$) shows an implementation based on external acquisition/actuation components. An example for this approach would be an ECU which relies on external components in its control loops where the microcontroller is more focused on processing data. Instances in this direction can be an external analog to digital converter (ADC) and smart actuators (e.g., intelligent "H bridge" separate controllers which incorporates dedicated extra logic and control functionality). In this case, some functionality/signal conditioning of the ECU is realized outside the microcontroller component by separate physical devices (sometimes also packed with additional ECU level services like power supplies, watchdogs, etc.).

Another approach is ($i_2 - o_2$) to use a microcontroller, which contains dedicated peripherals integrated on the chip. For example, many modern microcontrollers are having integrated ADCs with specific advantages for system partitioning (e.g., footprint, costs, etc.). Even in this approach, there is a need to provide the basic system functions like supplies, network connections, etc.

2 ENVIRONMENT CONSTRAINTS FOR AUTOMOTIVE EMBEDDED SYSTEMS

The shift towards more and more electronically provided functionality in modern cars is creating a whole set of new and specific constraints.

2.1 TEMPERATURE

By replacing the old mechanical/hydraulic system with new, modern mechatronic systems, the auto industry is creating complex electronic systems with special environmental needs and requirements. Particular to these types of mechatronic systems is a generic need to place the ECUs closer to the acquisition and actuator parts. This placement is a requirement deriving from cost and electromagnetic interference (EMI) problems deriving from long wiring designs. However, by placing the electronics near the controlled mechanical parts (e.g., combustion engine) the environmental stress to electronic components and assemblies increases highly.

While the "normal" electronic components are qualified in their vast majority for an industrial range of −40°C to 85°C, the ones intended for automotive usage are having much higher requirements deriving from their harsh environmental needs. Producing these components, and especially highly integrated and sophisticated modern microcontrollers is a challenge to the semiconductor vendors. In Fig. 4, some high temperature spots in a car are shown together with some informative (as these limits highly depend on car type) temperature values.

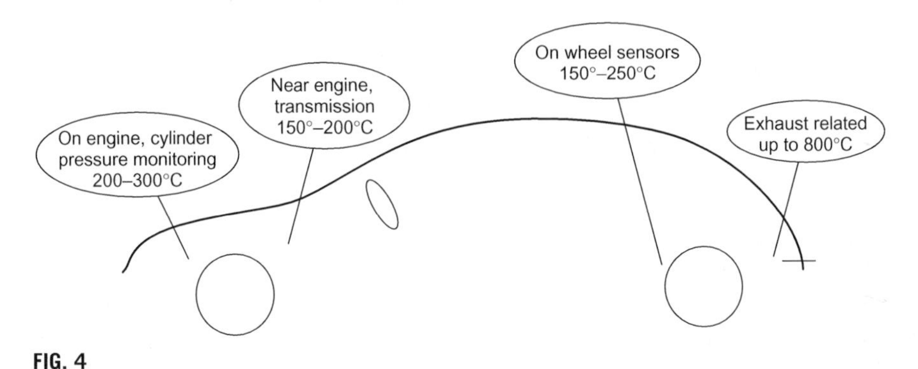

FIG. 4

High temperature ranges for automotive applications (informative values).

Automotive systems using ECUs in the proximity of the combustion engine or transmission units (e.g., gearbox) with demanding high temperature requirements are typically engine control units, electronic fuel injection, knock control systems, transmission control units, cooling fan control, etc.

In general, the temperature requirements for vehicles (highly dependent on the manufacturers and models) are falling under some generic classes like:

- Car interior (ECUs in the dashboard and other electronics inside the driver compartment) have an operating temperature profile of −40°C to −30°C, to 85°C;
- Underhood mounted ECUs (depending on their placement with respect to the hot sources like the engine) need an operating range of −40°C to −30°C, to 105°C to 125°C;
- On-engine mounted ECUs are operating in a range of −40°C to −30°C, to 150°C;
- Specific combustion, in-transmission and exhaust electronics need a higher limit of operating range which exceeds 175–200°C (for some sensors even higher than 600°C).

Temperature requirements commonly exceeding 85°C need special attention and especially components with operating temperatures above 125°C are critical for vehicles [4]. In this case, for semiconductors, the challenge for high temperature profiles can be divided, also considering the effects of high temperatures, between two main contributors:

- Silicon (die) related effects
- Package influence

The silicon effects induced by high temperature are diverse going from increased leakage, decreased carrier mobility, bandgap effects to reliability, and negative effects like electromigration and reduction of time dependent dielectric breakdown (TDDB). The functional impact caused by these effects is highly affecting the semiconductor performance. Effects like decreasing switching speed, increasing power consumption, decreasing noise margins, modifying thresholds and increasing offsets, are to be taken into account by the designers when designing high operational temperature ECUs.

One of the most important semiconductor effect with the temperature increase is the considerable (substrate) leakage current increase. Accounting for approximately doubling the leakage current for each 10°C of junction temperature increase for CMOS technologies, this effect is becoming important at high temperature. This significant increase in the leakage current leads to an increased power dissipation and may cause strong negative effects like latch-up. Additional, leakage effects are affecting the memories reducing the time of the correct charge per memory cell. This charge leakage is particularly important in the flash memories that are widely used in automotive applications (very often integrated directly on the microcontroller). Modern considerable smaller structures in silicon and epitaxial CMOS are contributing to lower latch-up risks and to produce silicon with improved high temperature behavior. Furthermore, modern structures like silicon on insulator (SOI) help to further reduce leakage thus allowing an even higher operating temperature (e.g., 200°C and above). These technologies are however not yet used on a large-scale mainly due to the higher cost associated.

The bandgap effects are also an important factor negatively affecting the semiconductor performance with temperature increase. At higher temperatures, due to intrinsic carrier concentration increase, the silicon p-n junctions are changing their properties (diode behavior migrate towards resistive behavior) and the transistors lose their switching capability. The solution of increasing the doping levels in order to minimize this effect may result in possible breakdown of devices due to increased electrical fields inside the silicon.

The reduced reliability of semiconductors exposed to high temperature is also an important problem in the automotive domain. One of the factors of this reliability reduction is electromigration. The electromigration effect consists of the position change of atoms due to high-energy electrons which impact them. This atoms shifting position along a wire will cause in time a narrowing of the wire hence increasing the current density in the respective portion of atoms displacement. The result of this effect in time is the increasing of the wire resistance which may lead to timing violation especially in high frequency switching logic. With time, this effect will continue until the respective wire (connection) completely breaks inducing a structural failure of the particular current path. The dependency to temperature of the electromigration can be evaluated with the following formula:

$$\mathrm{MTTF} = A_{\mathrm{w}} J^{-n} e^{\frac{E_{\mathrm{a}}}{kT}}$$

where “MTTF” is the mean time to failure, “A” a constant dependent on the cross-section of the respective wire, “n” is a constant scaling factor (usually 1–2), “J” is the current density, “E_{a}” is the activation energy (0.5–1 eV), “T” is the temperature, and “k” is the Boltzmann constant. The resistance to electromigration is realized by addition of copper to the aluminum conductors or, following the current trend, to replace it altogether with copper.

Another temperature dependent factor which contributes to a decreased reliability of semiconductors is the TDDB. The increased effect of this isolation defect with the temperature may also cause early failures in the silicon hence significantly reducing the availability of such components.

Special design measures related to high temperature, semiconductors can be used to 150–165°C (epitaxial CMOS goes up to 250°C and SOI up to 280°C). For higher temperatures, special emerging technologies, like silicon carbide (SiC) may be used up to 600°C.

The package of the semiconductors is also having a strong temperature dependency. While in general the plastic packages using in general small ball grid arrays or quad flat-packs are suitable with special measures up to around 150°C, for higher temperature ceramic packages are required (coming however with an increase cost). Special constraints for the die-attach in package methods and materials are also coming from high temperature due to different coefficient of thermal expansion which may cause significant mechanical stress to the die. Bonding also needs special attention due to mainly chemical effects coming with the increase of the operating temperature (also enabled by corrosion from mold compound components). To mitigate this effect, monometallic bonding is preferred, in which the bond wire and the bond pad are made of the same material. Also assembling these packages on the printed circuit board (PCB) has its own challenges when the operating temperature is increasing. As the maximum recommended operating temperature shall be around 40°C below the soldering melting point and the most common soldering alloys have a melting point below 200°C, special high melting point soldering alloys need to be used and special care shall be taken that the components can withstand the assembly temperature.

For modern complex semiconductors, the failure rate (frequency with which a component fails) is strongly dependent on temperature: high temperatures consistently increase the failure rate (in other words the probability of a component to fail). For example [2], for a microcontroller the increase of junction temperature from 90°C to 150°C is equivalent to a 13 times failure rate increase. Such a considerable failure rate creates an additional risk for mission critical systems and needs additional dedicated safety measures for mitigating the resulting effects.

2.2 VIBRATION AND OTHER ENVIRONMENTAL FACTORS

Another important aspect of the automotive electronics is the stress coming from vibration. In general, the vibration can be divided into harmonic (single frequencies) and random where a continuous spectrum of frequencies are affecting the electronic components. For the harmonic vibrations, the frequency range varies (depending on the placement of the ECUs) from around 20–100 Hz up to around 2 kHz with accelerations in the range of $15–19 \times g$. For random ones, high levels of vibrations, with values around 10 Grms,[2] are present in electronics mounted directly on the engine and on the transmission paths.

One other aspect of mechanical stress is shock. Without considering the requirements for assembly line (e.g., drop tests), during the operation life, the automotive electronics shall withstand mechanical shocks in the range of $50–500 \times g$.

The effects and failure modes for vibration are mainly affecting the mechanical assemblies of the electronic components on the PCB inside of the ECUs. The connections of the electronic components (e.g., soldering, connectors) are subject to fatigue due to vibrations and the failure modes result, generally, in loss of contact. Nevertheless, vibrations and shocks can also lead to damage inside the integrated electronic components resulting in failures like:

- Substrate cracks;
- Cracks and chip peeling for die bonding;
- Wire deformation, wire breaking, and wire short-circuits for wire bonding.

For automotive electronic components, special design methods and technologies are used in order to mitigate these effects. Similarly, the characterization (and testing) of automotive electronic components is following strict standards in which the

[2]Grms: root mean square acceleration—value showing the overall energy for a particular random vibration.

environmental parameters which may affect the operational life of a component are tested (e.g., thermal, mechanical, electrical stress). As an example [3], the mechanical stress is tested for shocks with five pulses of $1500 \times g$ acceleration, a harmonic vibration test covers 20 Hz to 2 kHz in each orientation with a $50 \times g$ acceleration peak, a specific drop test on all axes is performed, etc.

One other important environmental factor specific to automotive industry is the fluid exposure. The ECUs exposure to fluids is considered harsh, especially for ECUs mounted outside of the driver compartment, due to the possible exposure mainly to humidity and salt water. Depending on the particular position of each ECU, exposure to fuel, oil, brake fluid, transmission fluid, and exhaust gases is also possible.

2.3 SUPPLY AND POWER CONSUMPTION

Power consumption is a fundamental recurring thematic in automotive electronics. The push of the industry towards reducing the power consumption becomes even more critical with:

- Aggressive targets for emission reduction and
- Constant increase of electronic functionality in the modern vehicles.

The emission reduction targets have a multiple influence on the modern ECU architectures. On one side in order to achieve the required emission levels (like particulates, NOx, etc.) there is a strong need for high performance computation in the engine control. This computation power, which for modern standards like EURO 6, needs the power of multicore microcontrollers and sophisticated algorithms (the need for the main microcontroller for powertrain applications needs more than 1 Gips), comes with corresponding higher power consumption. On the other side, the need for CO_2 reduction implies a stringent need to lower the power consumption of the vehicle internal systems. Knowing that 100 W need in electrical supply equates to around 0.1 L of fuel increase consumption, the need of electrical power reduction is evident. Similarly, the equivalent of 1 g CO_2 is around 40 W electrical power need (also equivalent with around 20 kg weight increase). These targets are extremely important and cause a huge pressure on the automotive industry, as they are more and more difficult to reach. As an example [6,8], in the EU, the average target for CO_2 emission is gradually decreasing from 130 to 95 g CO_2/km in 2020 in the context of a dramatic increase in electronic functionality required by modern cars.

The modern need for more and more functionality covered by ECUs creates also a pressure on the power supply budget. With more than 70–100 ECUs in today's middle-high end cars (trend continues), each having an average consumption of around 200 mA, smart power saving strategies are needed.

One important constraint deriving from power consumption in automotive domain is the standby consumption. More and more ECUs need certain functionality even outside of the driving cycle. Depending on the original equipment manufacturer (OEM), requirements for standby consumption are varying between max 300 to max 100 μA. This target is difficult to achieve due to strict requirements for cyclic wake-up:

- Periodic wake-up for analog inputs acquisition (also watchdogs);
- Wake-up on external events (e.g., switch detection);
- Very fast complete wake-up if needed for operating actuators (e.g., door unlock);
- Periodic communication for antitheft units;
- Blinking LEDs, etc.

The ECUs are highly interconnected in modern cars. Usually CAN, FlexRay, LIN, MOST and most recently, Ethernet are the backbones of these complex interconnections. While there is a certain support for global wake features (all ECUs are waked up by specific messages on the bus), network architectures in which only some regions of the bus are powered down are not yet mainstream. Concepts like "partial networks" and "pretended networks [5]" are addressing these power requirements allowing certain "degraded" functioning modes in which only a subset of functionality is available at a given moment in order to save power, but the full functionality can be available depending on the system needs—the network can wake up the complete functionality of ECUs on demand.

2.4 PLATFORMS, LIFECYCLES, AND MAINTAINABILITY

Despite the relatively high numbers of passenger car models available on the market today, the high cost pressure forces OEMs into platform development where basic engineering design is (re)used across multiple models or family of vehicles. According to Ref. [9], in 2015, around 20 platforms account for around 45% cars launched globally. These platforms are cross models for a specific brand but it is common to have cross-brands platforms. This approach also allows the sharing of components to reach, depending on brand and models, more than 60% of common components. While the platform based development brings big benefits like cost reduction, standardization (in many times industry wide), maintainability, etc., it also comes with a certain risk concentration and important constraints challenging the engineering process used to develop these platforms. In the electronic domain, designing ECUs with a relatively high degree of variability, fitting a complex and wide range of requirements poses specific challenges to the design, implementation, and test processes. As the functional requirements are extending to fit more and more variants of functionality, the ECU design gets more and more complex and smart compromises are needed for fitting them. All these requirements are coming on the top of already stringent ones regarding reliability, safety, security, etc [7].

For example, from the semiconductor content of the platform, a balance between extra functionality needed by some particular functions and the overall power budget have to be considered. Enabling hardware functionality when needed and allowing for different configurations may be a solution, which on the other side, complicates the testing and maintainability of the final implementation. Similarly, an intelligent partitioning between HW and SW functionality is needed for modern ECUs. Also this partitioning comes with a considerable complexity add-on (in general from the SW side) which generates also risks for the automotive applications.

3 FUNCTIONAL CONSTRAINTS

In the current automotive landscape, the tendency of the electronic systems is towards achieving a high degree of dependability. While all the properties of dependability defined in Ref. [10] are needed for modern car electronics, extra focus may be allocated to requirements regarding safety, security, reliability, and availability. An example of such dependability vectors and their high level meaning is shown in Fig. 5.

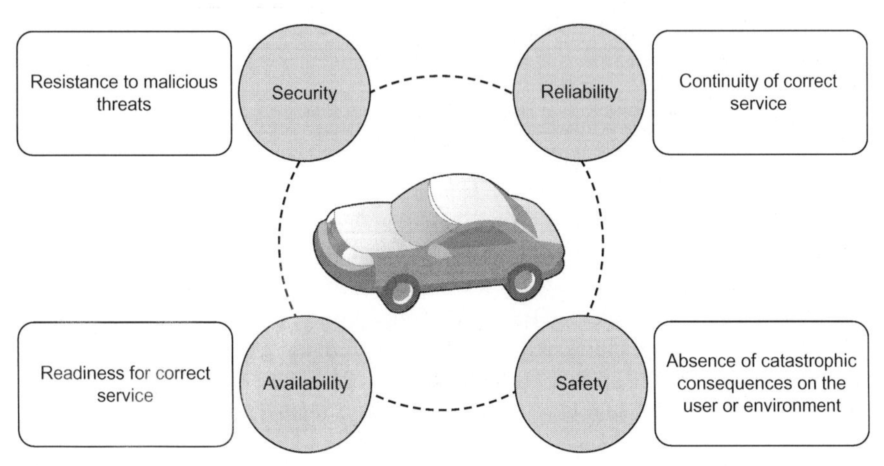

FIG. 5

Some important dependability vectors of a modern car.

Some strong constraints for the actual ECUs and electronic devices are derived from such stringent requirements.

3.1 FUNCTIONAL SAFETY

Starting with the introduction of seat belts, automotive safety has achieved a long road in reducing casualties due to car accidents. From 1971 seat belts (mandatory only from around 1984), continuing with antilocking braking (ABS) and (driver) airbags at the end of the 1970s, and with side airbags in the 1980s through around 2000 when the vehicle stability control started to be implemented in passenger vehicles, to today's active lane keeping assist and partial autonomous braking systems, automotive safety made huge progress and continues to be one of the main target for future automobile developments. With the perspective of the autonomous driving, the safety aspects become even more stringent influencing the whole development of the modern automotive systems and in particular electronics.

Traditionally, the automotive safety domain is divided into two big categories: passive and active safety systems. The passive safety systems like crash cells, laminated glass, collapsible steering column, seat belts, airbags, etc. are helping to reduce the negative effects of traffic accidents (mainly by means of absorbing or diverting the crash energy) and have lately reached a good maturity and stability in the actual car industry. On the other hand, the active safety elements (like antilock brakes and stability control), which are presumably the big current innovation area, are focused on preventing the potential accidents before their occurrence.

Due to the high potential of human injury in case of system fail, the automotive industry has derived its own functional safety standard, published in 2011. Derived from IEC61508, the new standard, ISO 26262 is applicable to all automotive safety-related systems including electric/electronic systems in passenger cars with a maximum gross weight up to 3.5 tons. In this context, functional safety has the meaning of "absence of unreasonable risk due to hazards caused by malfunctioning behavior of electrical/electronic systems" [11]. Specific to this standard is the risk-based approach in which the risk levels (classes) are derived combining the severity (consequences), exposure (probability of occurrence), and controllability (how good can an average driver deal with the results from a malfunction) of the considered hazard. Depending on this risk-based analysis, different classes of "safety integrity" can be defined, each with its own clear requirements for all phases of development cycle. In general, the notion of safety integrity relates to the probability of a safety-related system to perform its required/specified safety functions in a given time under all stated conditions. The higher the identified risk is for a given system, the higher safety integrity is required for its functionality.

This safety standard addresses a full set of aspects related to the whole safety lifecycle like:

- Requirement engineering
- Architectural aspects on different levels
- Functional aspects
- Procedural aspects (including safety lifecycle)
- Fault classification, definition, metrics, etc.
- Verification and confirmation aspects
- Safety management aspects

ISO 26262 is specifying four levels of automotive safety integrity (ASIL) going from the lowest "A" to the highest "D." Each of these levels has its own targets with respect to the probabilistic fault metrics which needs to be reached and in the stringency of the methods used to define, design, build, and test the product. Some examples of each category are:

- ASIL A (lowest): Loss of rear lights (both sides);
- ASIL B: Speedometer not available, loss of brake or driving lights (both sides), instrument cluster, etc.
- ASIL C: Inadvertent braking, unwanted vehicle acceleration
- ASIL D (highest): Inadvertent airbag system deployment, self-steering, unintended full power braking, stability systems, etc.

For the electric/electronic safety-related systems and in particular for the microcontrollers largely used in the context of functional safety, the notion of failure is of a crucial importance. The failures in the electronic components in general which have a high relevance in microcontrollers can be divided roughly into two main categories: systematic failures and random failures.

The systematic failures in semiconductor devices are an important contributor to the system safety integrity. Especially microcontrollers systems, which incorporate complex logic, protocol engines, and specialized processing units may suffer from such failures. In general, the systematic failures are resulting from failures in the design activity or manufacturing process for an electronic component. The design systematic failures are very difficult to find as they are assumed to be part of the correct intended functionality and therefore not likely to be found during the verification phase. A generic method against systematic failures in design is to implement diverse redundancy for the target system (or at least for some subfunctionalities or parts of design). The assumption for this approach is that the probability of a diverse design (done by different people using different methods) to have the same failure mode is highly reduced. A drawback of this method, besides the significant cost add-on needed, is that it will not be efficient against specification faults—a diverse implementation starting from a wrong specification (or assumptions) will not be correct.

Another type of systematic failures for semiconductor devices results from the manufacturing process. In these cases, a failure during the production process will affect all the components with the potential to produce "hidden" failures in the field application. Resulting in general from a process violation, these type of failures can be reduced only via continuous and rigorous process improvement in the production facilities (these constraints add also to the costs needed to produce high integrity semiconductor components).

The other fundamental failure type affecting the electronic devices is the random failures. These failures are inherent to physical semiconductors and results from production process and usage conditions. One problem with the random failure rate is that, in general, this rate cannot be reduced; therefore, the functional safety domain is focusing on detection and handling of this type of failures.

The origin of these failures are faults which can be allocated to a variety of causes like: EMI resulting in electrical overstress (EOS) of the semiconductor as well as different other types of electrical disturbances, physical defects of the semiconductor (arising during the functioning time), faults caused by the effect of electrically charged particles (alpha particles, neutrons) affecting the logical elements (e.g., flip-flops) of the implemented functionality, etc.

A further classification of these faults can be done with respect to their time behavior:

- Permanent faults: Persistent faults which affects continuously the hardware functionality. These faults can be caused by production faults, temperature, humidity, vibrations, EOS, etc.
- Transient faults: Faults which are nondestructive to the semiconductor structure but are nevertheless affecting the correct functioning of the devices. These type of faults are usually caused by charged particles, EMI, clock disturbances, etc. These faults will disappear once the affected logical region is re-initialized.

- Intermittent faults: Faults which are usually permanent faults in their first stage of manifestation. These faults are usually physical faults whose early development is strongly dependent on a physical parameter, e.g., temperature. With time, these types of faults are usually transformed to permanent faults.

The problem is important in the modern semiconductor technologies as the related failure rates are important. While the permanent faults, associated with "hard-errors" type of random hardware for high complex modern automotive processing units (e.g., microcontrollers) failure rate is usually in the range of 500 FIT, the failure rate associated with transient "soft-errors" is significantly higher (usually in excess of 2000 FIT). FIT (Failures In Time) is the unit for failure rates and is defined as the number of failures that can be expected in one billion (10^9) device-hours of operation (also interpreted as, e.g., 1 FIT = one failure in one billion hours or one million devices for 1000 h).

As the targets for the resulting failure rates applicable to high safety integrity systems are substantially lower as the failure rates resulting from current technologies, special functional safety mechanisms and measures are needed in order to achieve these constraints. For example, according to Ref. [11], the total failure rate of an ASIL D (highest safety integrity) system shall be smaller than 10 FIT which is at least two orders of magnitude smaller than the achievable technology results. In this case special dedicated safety measures are needed to reduce the raw failure rate from some thousands to less than 10 FIT.

Depending on the targets of the safety strategy, the automotive systems can be roughly divided into:

- Fail-safe systems in which in case of faults, the systems detects these faults and transitions to a "fail-safe" state in which no harm can occur. For these systems, the fault detection capability and the transition to the safe state in the given time interval are of highest importance. Special mechanisms are also needed for fault containment which prevents a fault to propagate. This type of systems are the majority of today's automotive safety-relevant systems.
- Fail-silent systems are a special subset of fail-safe systems in which a potential fault, once detected, is preventing the system of producing any output and hence limiting the possibility of a wrong output.
- Fail-operational systems are systems, which shall continue to operate correctly at their outputs despite the existence of faults. These systems are usually needed where a safe state is not possible (e.g., angle of wheels from steering) and are extremely important in the next automotive developments related to autonomous driving.
- A subset of fail-operational systems are fail-degraded systems in which the primary function is no longer provided in full quality in the presence of faults but a reduced correct functionality is delivered at the border of the system.

In principle, for all these strategies in dealing with faults, redundancy plays a very important role and is the usual way to implement effective safety-relevant systems.

However, redundancy alone will not be able to properly cover all the fault types (e.g., a SW fault will not be detected by running the same piece of code multiple times) and special care shall be taken for avoiding common cause faults (which may affect in the same way both redundant channels). This is why, the usual redundancy is combined with certain degree of diversity (different ways to implement the same functionality) resulting in a better coverage of the entire fault model. For example, for safety-related automotive microcontrollers such strategies may include:

- Time redundancy:
  - Same piece of code (SW) is executed at different moments of time (or on parallel execution threads) and the results compared;
  - This redundant execution is not efficient against systematic SW faults and causes a significant performance impact;
  - Common causes affecting both executions together with the comparator code shall be analyzed;
- Diverse algorithms:
  - Different hardware/software implementations are executed, e.g., integer/float calculations, plausibility checks, challenge-response, etc.
  - This method has the advantage to cover also systematic faults which may affect the implementation;
  - The added complexity of such implementations is however adding a high cost to the system and substantially complicates the verification;
- Hardware redundancy/diversity:
  - Different hardware blocks are executing the same functionality;
  - Similar hardware blocks are executing in parallel with special measures against common causes, e.g., lockstep CPUs;
  - Diverse hardware is used for safety-relevant purposes, e.g., signal acquisition using two different types of ADCs or using analog and digital sensors for signal acquisition;
  - This method adds costs to the system but by keeping the complexity in a manageable area represents usually a good compromise in today's automotive systems.

An example of the above-discussed techniques used for microcontrollers in state-of-the-art automotive safety-relevant applications is the AURIX© family from Infineon Technologies. A platform block diagram for this architecture is shown in Fig. 6.

In order to fit the required application level safety-related requirements, the modern automotive microcontrollers have a considerable amount of built-in safety mechanisms. In general, these architectures shall provide means to allow the application level to execute plausibility checks for the intended functionality. These checks are allowed mainly via module diverse redundancy (e.g., with diverse implementation of ADCs), redundant computation and a high number of monitoring capabilities.

The extensive monitoring capabilities of such modern microcontroller are covering all the relevant functionality which may affect the correct behavior and shall

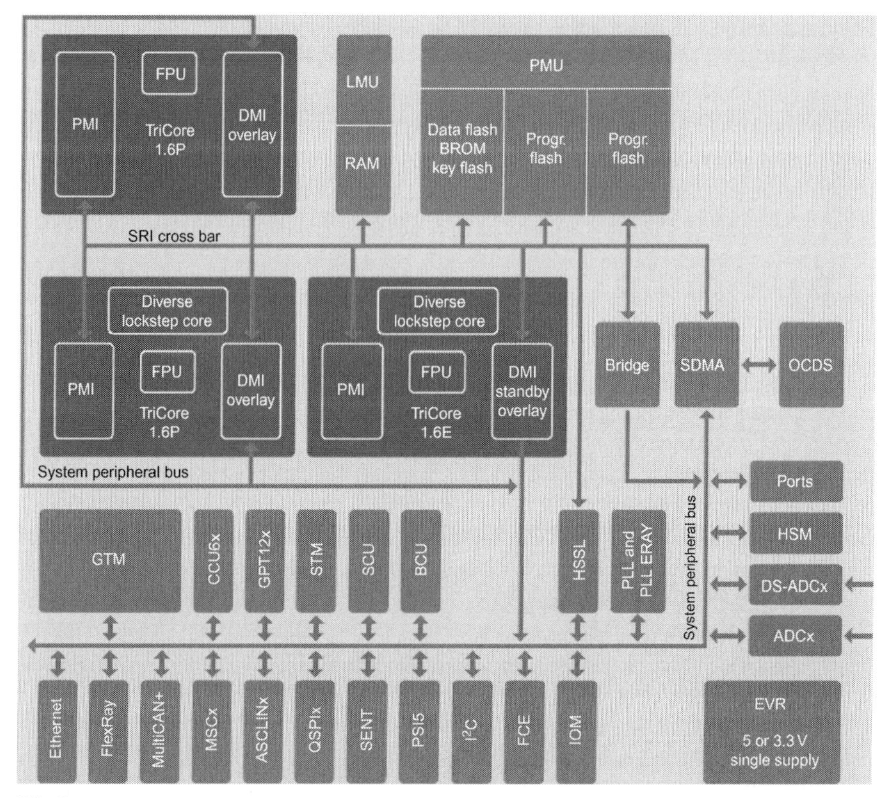

FIG. 6

AURIX© Platform—courtesy of Infineon Technology.

detect with high accuracy the presence of faults and trigger the corresponding fault reaction. Examples of monitoring capabilities are:

- Complex clock monitoring covering the full range of potential clock failures including the complete loss of clock in the system;
- Voltage monitoring for all the supplies in the system, with separate thresholds allowing complex fault scenarios based on different supply failure scenarios;
- Safe transfer through bus systems and direct memory access (DMAs);
- Memories (Flash and SRAM type of memories) error correction codes (ECCs) with enhancements to detect also multibit failures;
- Memory protection mechanisms for memory areas and for peripherals;
- Input/output monitoring capability;
- CPU replication (lockstep) and self-tests;

One of the main safety-related mechanisms in modern microcontrollers is the CPU lockstep. This hardware mechanism allows the double execution of the software via a "shadow" redundant core and is able to detect any difference between the two CPUs by comparing (also in hardware—so no software overhead) internal signals during the execution. This duplication (e.g., the "diverse lockstep core" in Fig. 6) is built in order to avoid common causes, which may affect the execution in both redundant paths (e.g., power supply glitches or other electrical effects) and has a very good coverage of faults with potential to affect the correct computation. The modern safety-relevant applications need such "safe computation" platforms in order to achieve the stringent metrics with respect to hardware faults.

A different scenario relates to the safety-relevant software implementations. As software has only a systematic fault model, i.e., faults deriving in the process of defining and implementing the software, different techniques are needed in order to achieve higher safety integrity levels. In general, process measures are required and stringent methods for requirement, design, implementation, and verification are required by the safety standards (e.g., [11]). From technical point of view, higher safety integrity levels can be reached only by a given redundancy including diversity, as homogenous redundant software is not efficient against systematic faults.

3.2 SECURITY

As opposed to safety, security is relatively new to the automotive standard applications but gains a significant momentum mainly due to the modern context in which the connectivity is becoming an important aspect of the vehicles. In addition, different from safety perspective where the hazard source is a malfunction of a system due to systematic hardware and software faults or random hardware faults, security hazards are caused by intelligent and malicious attackers. Even though the security aspects linked for example to tuning protection are relatively old, the security issues are getting a high attention with the increased opening of the car to data streams from external sources.

Fig. 7 shows some typical connectivity types for a modern car. While these communication channels are changing the driving experience highly improving the comfort and the information flow for the driver and passengers, they are at the same time potential attack targets for hacker attacks. In general, there are two main classes of communication for vehicle networks: on-board communication in which the ECUs are communicating using the internal networks of the vehicle (CAN, LIN, FlexRay, Ethernet) and off-board communication where the vehicle is connected to external network(s). Both domains are however linked with gateways which allows the potential attack from outside also on internal communication channels.

The benefit and motivation for these potential security attacks on vehicles are multiple:

- Activate unpaid features, use "nonofficial" software in vehicle, manipulate vehicle recorders; this can have a financial impact but also can cause a safety risk;
- Tuning of existing hardware, use third-party "nonofficial" ECUs, refurbishing components, etc.; financial impact but also safety and customer trust from OEM side;

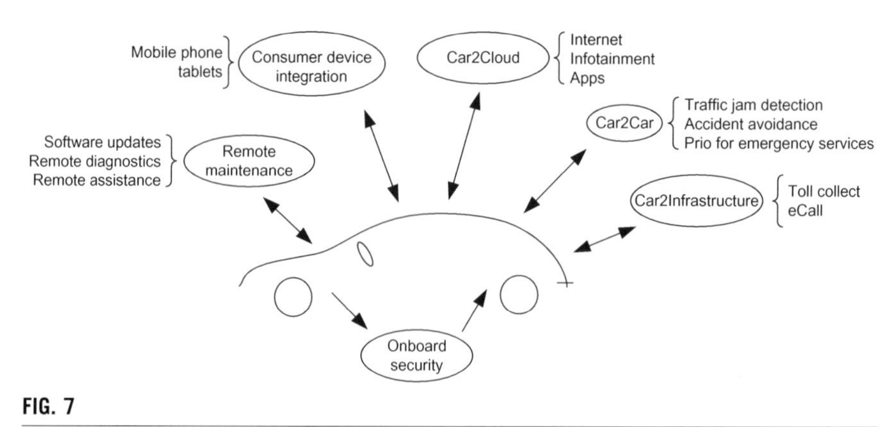

FIG. 7

Some examples for a modern connected car.

- Steal personal data, unauthorized user profiles, malware installation, etc.; hacker typical attacks with impact on privacy, customer trust, etc.;
- Vehicle remote control, large-scale accidents, attacks on traffic infrastructure, deactivation of safety features; terrorist/criminal type of attacks with potential high impact on safety at bigger scale;
- Intellectual property stealing from existing designs; financial and competitiveness impact.

The measures against the security attacks have a big range of methods ranging from complex software solutions to hardware-software solutions. For the modern microcontrollers used in automotive applications, the security threats are translated into security requirements, which are generating security mechanisms implemented at chip level.

In general, offering ineffective and precarious approaches, "security by obscurity" approach is no longer considered as a viable alternative for the automotive security requirements today. Instead, security algorithms and specifications are publicly available and the resulting systems shall offer the required robustness against the specified attacks via a correct and well-designed implementation.

One the first levels of security protection needed on current microcontroller architectures is the debugging protection. This type of protection is used to forbid the physical access of a debugger to the flash type memories once the ECU is leaving the assembly line. This protection will prevent a subsequent debug access and hence the unauthorized reading of the code and data from inside a microcontroller once in the field. Depending on the implementation, this protection goes from blocking the debugging interface to complex protection of the memories content preventing any unauthorized data reading.

Other current implemented methods requiring hardware support, which are today quasistandard in automotive microcontrollers, for limiting the potential of a malicious attack are:

- Secure boot: Startup check that only the correct software is allowed to start and no modified copy of the software can boot the device;
- Secure flashing: Check that only authorized flash software images are allowed for the software updates (prevent a modified software image for being loading in an ECU);
- Runtime tuning detection: Prevent modified software parts (executable or configuration data) to be changed or used during runtime of the ECU.

Depending on the vendor, some modern automotive microcontrollers are including advanced hardware elements that allow the industry to provide secure ECUs for automotive applications. In general, these hardware security dedicated parts are at the minimum hardware accelerators for cryptography purposes (e.g., for symmetric block ciphers, cipher block chaining, etc.) and some protected and encrypted memory area(s). In addition, a true random number generator is usually included as specialized hardware implementation. The more advanced hardware implementations are additionally including a complete processing unit (in many cases, modern 32-bit architectures) including its own protected memories (e.g., flash, SRAM). This whole construct, including the hardware accelerators, are placed in a special separate shielded domain with a strong firewall to rest of the microcontroller. Based on this separate computation unit, different dedicated software for authentication, encryption/decryption, key management, etc. can be quickly and efficiently executed allowing the runtime usage of complex security algorithms.

In the automotive domain, some specification efforts have meanwhile reached a certain maturity allowing them to be established as de facto standards. Some examples are:

- Secure hardware extension (SHE) specification developed by the German OEMs HIS (Hersteller Initiative Software) [12] organization. This specification defines a set of application programming interfaces (APIs) allowing a "secure zone" to coexist within a vehicle ECU. Some important features included in this specification are storage and management of security keys, API access to authentication and encryption/decryption, etc.
- EU funded project EVITA [13] developed guidelines for design, verification, and prototyping for automotive ECU architectures.

3.3 AVAILABILITY

Being always in the first 2–3 customer criteria for selecting a new car, the vehicle quality/reliability is one of the most important targets for the automotive industry. With electric/electronic failures accounting for more than 50% (or much higher depending on the statistics) of causes for vehicle unavailability, the drive for electronic reliability is a highly important driving factor for the component producers.

On one side, we have the stringent constraints related to the semiconductor production process allowing components to survive in the required harsh environments related to temperature, vibration, humidity, etc., leading to sophisticated quality assurance programs being able to deliver the required "zero-defect" quality targets. On the other side, the design of the components shall allow a longer life in the field and allow for designs maximizing the reliability of ECUs and hence the availability of the vehicle.

Some of such requirements generate functional requirements on the level of microcontroller and require a different design as for their consumer electronics counterparts. In addition, the integrator shall carefully design the ECU level with measures for mitigating effects of potentially reducing availability factors. Examples of such measures are:

- Power budget: Ensure that the microcontroller meets its power envelope. Usually, supply monitors are used with progressive thresholds which are generating corresponding reactions on ECU level in case the power consumption/supply is getting out of specified values.
- Temperature management: Die temperature sensors (built in the microcontrollers) are permanently monitoring the die-junctions temperature. Together with the ECU temperature sensors, these sensors are able to determine a temperature profile inside the electronic system and trigger an appropriate reaction in case of violation of prescribed limits. These monitoring systems are usually working together for complex diagnostics; e.g., in case of junction temperature out of specification increase, the current consumption will also dramatically increase thus signaling also an out-of-bounds power consumption.
- Clock integrity: Complex clock monitors are usually available for all relevant modules inside modern microcontrollers. In case of clock problems, reactions are triggered depending on the importance of the deviation and the considered application. In addition, the modern designs are also providing backup clocks able to clock the device even in the case of a complete failure of the external provided clocking.
- EMI regulations: These requirements are important in automotive domain as there are strict regulations related to the maximum allowed EMI[3] levels. On the other side, special design measures are taken for reducing the EM influence on the correct functioning of the whole chip(s). Different clock domains, phase-locked loops (PLLs), as well as special filters for pin inputs are part of modern architectures.

In general, the availability requirements are contradicting the safety requirements. While from availability point of view, the continuous service delivery is necessary, from the safety domain, once a fault occurs, tries to transition the system into its "safe-state" mode, which, normally, is making the vehicle no longer available. Depending on the particular fault type, the transition to safe-state, although "safe," may not be always the best solution for the driver. In many cases, a degraded mode, in which only a certain type of functionality is disabled in the presence of a fault and the vehicle is still operational in a restricted way (e.g., for a limited maximum speed,

[3]EMI: electromagnetic interference.

or with electronic stability program switched off), is preferable to the complete immobilization of the vehicle.

In order to support this degraded modes, "limp home," the microcontrollers used in modern ECUs, have to incorporate special features like fault containment regions, reduced functionality with lose of redundancy, restarting of (parts of) application during runtime, shifting tasks on other processing units inside the same microcontroller or even on other physical processing unit over the connection buses, etc. While many of the techniques are usually implemented in software, the underlying hardware shall also provide means to implement such techniques. For example, the possibility to reset only a part of the microcontroller (e.g., one core) leaving unaffected the rest is a requirement allowing for a substantial availability increase. Memory partitioning together with the ECCs allows for a fault tolerant behavior and extend the functioning of the device; depending on the criticality of the application, this fault tolerance threshold can be set accordingly.

4 CHALLENGES FOR MODERN AUTOMOTIVE SYSTEMS

4.1 NEW FUNCTIONS FOR MODERN CARS

Another trend in automotive industry is represented by the opening of the car to the existing connectivity potential: the connected car. Even though this domain is just beginning, there are already, especially in the high segment area, connected cars on the road. According to market forecasts, by 2018 at least 20% of the cars will be "connected cars."

Starting more than 20 years ago, CAN is one of the most traditional communication backbone for vehicles. Due to the exponential increase of the communication needs between the ECUs in a modern car, the automotive industry is shifting towards other communication means like FlexRay, CAN FD, and Ethernet (currently 100 Mb but also shifting in the direction of Gb Ethernet).

This increase in communication needs has many reasons related to the permanent complexity increase of the modern cars but some modern developments are putting yet more pressure on the communication infrastructure:

- The need to connect the personal communication: This usually means—"bring your own device." Consisting on mobile phones, tablets, etc., the modern cars shall allow the integration of these devices into the vehicle infrastructure. This integration comes with stringent requirements for security (privacy) and requires a certain market standardization deriving from the different product cycles consumer—automotive.
- Automated driving: In this case, the needed information aggregation between the many environmental diverse sensors (sensor fusion) requires an important available bandwidth as the data quantities, which needs to be processed, is enormous. With the requirements resulting from fail-operational type of operation, the possibility for re-routing information flows to redundant structures

becomes also a requirement for such systems. One more information flow, which needs to be considered for automated driving, is the vehicle-to-vehicle communication, which has also very specific requirements in terms of bandwidth but also in terms of communication security.

– Infrastructure data flows: Also known as "car-to-x," this type of connection connects the modern vehicle with available infrastructure or "cloud" type of services. One example here is the remote maintenance and the "update over the air" which allows the distribution of updates and enhancements without the physical need to have the car in a specific garage.

As already mentioned, the modern cars are complex systems and complex networks of processing units with increasing complex requirements. In the past, each needed vehicle electronic function was clearly allocated to an ECU responsible to execute this function. There was a certain communication between ECUs on the vehicle networks in order to change required information regarding the required actions, configuration, etc., but in general, this communication was not changing the initial functionality allocation. In the modern vehicles, this system view is changing towards a more integrated view decoupling to a certain extent the functionality from the physical topology. The practical need for this shift, with massive implications on the modern network architectures is coming from several new approaches like providing new functionalities without adding new ECUs and shifting software components.

New functionality resulting from aggregating functions and information from multiple ECUs consists of a new functionality of the system being available without the necessary need to add another physical ECU to the system (idea shown in Fig. 8). Such add-on functionality, made available with integration of multiple existing system capabilities is for example an adaptive cruise control in which the resulting

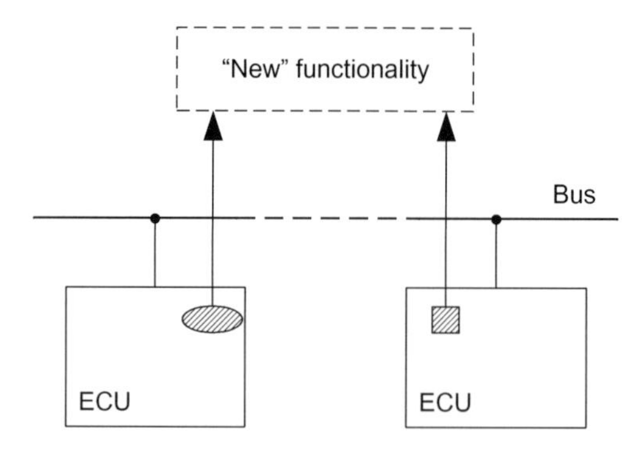

FIG. 8

New functionality resulting from aggregation.

functionality is a "software function" integrating results from wheel speed sensors/ABS controllers, trajectory sensor fusion, motor control, gearbox, etc. Sensor fusion is another example in which information is available based on the diverse amount of data available form a multiple set of sensors.

Shifting function execution (example in Fig. 9) consists of delegating the execution of a specific piece of software to other physical execution unit (e.g., shifting the execution of a software component from an ECU to another ECU via the connection network). While allowing for "smart" load balancing scenarios (in which software components are allocated to processing units depending on the current processing load), this execution shifting becomes important in the context of fail-operational systems (e.g., for autonomous driving). In this case, the possibility to execute an algorithm (or parts of it) on another processing unit assuming the malfunction of the initial processing unit is allowing the continuation or at least a degraded continuation of the control function. In addition, this capability is one of the building blocks of the future adaptive/self-healing system capable of self-reconfiguration from a pool of redundant available resources.

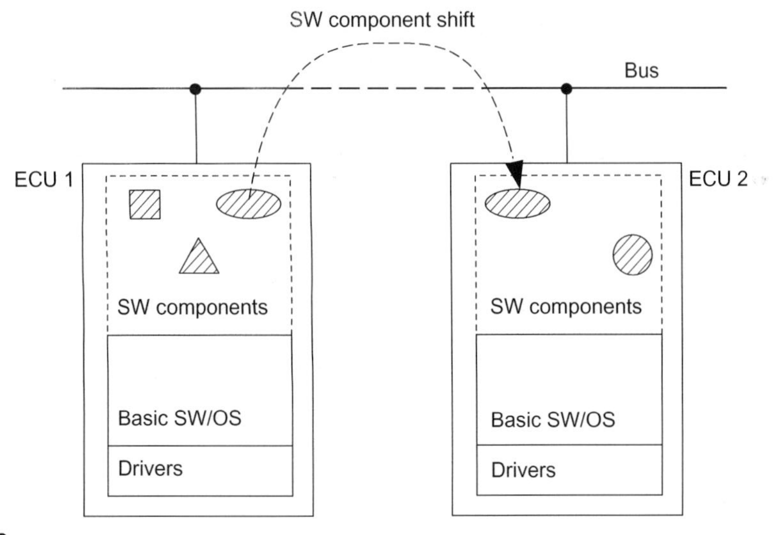

FIG. 9

Shifting of software components execution over the network.

Considering the "opening" of the car towards external connectivity and the new system requirements deriving from modern fail-operational applications, the challenges for the automotive networks are increasing. Properties like dependability, security (with a strong privacy component), performance, real-time, low latency, fault tolerance, dynamic reconfiguration, etc. are and most probably will be the driving factors for the electronic embedded systems in modern vehicles.

4.2 AUTONOMOUS VEHICLES

Self-driving cars are today one of the important trends of automotive industry. Most of the major car manufacturers have elements of assisted driving already on the market and on their roadmaps. In addition, prototypes of highly autonomous vehicles are intensively tested worldwide with constantly increasing confidence in their constantly increasing capabilities. In a certain way, the self-driving cars seem to be "the future of car industry."

The biggest benefits for the autonomous driving are assumed today in avoiding traffic congestion, highly increasing the safety of both car occupants and other traffic persons (e.g., pedestrians) and in reducing pollution by "intelligent" resource usage. All these enhancements make autonomous vehicles a big promise of the near future.

While totally self-driving cars are the target of the current research (and some implementations are already available), an important part of the industry is taking a gradual approach starting with simple driving assistance features and making the vehicle more and more "autonomous." In order to differentiate between the different automation levels, international classification was developed with similar categories [14,15]. A short overview of these levels based mainly on [14]:

(1) Driver only (Level 0): On this level, all the vehicle control is done by the human driver who is completely controlling both the longitudinal and lateral dynamic of the vehicle. At this level, there is no vehicle system, which interferes with the driving activities. However, some systems that can help the driver to perform better his activities like "lane change assistant," "park distance control," or "forward collision warning" may be available at this basic level.
(2) Assisted (Level 1): The vehicle systems perform one of the longitudinal or the lateral vehicle control for given use cases while the driver performs the other type of control. However, it shall be noted that at this level, the driver is supposed to monitor permanently the vehicle behavior and the driving environment. For this level, the assistance system is not capable of recognizing all of its performance limits. Some examples for the assisted level are "adaptive cruise control (ACC)," "stop-and-go," "park steering assistant," etc.
(3) Partial automation (Level 2): The vehicle systems perform both the longitudinal and lateral dynamic of the vehicle. Also at this level, the driver shall monitor permanently the vehicle behavior and the driving environment and the driver is not allowed to have nondriving related activities. When the assistance system recognizes performance limits, the driver is immediately requested to take over the control. Example of systems at this level can be "parking assistant" (driver monitors the behavior and can interrupt the maneuver), "lane change assistant" (limited to motorways, driver initiates the lane change and monitors permanently), or "construction site assistant" (also limited to motorways, driver monitors permanently).
(4) Conditional automation (Level 3): Similar to level 3, the vehicle systems perform both the longitudinal and lateral dynamic of the vehicle but the driver does not need to monitor the driving dynamics at all times hence allowing some

limited nondriving related activities. When the automated systems recognize reaching its performance limits, a request to resume vehicle control is issued to the driver with a sufficient time margin. Some of the implementation at this level are "highway congestion pilot/traffic jam chauffeur," and "overtaking chauffeur" both limited to motorways and allowing some nondriving activities from the driver (which may however be requested to take over the control of the vehicle if necessary).

(5) High automation (Level 4): The vehicle systems perform both the longitudinal and lateral dynamic of the vehicle and the driver does not need to monitor the driving dynamics and can perform nondriving related activities at all times during an use case. This behavior is limited to certain defined use cases and the system recognizes its own limits and can cope with all situations automatically until the end of the use case. After the finish of the use case, the driver is requested to resume vehicle control. Examples of systems at this level are fully automated "parking pilot" in which the car parks fully automatically completely without the need of a driver and "motorway pilot" where the system fully controls the vehicle during the full motorway trip.

(6) Full automation (Level 5): As in Level 4, The vehicle systems perform both the longitudinal and lateral dynamic of the vehicle and the driver does not need to monitor the driving dynamics and can perform nondriving related activities at all times during the entire journey. The system is able to recognize its limits and cope with all situations during the entire journey. In other words, no driver is required. Even though prototypes for this level do exist, regulatory problems are still to be defined before wide acceptance of this automation level.

In some other classification schemes (e.g., [15]), the high and full automation levels are merged in one "self-driving automation" level.

In order to implement such systems, a three staged approach is usually used for reference:

- Sensing stage in which a model of the vehicle environment is built and maintained,
- Planning stage in which, depending on many preset factors and to a certain extent also depending on human interaction, a driving strategy and a trajectory is calculated, and
- Actuating final stage, where the vehicle is controlled according to the established trajectory.

For computing an "environment model," an internal representation of the space and traffic participants around the vehicle, a consolidated input from environmental sensors is needed. As shown in Fig. 10, there are many sensors mounted on the vehicle, which, through data fusion, contribute to the creation of this model. Starting from Radar sensors (distance ~50–200 m, detection angle ~20–70 degrees), these sensors can also be video cameras (distance ~50–150 m, detection angle ~50–70 degrees), ultrasonic sensors (distance ~2–5 m, detection angle ~30–50 degrees), Lidar sensors (distance ~10–150 m, detection angle ~10–30 degrees), etc. The high amount of data coming from these sensors (e.g., only one camera in high-end vehicles can deliver a raw data flow in the order of 250 Gb/h) needs processing and consolidation: due to different sensors characteristics (e.g., a camera detection accuracy decreases with degrading light conditions), a certain redundancy in the acquired environment data is also needed.

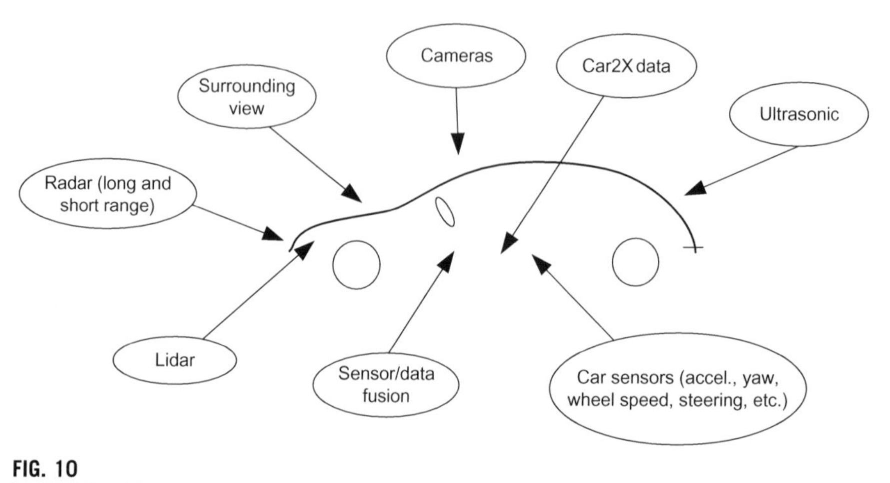

FIG. 10

Example for sensors and data input for assistance systems.

The need for such high computation power requires special approaches for the embedded systems used for self-driving vehicles. New computing structures, combining heterogeneous mixes of traditional automotive embedded CPUs with modern GPUs are one possible solution to address these needs. Due to the already discussed constraints in the automotive environments, this processing need shall be passive cooled therefore adding another constraint on the power consumption of such computing structures.

The next stage for the autonomous modes is the maneuver planning or trajectory calculation. In this stage, the planned future trajectory of the vehicle is computed based on the environment model and the predefined set of rules and/or driver interaction. At this stage, multiple trajectories are calculated and evaluated using complex optimizing algorithms, often including solving of complex differential equations. One constraint added on the embedded computing systems in this application use case is deriving from the fail-operational type of functionality mandatory at this stage. High integrity systems are able to detect failures and switch to a "safe-state" with high confidence; in these systems however, there is no safe-state and therefore, high reliable architectures with optimized availability characteristics are needed. Different solutions for reaching required availability targets are used ranging from full redundancy to degraded functionality in which the redundant channel offers only a subset of the main channel features. The design of such high availability systems while fulfilling the safety requirements at their high levels required by the automotive industry and, in the same time, keeping the system under the required cost envelope is a challenge of the near future.

The last stage of this model is the vehicle control. This part of the control chain is responsible for the control of the motor, steering, and braking of the vehicle in order to maintain the calculated trajectory. To a certain degree, such systems are also present in the lower automation levels like adaptive cruise control, stop-and-go, etc. The constraints on these systems when used in higher level of autonomous driving are coming also from the fail-operational requirement, which assumes a certain level of availability for the systems controlling the vehicle. As for the rest of the systems involved in self-driving vehicles, possible redundancy scenarios including degradation can be considered.

One of the important issues with self-driving cars (besides the needed adaptation of international regulations regarding vehicle traffic) is the testing of such systems. Some known achievements [16] like the Audi high automated car running in the United States from west to east-coast or Google cars which already have a good record of autonomous driving on real streets (high automated without steering controls) are demonstrating some of these technologies potential. However, the testing of such autonomous systems is still a challenge as the real traffic situations are highly complex and unpredictable. The actual testing, which is an important part of the automotive development lifecycle, is driven by clear standards and specifications and may be not fully appropriate without adaptation to detect with the required coverage border conditions (e.g., misinterpretation of objects or traffic conditions). In order to test self-driving cars, special methods may be needed, e.g., for dealing with the huge number of relevant environmental configurations and scenarios, the complexity of such systems taking into consideration the different possible configurations, the unpredictability of a real traffic, etc. Most probably, the testing of autonomous vehicles will result from a combination of scenario based testing and system based testing also gradually improved with incremental introduction of driving automation.

REFERENCES

[1] Industry Overview—The Automotive Electronics Industry in Germany 2014–2015, ©Germany Trade & Invest, February 2014.

[2] Siemens Norm SN 29500–2/Edition 2010-09, Failure Rates of Components, Part 2: Expected Values for Integrated Circuits.

[3] AEC-Q100-Rev-H, Failure Mechanism Based Stress Test Qualification for Integrated Circuits, Automotive Electronics Council, September 2014.

[4] R. Wayne Johnson, J.L. Evans, P. Jacobsen, J.R. Thompson, M. Christopher, The changing automotive environment: high-temperature electronics, IEEE Trans. Power Electron. 27 (2004) 164–176.

[5] T. Liebetrau, U. Kelling, T. Otter, M. Hell, Energy Saving in Automotive E/E Architectures, Infineon, Neubiberg, 2012.

[6] Regulation and competitiveness of the EU automotive industry, Final Report, FTI Consulting, June 2015.
[7] M. Brylawski, in: Uncommon Knowledge: Automotive Platform Sharing's Potential Impact on Advanced Technologies, Hypercar, Inc., 1999.
[8] Regulation (EC) No. 443/2009 of the European Parliament and of the Council of 23 April 2009 setting emission performance standards for new passenger cars as part of the Community's integrated approach to reduce CO_2 emissions from light-duty vehicles.
[9] Platform Strategy Will Shape Future of OEMs, Flexibility to Drive Growth, Evaluserve, Schaffhausen, 2012.
[10] A. Avizienis, J.-C. Laprie, B. Randell, C. Landwehr, Basic concepts and taxonomy of dependable and secure computing, IEEE Trans. Dependable Secure Comput. 1 (1) (2004) 1–23.
[11] ISO 26262:2011, Road Vehicles—Functional Safety—Part 1–10.
[12] Hersteller Initiative Software (HIS): http://www.automotive-his.de.
[13] EVITA—E-Safety Vehicle Intrusion Protected Applications: http://www.evita-project.org.
[14] VDA—Verband der Automobilindustrie—VDA Position Automated Driving, 15 Mai 2014.
[15] NHTSA—National Highway Traffic Safety Administration, Preliminary Statement of Policy Concerning Automated Vehicles, 2013.
[16] Hochautomatisiertes fahren auf Autobahnen—Industriepolitische Schlussfolgerungen, Fraunhofer-Institut für Arbeitwirtschaft und Organization (IAO), Stuttgart, 2015.

CHAPTER

Harsh computing in the space domain

J. Abella*, F.J. Cazorla†

*Barcelona Supercomputing Center, Barcelona, Spain**

IIIA-CSIC and Barcelona Supercomputing Center, Barcelona, Spain†

Computing systems can be broadly classified into mainstream systems and real-time (embedded) systems (RTES). While the former are usually optimized mainly to improve average performance, the latter care about worst-case performance. Further, the correctness of RTES does not only rely on their functional correctness—as it is the case for most mainstream systems—but also on their timing correctness. RTES have to meet certain timing restrictions, which are accomplished by providing evidence that system's tasks fit their assigned timing budget (i.e., execute before a given deadline).

Another differentiating feature of RTES is that they are often critical, meaning that their unexpected misbehavior can result in damage for equipment or people. The potential damage determines the degree of criticality of the system which broadly covers safety, security, mission, or business aspects. For instance, misbehavior (either on its functional or its temporal domain) of a mission-critical RTES may cause a failure to accomplish the mission or goal of the RTES. Likewise, misbehavior in safety-critical RTES may involve human injury or fatalities. In the space domain, misbehavior in the spacecraft guidance and navigation control may take the spacecraft into a wrong position or orbit and the potential loss of the spacecraft.

Critical RTES can be found in different application domains including automotive, railway, avionics, and space. Among those, RTES in space are subject to specific performance, verification, and reliability requirements different from those of RTES in other domains.

Space missions comprise a number of segments with different requirements and conditions. Those segments can be typically classified into: the launch segment (launchers), the ground (segment), and space segment (spacecraft).

- Launchers transport the spacecraft into space during the launch itself. Therefore, the electronic systems of the launchers are subject to harsh conditions due to extreme temperature and vibration during the launch, but this process lasts in the order of some minutes at most.

Rugged Embedded Systems. http://dx.doi.org/10.1016/B978-0-12-802459-1.00009-9

- Ground facilities control the launcher and the spacecraft, monitoring them and collecting the payload data of the spacecraft, which is also processed in ground facilities. Their lifetime can be in the order of dozens of years to keep control of the spacecraft until it is decommissioned. However, electronic equipment in ground facilities, in general, is not subject to particularly harsh conditions and their size, power, and weight are not severely constrained. Furthermore, faulty components can be replaced if needed.
- The spacecraft is the most critical element of a space mission due to the—many—challenges that need to be faced in its design and verification.

Among those challenges in the spacecraft design and functioning, in this chapter, we focus on those related to its electronic equipment and, more specifically, those affecting its microcontrollers. We identify the following main challenges, summarized in Fig. 1:

(1) Size, power, and weight limitations are severe for the spacecraft, thus requiring the use of as few electronic systems as possible able to provide the required functionality.
(2) The lifetime of the spacecraft can be easily in the order of dozens of years and faulty components cannot be replaced since the spacecraft is not at reach.
(3) The spacecraft is subject to harsh operating conditions that include extreme temperatures, vibrations, radiation, electromagnetic interference (EMI), etc. In particular, extreme radiation conditions out of the Earth atmosphere represent the most critical challenge, since it affects microcontrollers during the whole lifetime of the spacecraft in a way that differs from what can be found in other systems on the Earth.
(4) Due to the long lifetime of many missions, software is typically updated during the lifetime of the mission from the ground facilities. This challenges the

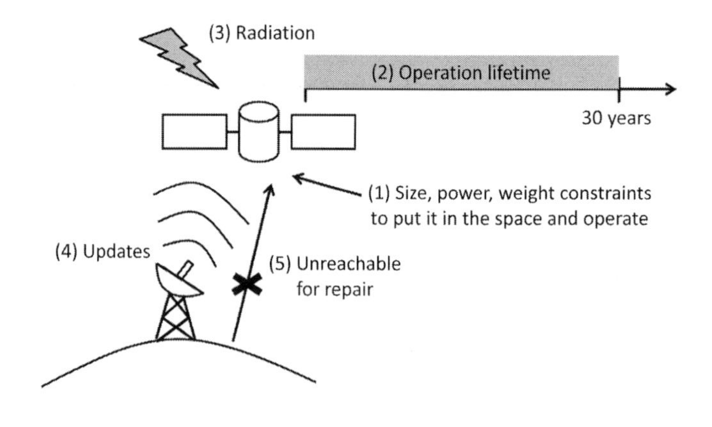

FIG. 1

Challenges in the design and verification of spacecraft electronic systems.

functional and timing verification of the electronic systems since they must operate correctly with existing and future functionalities.

(5) Systems cannot be generally tested in the target environment, thus needing some degree of flexibility to adapt to the unknown.

In this chapter, we first review the main on-board functionalities and characteristics of spacecraft microcontrollers identifying the key environmental conditions in space, the performance demands of future missions, and the type of microcontrollers that are already being developed and will be needed in spacecraft. Then, we put those designs in the context of the verification process of space systems identifying the most critical challenges. Finally, we present a new approach that relies on randomizing structural dependences on hardware and software for the seamless design and verification of microcontrollers in future space missions that allows the use of high-performance features, the consolidation of mixed-criticality functionalities onto a single hardware platform, and—functionally and timely—correct operation on top of potentially faulty components.

1 ON-BOARD COMPUTING

On-board spacecraft functions, which are the main focus of this chapter, can be classified into two main groups: *platform* that covers those components and the related functionalities required to allow the satellite to function in space (e.g., power management, communication, guidance, and navigation) and *payload* functions that deal with the specific mission of the spacecraft, such as infrared detectors, and telescopes.

Platform functions are quite standard across missions since they deal with space conditions that, in general, do not change much across different missions. This results in relatively high similarities among platform functionalities in different spacecraft. Conversely, payload functions can significantly vary since they are specific for each mission. For instance, payload may include a wide variety of functions including communications, data acquisition, experimental, weather, human manned, etc., and even within each type differences across missions can be huge due to the distance, atmosphere, light, and electromagnetic noise of the target of the mission.

Next, we review the specific—harsh—conditions under which on-board computing systems operate, their functional characteristics, their performance needs, and how such performance can be achieved in a reliable way under space extreme conditions. As explained later, the most distinguishing matter of spacecraft with respect to RTES in other domains is the extremely high radiation. Microcontrollers are particularly sensitive to radiation and hence, they are the focus of this chapter. Other semiconductor electronic components in the system (e.g., memory chips) are not discussed explicitly, but most of the discussion in this chapter also holds for them. The impact of space harsh conditions on data buses (or other communication links in the payload or platform data systems), sensors, and actuators is left out of this chapter.

1.1 THE SPACE ENVIRONMENT

Several parameters determine the operation conditions of spacecraft in space. Most of them affect similarly other systems on the Earth surface as it is the case of extreme vibration, humidity, EMI, and temperature conditions, which can already be experienced by systems in cars, trains, etc. However, microcontrollers in spacecraft are subject to radiation levels far higher than those that can be experienced on Earth.

It is not the purpose of this chapter making a detailed review of the existing types of radiations and their characteristics. Instead, we identify those radiation sources whose impact exacerbates in the space with respect to their potential impact on the Earth surface.

Radiation-induced failures are caused by strikes of high-energy particles (typically above 50 mega electron volts, MeV) including neutrons, protons, and heavy ions. Interestingly, radiation failures are more abundant in the space than on the Earth since the atmosphere acts as a filter for high-energy particles. For instance, even still within the atmosphere, it has been proven that soft error rates due to radiation grow by a factor of around 650 × when moving from sea level to 12,000 m of altitude [1]. The frequency of those single-event phenomena and their charge depend on a number of factors including the following:

(1) The particular position of the spacecraft with respect to the Earth. On the one hand, whenever the spacecraft is close enough (i.e., orbiting around the Earth), the distance to the Earth determines the degree of filtering exercised by the atmosphere. On the other hand, the radiation around the poles is higher than far from them. Analogously, if the spacecraft is orbiting or landing in another celestial body (e.g., another planet), its atmosphere will also act as a filter.

(2) The distance to the Sun that determines the intensity of filtering of solar radiation due to the space background.

(3) Whether the spacecraft is under the influence of the radiation produced by a solar flare, whose intensity is far much higher than regular radiation. Moreover, the solar cycle, whose frequency is around 11 years, leads to oscillations in the number of solar flares of several orders of magnitude.

(4) The influence of the radiation belt, since heavy ions get trapped within the magnetosphere of the Earth.

(5) Other sources of radiation may exist in spacecraft due to particle emissions of radioactive elements in the spacecraft itself.

While extreme temperature conditions in the space have also some influence on radiation effects, such influence is less relevant in comparison with the factors outlined above. For instance, some (weak) dependence has been reported between temperature and radiation. In particular, colder devices suffer lower transient fault rates due to particle strikes with an approximate increase of the rate of 0.2% per temperature Kelvin degree [1]. Therefore, even a 100 K degree variation would only lead to a radiation impact variation of around 20%, thus far from those other factors that may increase radiation impact by several orders of magnitude.

Particle strikes can directly interact with the semiconductor devices of the microcontrollers in the spacecraft or, in the case of neutrons, can lead to secondary strikes after a first collision. Regardless of whether particle strikes are direct or come from secondary collisions, they may lead to a number of effects depending on the energy of the strike and the charge of the particle. Those single-event effects (SEE) can be directly related to specific types of faults in the microcontroller [2]:

(1) Single-event upsets correspond to those SEE, where the state of a sequential device (e.g., a memory cell) changes or a combinational device suffers a glitch that can be latched in a sequential device. Although these events may alter the state of the circuit, they are transient and cause no permanent damage.

(2) Single-event latchups correspond to those SEE, where the state of a device gets statically changed due to a high current above device specifications that can only be removed by resetting the power of the circuit. This is a form of long transient fault that causes no permanent damage.

(3) Single-event burnouts, gate ruptures, and gate damages correspond to those SEE able to create a permanent damage in the device that leads to its degraded operation or simply a complete malfunction.

All in all, increased radiation levels in the space may increase SEE rate several orders of magnitude in frequency and intensity, resulting in much more frequent transient and permanent faults than those experienced on the Earth surface.

1.2 PERFORMANCE REQUIREMENTS

Software complexity in spacecraft has grown steadily since the very first space missions, and it is expected that future applications will demand increased performance to provide computation-intensive value-added on-board functions. This trend can be observed in Fig. 2 that reports the increasing number of lines of code in space

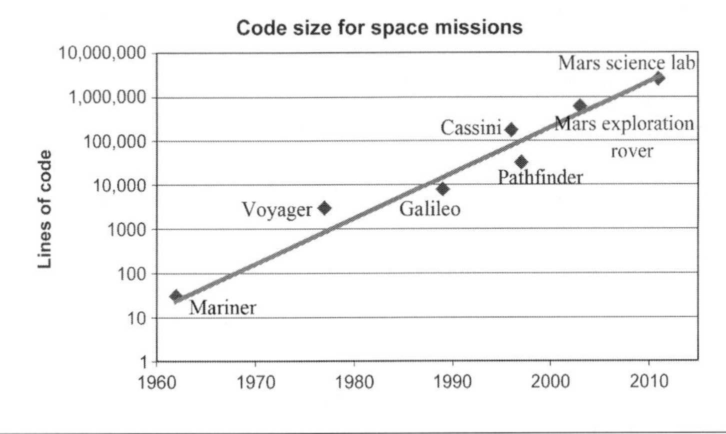

FIG. 2

Code size of several space missions [3,4].

missions: from a few tens in early 1960s to a few millions 50 years later, thus by a factor of around 10× every 10 years.

Increased on-board performance requirements affect both platform and payload functions. On the platform side, despite many functions do not change across missions, increased performance is needed to deal with the advent of more autonomous systems and the use of new materials. Advance autonomous systems are becoming pervasive in all critical RTES domains, such as automotive, railway, or avionics. In the space domain, increased autonomy (1) allows mitigating issues related to latency transmitting/receiving commands to the satellite—in the event of phenomena with time scales shorter than the communication latency between the spacecraft and Earth (which increases as missions go further and further); (2) allows improving efficiency and making spacecraft more resilient by self-diagnosis and correction of errors; and, even in manned missions (3) helps managing increasingly complex spacecraft systems. The other side of the coin is that higher autonomy levels increase the demand of reliable computing performance.

On the payload side, we also observe a significant increase in the need for performance as missions become more complex and ambitious. For instance, the use of more complex imaging applications creates orders of magnitude larger data than what can be transmitted with satellite communication links. This calls for on-board computation capacity able to process (on-board) these data reducing the pressure on the satellite communication links. Further, the number of payload functions due to the large number of on-board instruments (typically around 10 different types of instruments) poses a significant performance requirement for space systems.

1.3 ACHIEVING PERFORMANCE REQUIREMENTS

Simple single-core microcontroller designs cannot scale up to the challenge to meet the performance requirements of future spacecraft. Instead, new microcontroller architectures are considered, not only in the space domain, but in most critical RTES. The main path chosen in other industries relies on the use of multicore microcontrollers, cache hierarchies, and higher performance core designs. For instance, automotive industry considers processors such as the Infineon AURIX, a 3-core multicore, and avionics and railway domains are also evaluating some ARM and Freescale multicore processors (e.g., Freescale P4080), with multilevel cache hierarchies and increasingly complex pipelines. In this line, space industry is also moving toward more powerful—yet more complex—multicores with multilevel cache hierarchies for their future spacecraft.

1.3.1 High-performance architectures

From a hardware perspective, multicore microcontrollers with multilevel cache hierarchies offer an excellent solution to improve performance without significantly increasing processor complexity. Instead of striving to increase single-core performance by deploying complex designs that exploit instruction-level parallelism or increasing CPU clock frequency, system performance is scaled up by deploying simpler cores on a single chip with multilevel cache hierarchies. Those microcontrollers offer a better performance/Watt ratio than a single-core solution with similar performance, while reducing the average execution time for parallel/multitask workloads.

All these advantages of multicore microcontrollers equipped with caches have motivated their adoption in both space and nonspace industries. In the case of the European Space Agency (ESA), it has used recently in space missions LEON2 [5] and LEON3 [6] processors, which implement deep pipelines and cache memories. In this line, ESA is currently funding the development of the LEON4-based Next Generation MultiProcessor (NGMP) [7], an even more complex microcontroller implementing a 4-core multicore with a two-level cache hierarchy for increased computing power needed for future missions.

Integrating several high-performance cores per chip allows consolidating several functions into a single hardware platform, thus reducing the overall weight, cost, size, and power consumption of the system. These parameters are of prominent importance in spacecraft due to the cost of space missions, the challenges associated with launching large and heavy spacecraft, and the limited energy available in the outer space. However, as detailed later, the concurrent execution of several applications challenges verification due to unintended interactions in shared hardware and software resources, especially in the context of mixed criticalities, where a low criticality function could compromise the integrity of the spacecraft by interfering with the functionality or timing of a critical platform function.

1.3.2 Semiconductor process technology

Multilevel caches and multicores require an increased number of transistors to be implemented into a single chip. These high-performance functionalities need to be packaged into chips with similar dimensions and power constraints as those for simple single-core processors due to the limitations of the equipment and technology used by chip vendors. This can only be realistically achieved by using smaller technology nodes where geometry is scaled down. The main advantages of smaller technology nodes include (1) an increased integration capacity, (2) reduced power consumption per device and so, reduced heat dissipation per device, and (3) increased on-chip storage and communication means that allow reducing the number of slow and power-hungry off-chip activities.

However, smaller technology nodes bring a number of challenges, some of which may jeopardize their advantages in the space domain. Among those challenges, we identify the following ones:

(1) Although power per device decreases, the number of devices per mm^2 increases exponentially, thus leading to an increased power density, and so increased temperature. While power limitations and temperature-related aging are key challenges in the space, they are often mitigated by simply powering chips with lower operating voltages. However, reduced voltage also means increased susceptibility to radiation that can easily produce SEE.

(2) The use of a larger number of smaller devices also leads to an increase of the FIT rate (Failures In Time—number of failures per 10^9 h of operation) because of the exponential increase of the device count.

(3) Fabricating devices whose dimensions are smaller than the wavelength used in the lithography process lead to devices whose actual shape differs from their nominal characteristics. These devices may probabilistically be more susceptible to SEE and suffer faster degradation, thus shortening their lifetime. Even if these process variations are made negligible, devices consist of fewer and fewer atoms, so the relative impact of degradation is higher and lifetime shortens.

Increased susceptibility to SEE and shorter lifetime are key concerns in spacecraft, where microcontrollers are exposed to high radiation and missions may last long. For instance, the Voyager 1 was launched in 1977 and is expected to keep working until 2025, so almost 50 years of operation. Further, by reaching interstellar space after being subject to the gravity effects of several planets, microcontrollers in the Voyager 1 have been exposed to a wide range of radiation levels, much higher than those within the Earth atmosphere. Several other scientific spacecraft and communication satellites have been operating for long periods of some dozens of years. However, if smaller semiconductor process technology is to be used, failures can become more frequent and lifetimes shorter. This is better illustrated in Fig. 3 for different technology nodes evaluated in the context of space environments. As shown, scaled geometries increase FIT rates and decrease lifetime exponentially. For instance, when moving from 130 to 90 nm lifetime decreases by a factor of $7\times$ and FIT rate increases by a factor of $2.2\times$.

It is obvious that space industry uses specific gate designs, so that much higher lifetimes and lower FIT rates are obtained in comparison with those of microcontrollers fabricated for consumer electronics or other RTES domains. Still, the FIT rate

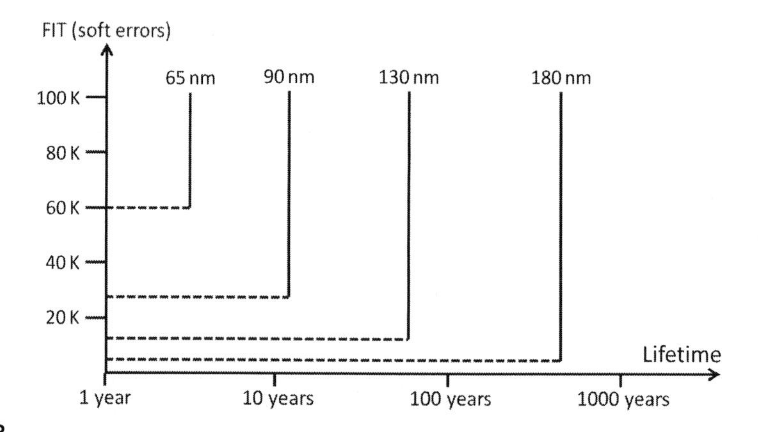

FIG. 3

FIT rate for radiation for STmicroelectronics DRAM technology in the context of an ESA study [8] and lifetime (in logarithmic scale) of technology in the context of an NASA study [9].

increase and lifetime decrease across technology hold and can only be mitigated at the gate design level by using larger gates that may completely jeopardize the gains obtained by shifting to smaller device geometries.

It can be concluded that scaling semiconductor process technology to fit more powerful architectures leads to increased failure rates and shorter lifetimes. Thus, *space industry faces a dilemma finding a good balance between performance and reliability*. This is a complex challenge for the design and verification of spacecraft microcontrollers. How to deal with the verification process in this context is analyzed in the next section.

1.4 BEYOND PERFORMANCE AND RELIABILITY

While performance and reliability are the main concerns in space microcontrollers, other concerns are also becoming of prominent importance in spacecraft. These include cost and security.

1.4.1 Cost

The use of domain-specific microcontrollers for a low-volume market such as the space domain increases the nonrecurring engineering cost per microcontroller unit to amounts that make them often too expensive. Therefore, commercial off-the-shelf (COTS) microcontrollers are desirable in the space domain due to their lower cost and immediate availability in the market. However, they need to be fabricated with radiation-hardened long-lasting devices to meet space requirements or, at least, stay close enough to them. This is still a hard-to-remove constraint in the use of COTS microcontrollers, but still allows using less expensive microcontrollers and shifting to more powerful microarchitectures faster than by using domain-specific microcontroller designs.

As it is extensively covered in Chapters 2 and 5, a specific form of COTS microcontroller consists of using designs implemented with FPGA, where those FPGA are also built using radiation-hardened long-lasting devices. Given the specific functionalities that may be needed in the space for the efficient collection and processing of large amounts of data—that need to be processed on-board due to communication bandwidth constraints—application-specific microcontrollers may be needed in the space domain. This is particularly true for payload applications where data collected with instruments may need to be filtered, processed, and/or compressed on-board. The use of FPGA allows implementing those functionalities more efficiently than processing data purely by software means on general-purpose microcontrollers. Further, since functionalities may be updated in ways that are hard to predict before the launch of the spacecraft, using FPGA provides the flexibility needed to implement future functionalities efficiently. For instance, the Voyager 1 was launched in 1977 and its mission was intended to last 5 years. However, it has been reprogrammed remotely and almost 40 years later it is still working and flying in the interstellar space, an operation scenario that was never conceived before its launch. Similarly, the Cassini spacecraft was launched in 1997, reached Saturn in 2004, and was expected to complete its mission in 2008. However, its mission was firstly extended until 2010 and secondly until 2017 by reprogramming the spacecraft accordingly.

Therefore, the use of COTS components (either microcontrollers or FPGA) is a key aspect in the space domain to obtain inexpensive, flexible, and scalable platforms able to support reduced development schedules.

1.4.2 Security

Security is a major concern in many critical RTES due to the risks associated with someone else taking control of the system, stealing data, etc. In the context of space microcontrollers, security is also a critical concern, especially when using FPGA-based microcontrollers. Due to the fact that missions may change over time to fix errors, improve functionalities, or extend missions beyond what they were planned to reach, means to program FPGA remotely need to exist. Unprotected FPGA programming could allow malicious reprogramming so that the operation of the spacecraft is changed, data are stolen, or the microcontroller gets unusable (or even destroys the spacecraft). For instance, a remote attack could reprogram the FPGA in a way that it performs high-power operations destroying the FPGA due to the induced heat, which in turn could produce serious damage in other systems; in a way that it starves other systems due to the high power consumption of the FPGA; or simply making it not to operate. Therefore, secure means to reprogram spacecraft systems need to exist, and those must include microcontroller configuration and FPGA programming.

In summary, while performance and reliability are the most critical aspects in the design and verification of space microcontrollers, other aspects such as cost and security cannot be neglected.

2 FUNCTIONAL AND TIMING VERIFICATION

Critical RTES undergo a strict validation and verification process to assess whether they will perform their functionalities during operation according to certain specifications. Operation in harsh environments, as it is the case of spacecraft, has specific implications in the functional and timing verification of critical RTES. It is not the objective of this section reviewing the verification and testing processes of electronic equipment in spacecraft, which are detailed in the corresponding standards. For instance, European space activities build upon a number of standards (more than one hundred, with thousands of pages in total) developed by the European Cooperation for Space Standardization (ECSS). In this section, we deepen into those aspects influenced by the use of high-performance architectures implemented with technology nodes subject to high transient and permanent fault rates. First, we describe those implications related to functional aspects of the system that depend on the characteristics of the microcontroller. Then, we analyze how timing verification is affected.

2.1 FUNCTIONAL VERIFICATION AND TESTING

Since microcontrollers in spacecraft are exposed to high radiation, they are typically implemented with technology particularly suited to tolerate radiation. This occurs at different hardware levels, spanning from electronics to architecture. The techniques related to material, fabrication process, and device sizing used in past and present microcontrollers in spacecraft need to be applied even more aggressively for future microcontrollers due to the drastic FIT increase and shorter lifetimes expected due to permanent failures. Also, techniques at the gate level need to be applied analogously. However, these methods are unlikely to cope on their own with the increased reliability issues of smaller technology nodes. Instead, the microarchitecture and the architecture of the microcontroller need to provide means to cope with increased reliability issues.

Fig. 4 sketches the main steps in the test of a microcontroller in the space domain based on the main European standards. As shown, the microcontroller is designed (or chosen) with specific characteristics to meet the performance demands of the particular mission. This includes both its architecture and the specific technology node for its fabrication. In parallel, based on the mission requirements, the radiation scenarios that can occur during operation are identified and properly described. Based on those radiation scenarios and the equipment available to perform radiation tests, test conditions are defined. Test conditions must account for discrepancies between radiation scenarios during the mission and those had in the test facilities, duration of the mission, safety margins due to inaccuracies in the different elements of the process (e.g., definition of radiation scenarios, mission duration, intrinsic inaccuracies of the test equipment, inaccuracies during data collection, etc.) among other parameters. Finally, the test campaign occurs and results are analyzed to assess whether the requirements of the mission that apply to the microcontroller are fulfilled. If this is not the case, then the design needs to undergo through modifications until requirements are met.

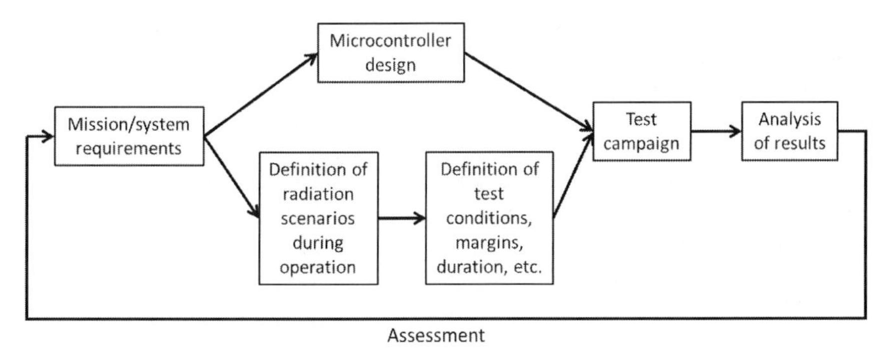

FIG. 4

Test process for microcontrollers for spacecraft [10,11].

In the context of spacecraft microcontrollers, conventional countermeasures at electronics, circuit, and gate levels can be applied. However, the microarchitecture and architecture will be naturally exposed to higher failure rates when using smaller technology nodes. In this context, safety measures need to be put in place at higher abstraction levels. As usual, those measures build upon redundancy, either in time or in space. While space redundancy is effective to deal with both permanent and transient faults, time redundancy can be regarded as effective only for transient faults and, at best, first occurrences of permanent faults, that typically manifest in the form of intermittent faults until damage is large enough so that faults become permanent.

Microcontrollers implementing multicores and large caches arranged in multilevel setups already account for high redundancy by construction, far beyond the typical redundancy one may have by using error detection and correction (EDC) codes. For instance, single error correction double error detection (SECDED) codes are common in storage in microcontrollers, and will remain in place in future microcontrollers, as well as other EDC codes with higher detection and correction capabilities. For instance, the Cobham Gaisler Leon3FT (Fault-Tolerant) is one of the microcontrollers used by the ESA in its missions [12]. It implements EDC codes able to detect four errors in 32-bit words in caches where redundancy exists (e.g., in main memory) so that, on a fault, the word or the full cache line can be invalidated and fetched again. If redundancy does not exist, as it is the case for register files, then it implements EDC codes able to *detect and correct* four errors in 32-bit words also transparently for the application.

Still, the increased permanent fault rates need further means to keep correct operation in front of faults not to allow lifetime to fall below the needs of long-lasting missions, which can be in the order of dozens of years, and are often extended beyond their initial expectations. Therefore, redundancy needs to be exploited in different forms since EDC codes cannot scale efficiently to arbitrary error counts. Multicores offer intrinsic redundancy in nonshared resources like those local to cores. This redundancy can be exploited for lockstep execution so that critical tasks are run simultaneously in more than one core and discrepancies in the results can be easily and quickly detected and corrected. While lockstep execution is a well-known solution for fault tolerance, the advent of multicores opens the door to implementing on-chip lockstep. This approach has already been exploited in other domains such as automotive, for instance, in the Infineon AURIX multicore, where five cores exist physically, but two of them are only used for lockstep execution with two other cores, so that there are only three cores visible to the software (two lockstep-protected and one unprotected) [13]. However, the increased permanent fault rate may make lockstep execution fail to provide the required fault tolerance during the target lifetime of the microcontroller due to unrecoverable faults.

Large and abundant caches in high-performance architectures are intrinsically fault tolerant due to the fact that they are in place to provide higher performance, but not needed to provide correct operation. For instance, some parts of a cache memory could be permanently disabled while still providing correct operation despite the degraded performance. Based on the fact that cache memories easily occupy most of the area of the microcontroller and they are implemented with the smallest transistors available—also in the particular radiation-hardened technology used—it is expected that most permanent faults will occur in caches. Hence, designing them properly allows, not only detecting and correcting errors, but also diagnosing fault types and reconfiguring them properly, so that permanent faults do not trigger EDC cycles, which are time and energy consuming [14].

Other shared components may still need some form of redundancy or additional fault tolerance measures if they are not replicated or intrinsically fault tolerant, as it is the case of on-chip interconnection networks and memory controllers, since they are single points of failure. A failure in any such component could deem the whole microcontroller unusable. Replicating them or making them radiation-hardened may increase their area and power noticeably, but given that their total contribution to the area and power of the chip is relatively low in general, those solutions have a relatively low cost with respect to the whole microcontroller.

In summary, multicores and large multilevel cache hierarchies enable a number of safety measures that can be taken into account during the microcontroller design, so that transient and permanent faults are naturally tolerated and so, the results of the test campaign (see Fig. 4) meet the requirements posed by the mission on the microcontroller. However, while these safety measures may suffice to provide correct—functional—execution, they may alter the execution time of the tasks due to the error recovery, diagnosis, and reconfiguration activities conducted by hardware, as well as due to the degraded operation of the hardware if some components (or parts thereof) are deactivated due to the presence of permanent faults.

2.2 TIMING VERIFICATION

Timing verification of critical RTES is a required step to provide evidence that tasks will execute by their respective deadlines. Timing analysis methods are needed to prove with sufficient confidence that tasks will execute within their timing budget. Those methods aim at tightly upper-bounding the worst-case execution time (WCET) of the task under analysis. As discussed later, a number of techniques exist for that purpose. However, the use of multicores and multilevel cache hierarchies challenges the ability of timing analysis techniques to obtain reliable and tight WCET estimates. This occurs because the timing behavior of applications running on multicores is much harder to analyze than on single-core processors due to contention (intertask) interferences when accessing hardware-shared resources like internal buses and shared caches. Further, how contention interferences affect timing analysis is a relatively young (and not fully mature) area of research, with implications on the reliability that can be obtained for WCET estimates of tasks running on multicores. While more powerful analysis techniques are developed, a way to deal with contention is by limiting resource sharing. For instance, in the 4-core NGMP to be used by the ESA in its future missions, the shared cache can be configured to allocate separated partitions (cache ways) to different cores, so that tasks running in different cores cannot alter the cache contents of other cores. Still, contention interferences may exist in other shared resources such as interconnection networks and memory controllers.

This situation is compounded by the fact that multicores allow co-hosting several applications with different criticality levels on a single MPSoC. In this context, low-criticality applications may run concurrently with high-criticality ones. Thus, means need to be put in place to prevent low-criticality applications to overrun their allocated access budget to shared resources to prevent them from affecting the functional and time predictability of high-criticality applications. That is, in addition to ensure that low-criticality applications cannot corrupt the state of high-criticality ones, it is needed to ensure that low-criticality tasks do not create unpredictable effects on the execution of critical ones (time analyzability).

In the space domain, several software and hardware methods to account for contention interferences in multicores have already been devised [15–17]. Those methods can also be applied to other domains since the problem of contention interferences in multicores is shared across different domains with critical RTES. However, harsh environments bring a number of timing issues that, in general, are not expected in other domains. Those issues relate to the frequent occurrence of transient and permanent faults during operation due to high radiation, and translate into timing issues due to the sequence of resolution phases presented in Chapter 2:

(1) Anomaly detection. Whenever an error occurs, sometime may be needed for its detection, thus increasing execution time with respect to fault-free computation. However, in general, error detection occurs transparently with constant latency regardless of whether an error occurs or not. For instance, this is the case of EDC codes, whose timing effect (if any) is constant regardless of whether an error occurs.

(2) Fault isolation. If fault diagnosis determines that a permanent fault occurred, hardware may need to be reconfigured by, for instance, disabling faulty components. In general, this is a quick step but it may also increase execution time by stopping or slowing down the execution of the program being run.

(3) Fault diagnosis. Once an anomaly has been detected and isolated, some form of hardware diagnosis may be needed to discern whether the problem was due to a transient or a permanent fault. Such diagnosis may also occur transparently or affecting execution time by, for instance, stopping the execution while diagnosis occurs.

(4) Fault recovery. Whenever an error is detected, some actions may be needed to restore a correct state and resume execution. In some cases such error recovery can also occur transparently, as it is often the case with EDC codes since EDC may occur for all data regardless of whether an error actually occurred. However, this is not always the case and, for instance, a previous (correct) state may be recovered to resume execution from there. The process to recover the state plus the time needed to reach the point in the execution when the error occurred may take some nonnegligible time, thus augmenting execution time in the presence of errors.

Upon recovery, execution can resume normally. However, some hardware may have been disabled (e.g., some faulty cache lines), thus increasing the execution time due to the reduced cache space (degraded operation).

Since faults occur frequently in high radiation conditions such as those for spacecraft, WCET estimates need to account for the timing effects related to error detection and recovery, fault diagnosis and reconfiguration, and degraded operation. Some of those actions may have easy-to-upper-bound latencies when safety measures are properly designed. This is the case of detection, recovery, diagnosis, and reconfiguration since their maximum latencies depend on the actual methods used rather than on the characteristics of the program being run. For instance, when some form of checkpointing and/or reexecution are deployed, detection latency is upper bounded by the latency between consecutive error checks triggered by the hardware/software platform. However, whenever hardware is reconfigured and programs run on degraded hardware, performance impact, and hence impact on WCET, can be high and unpredictable. This has already been shown in other domains for cache memories where, by disabling some specific cache lines, performance degradation can be huge and depend on the specific location of the faulty cache lines inside the cache memory [18]. In the context of critical RTES in the space domain it is required to identify the worst fault map that can occur during the lifetime of the microcontroller to obtain reliable WCET estimates. However, enumerating and producing all potential fault maps is often unaffordable. Further, even in those scenarios where this is doable, WCET estimates may degrade noticeably since the actual worst-case may be far worse than the average case.

All in all, dealing with frequent transient and permanent faults and operating on degraded hardware may have inordinate impact on WCET estimates, thus jeopardizing the advantages of multicores with multilevel caches. Hence, specific hardware designs and appropriate timing analysis methods are needed to obtain reliable, tight, and low WCET estimates for tasks running in harsh environments in spacecraft.

3 A PROBABILISTIC APPROACH TO HANDLE HARDWARE COMPLEXITY AND RELIABILITY

From previous sections it follows that the main challenges to face by on-board computing systems are to design high-performance, reliable, easy to analyze (qualify) complex microcontrollers on top of unreliable components to run complex mixed-criticality software.

In terms of timing, verification requires providing evidence that the application will perform its function in a timely fashion. Timing verification relies on methods to derive WCET bounds in the context of system-level task scheduling decisions. Two fundamental requirements on WCET estimates are tightness and reliability. The former talks about the need of sufficiently accurate estimates in order to avoid computer capacity to be unnecessarily reserved and hence wasted. The latter covers the fact that WCET estimates have to be trustworthy "enough" as determined by the domain-specific safety-related standards. Those standards, depending on the severity of the consequences of failure of a (sub)system, define the degree of rigor and depth for analysis to have confidence on the safety of the (sub)system.

Determining tight and sound (reliable) WCET bounds is a complex task [19]. The advent of sophisticated software running on top of high-performance complex and less reliable hardware negatively affects the scalability of current timing analysis approaches. On the one hand, the intrinsic increased complexity of high-performance hardware—comprising features such as several cache levels, deeper pipelined processors, and multicore technology—makes it more difficult for the WCET analysis technique to gather enough evidence on the timing behavior of applications running on that hardware. On the other hand, with few exceptions [20–24] WCET analysis techniques have been designed assuming fault-free hardware. However, the use of aggressive implementation technologies makes that hardware degrades throughout its lifetime, with its timing behavior becoming worse as it degrades.

3.1 TRADITIONAL WCET ANALYSIS TECHNIQUES

It is well accepted that timing analysis techniques can be broadly classified into three categories.

- *Static timing analysis* (STA) techniques derive WCET bounds for a program and target hardware architecture pair without executing the program. Instead STA constructs a cycle-accurate model of the system and a mathematical representation of the application code, which are combined to derive the timing behavior of the application.
- *Measurement-based timing analysis* (MBTA) techniques collect execution time measurements of the program running on the target hardware and derive WCET estimates by operating on those measurements. MBTA depends on the quality of the input data used to stress the hardware and program such that the actual WCET of the program is triggered in one run.
- *Hybrid timing analysis* (HyTA) techniques aim at getting the best from static and measurement-based techniques. In particular, static information on the program control flow is used to augment the measurement-based approach, gaining confidence on the derived WCET bounds.

Time predictability is a key feature required from the hardware and software platform to facilitate timing analysis techniques. Historically, *time determinism* has been understood as the only way to achieve time predictability. This is pursued by making every single hardware and software component affecting the execution time of a program in such a way that for the same input it responds on a fixed (or easy-to-bound) time. While this approach works well for relatively simple systems, it does not scale to more complex ones. This occurs due to the complex interactions and dependences among platform components in computing systems.

- For STA it is difficult, if at all possible, to model those dependences among platform components with a sufficient degree of accuracy in order to obtain tight WCET estimates. This complicates having a *causal* model to derive for every input (stimuli) the system receives, the state the system transitions to. In the absence of information on how components behave or interact, either pessimistic assumptions are made on the system state or several states are carried forward during the modeling, leading to a state explosion. Hence, STA techniques are intrinsically limited by the complexity of constructing a sufficiently accurate model of the hardware and modeling how the execution of complex software impacts the state of the hardware model.
- For MBTA, it is hard to define and perform test experiments that stress all possible state-transitions, or at least those leading to the WCET. In general, the user cannot make such an argument on whether bad system states have been covered. Therefore, the degree of confidence achieved on the WCET estimates strongly depends on the end-user skills to control software effects (e.g., execution paths tested), hardware effects (e.g., cache interactions), and combinations thereof.
- HyTA face similar problems to those of MBTA to handle state transitions.

Overall, while each platform component can have a perfectly deterministic behavior, it is the case that the timing behavior of an individual component is not solely affected by the inputs it is given, but also by the initial state and the complex interaction among components. This complicates the modeling of the timing behavior of the system as a whole. Further, complex interactions between programs may result in unnoticed abrupt variations in nonfunctional behaviors, which are called pathological cases [25].

3.2 HANDLING SYSTEM COMPLEXITY THROUGH RANDOMIZATION

Recently a new set of hardware designs based on the concept of smartly injecting randomization in the timing of specific components has been proposed to help dealing with system complexity [26,27]. The basic idea consists in making that hard-to-model components affecting program's execution time have a randomized behavior in their temporal part. This, on the one hand, allows building an argument on the probability of appearance of each response time of the component (including the pathological ones). On the other hand, it also allows deriving measures of *coverage* for the latencies of platform components when a given number of experiments (runs) are performed.

We denote source of execution time variability (or SETV) all those events—that can be generated in any hardware or software component—that can affect program's execution time. SETV include the execution path of the program, the memory location of data, and code that determines cache interactions, the initial cache state prior to the execution of the program, the input values operated in variable-latency functional units among others. The values that SETV take for a given experiment are referred to as the *execution conditions* for that experiment.

The basic idea behind the time-randomization approach is that, if events are truly randomized—and so probabilistic—the more execution measurements of the overall system are taken, the higher the probability to capture the impact of all latencies of the component. Making such an argument in a nonrandomized architecture is complex, if at all doable. In particular, the coverage on the response times of the component exercised during a set of experiments depends on the inputs provided by the end user, rather than on the number of runs performed. Therefore, the end user needs to control how the specific input data used for the program, the specific state of the operating system, the specific location of the program in memory, etc. will influence execution time in each experiment. Instead, time-randomized architectures aim at breaking as much as possible those dependences, so that a probabilistic argument can be built on the impact of most of the SETV relying only on the number runs are performed, rather than the particular inputs provided by the user in each of them. It is noted that for some SETV, instead of randomizing their response time, they can be made to work on their worst latency during the analysis phase. This approach provides satisfactory results when the jitter suffered by the resource is small, so that forcing it to work in its worst latency has a relatively small impact on the WCET estimate. Those SETV using time randomization and latency upper-bounding—typically hardware SETV—do not need to be controlled by the end user anymore, who can focus on controlling software-related ones (e.g., execution paths visited) that are much easier to control.

Time-randomized platform designs are accompanied by the appropriate probabilistic and statistical timing analysis techniques [26] to derive a probabilistic WCET estimate, like the one presented in Fig. 5.

With probabilistic timing analysis techniques WCET is represented with a curve in which each potential execution time the task can take is accompanied by

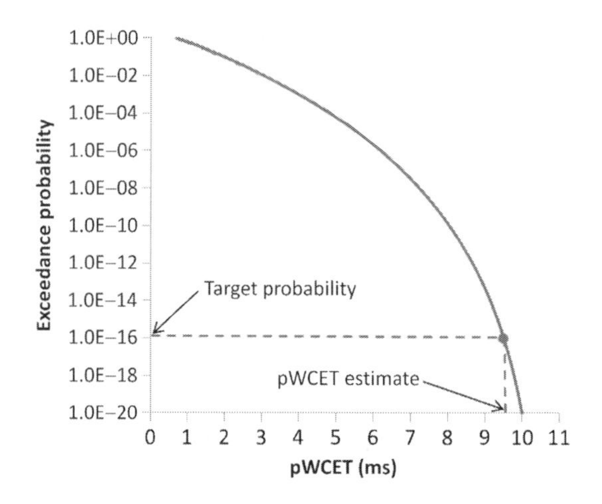

FIG. 5

Example of pWCET curve.

Table 1 Possible Taxonomy of Existing Timing Analysis Techniques

	Static	Measurement-Based	Hybrid
Deterministic	SDTA	MBDTA	HyDTA
Probabilistic	SPTA	MBPTA	HyPTA

an upper-bound probability for one run to overrun this execution time. For instance, in Fig. 5, the probability that the application takes longer than 9.5 ms is smaller than 10^{-16}.

It follows that, while timing analysis techniques working on time-deterministic architectures produce a single WCET estimate, timing analysis techniques meant for time-randomized architectures derive a curve of WCET estimates that relates each WCET value with an exceedance probability. Interestingly both families of techniques (deterministic and probabilistic) have their static, measurement-based, and hybrid techniques [28], resulting in the taxonomy presented in Table 1.

It is worth noting that each of these techniques builds on a set of steps to carry out the analysis as well as some information about the hardware and software analyzed to properly model them. While some steps can be easily taken, and enough information can be gathered, for some platforms, others are challenging, affecting the confidence that can be placed on the resulting probabilistic (p)WCET estimates. For instance, in SDTA (and SPTA) inaccurate timing models, from which the latency to the different resources is derived, or potentially unreliable manual annotations affect the reliability of the derived WCET estimates.

3.3 MBPTA

Measurements is the most common industrial practice for timing analysis [19] due to its simpler application steps and lower-cost to apply to new hardware. MBPTA, pursues the same goals as MBDTA while attacks some of its weaknesses that affect the reliability of MBDTA-provided WCET estimates.

MBPTA recognizes two phases in the lifetime of the system.

- First, during the analysis phase the verification of the timing behavior takes place. Measurements of the program on the target platform, under specific conditions, are taken.
- And second, the operation phase when the system becomes operational.

The goal of MBPTA is using execution time measurements taken during the analysis phase to derive a pWCET function of the program of interest that holds during operation. This carries the requirement that the execution conditions under which the analysis phase measurements are obtained, are guaranteed to represent, or upper-bound, the execution conditions that may occur during operation. Hence, it is required to have good control of all SETV.

Some SETV are well known and understood by CRTES industry. This includes the input vectors that determine the control flow (execution) path taken by the program in the measurement runs. We refer to those SETV as high-level SETV. In contrast to high-level SETV, we find low-level SETV, which arise due to the use of high-performance hardware. Among others, we find the mapping of program objects (e.g., code) to memory that determines how objects are laid out in cache and hence, the cache conflicts suffered accessing those objects and the resulting impact on execution time. In a multicore setup, low-level SETV include the parallel contention among tasks on the access to shared hardware resources, which impact tasks' execution time.

Interestingly the user has means to control high-level SETV such as execution paths. To that end tools, such as Rapita's Verification Suite (RVS) [29], help deriving the path coverage achieved in observation runs and whether this is deemed as enough according to the applicable safety standard. However, the user lacks cost-effective means to control low-level SETV.

- On the one hand, for some low-level SETV the user lacks support to monitor their timing behavior in different experiments and hence, understand whether execution time measurements can be related somehow to the WCET. For instance, in some cases the hardware provides no Performance Monitoring Counter support to track the contention delay suffered by each request when accessing a given shared bus (or other shared resource). This would difficult the work of the user to build sound arguments on whether the observation runs do indeed capture the jitter of that resource.
- On the other hand, other low-level SETV may be known by the end user, but it is no affordable in terms of experimentation requirements to design and execute a set of experiments that expose the effect of the jitter of that low-level SETV. For instance, some long-latency floating point units (FPU) have a variable response time depending on the values that they operate on. Managing their jitter would require the user to know the response time of the entire instruction set of the FPU in relation to the possible operands; and to force in test experiments a distribution of FPU operations that either cause it to operate with its worst-case profile or at least upper-bound the distribution that will occur during the operational life of the system.

Overall, the lack of control on the impact of low-level SETV challenges the confidence obtained with MBDTA. As a result, the end user may fail to provide sufficient evidence that the variability (jitter) that those SETV inject on the observed execution time does indeed upper-bound what may happen during operation.

In order to address this problem, MBPTA stipulates that the execution conditions under which measurements are collected during analysis lead to equal or worst timing behavior than determined by the conditions that can arise during operation [26]. This requires the identification of SETV and evidence that their timing behavior captured at analysis upper-bounds their timing behavior during operation by randomizing them or making them work on their worst latency. Those architectures for which this is achieved are referred to as MBPTA-compliant and build upon the approaches explained before: time randomization and latency upper bounding.

Interestingly, the degree of MBPTA-compliance for a given processor architecture depends on the architecture itself and the possibility of injecting randomization in its hardware design or by software means. For LEON3 processors, technology deployed in the space domain, the first MBPTA-compliant designs have reached the FPGA implementation level [30].

Fig. 6 shows the main idea behind MBPTA. Fig. 6A shows the probability distribution function (PDF) of the execution time of a synthetic program on an MBPTA-compliant processor. This represents the program probabilistic Execution Time (pET). From the PDF, we build the cumulative distribution function (CDF) and its complementary (or 1-CDF), which is referred to as the exceedance function. Both, CDF and 1-CDF are shown in Fig. 6B in decimal scale. When represented in logarithmic scale, Fig. 6C, the 1-CDF clearly shows the probability that one execution of the program exceeds a given execution time bound, which is what MBPTA aims at tightly upper-bound.

For a set of R runs, we could derive an exceedance probability at $1/R$ in the best case. For instance, for 1000 runs exceedance probabilities for execution times at

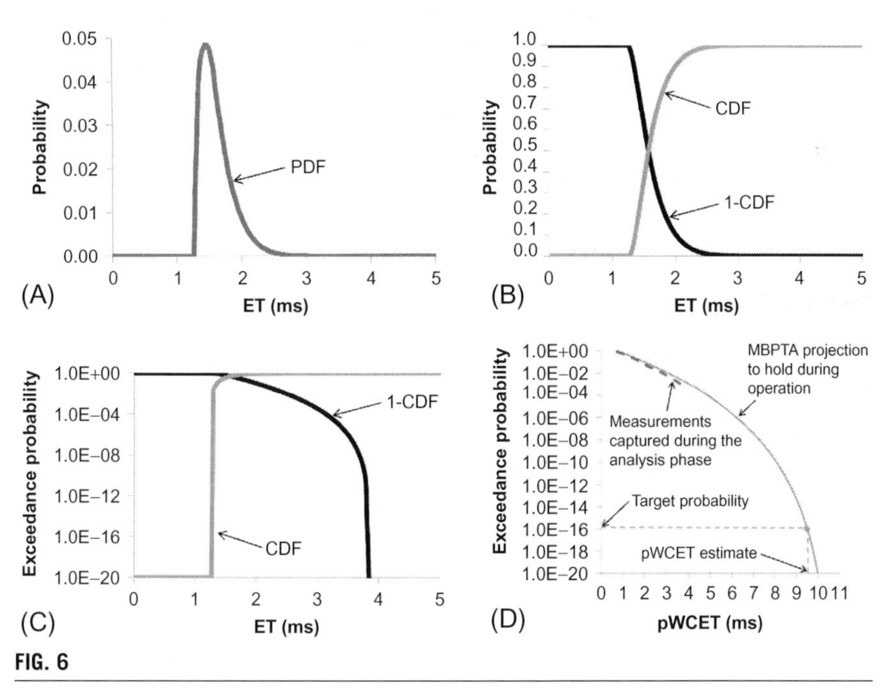

FIG. 6

Probabilistic distribution function or PDF (A), cumulative distribution function or CDF (B), and complementary CDF or 1-CDF (C) of a program in an MBPTA-compliant platform. CCDF of a set of measurements for that program and EVT projection obtained from those measurements (D).

round 10^{-3} could be derived. For smaller probabilities, such as 10^{-9}, techniques such as Extreme-Value Theory (EVT) [31] are used to project an upper-bound of the tail of the pET of the program, called pWCET. Fig. 6D illustrates the hypothetical result of applying EVT to a collection of $R = 1000$ measurement runs taken on an MBPTA-compliant architecture. The dashed line represents the 1-CDF derived from the observed execution times captured during the analysis phase. The continuous line represents the projection (pWCET) obtained with EVT.

Overall, MBPTA applies to resources with either time-randomized or time-deterministic behavior. MBPTA techniques compute a distribution function called probabilistic WCET (pWCET) whose tail can be cut at a probability of exceedance sufficiently low to make a negligible contribution to the overall probability of failure of the system according to the applicable safety standard.

3.4 MBPTA AND FAULTY HARDWARE

So far, in this section we have assumed fault-free hardware. However, as shown in Section 2, the use of smaller technologies increases both permanent faults caused by fabrication defects and transient fault susceptibility caused by the low charge required to alter a transistor's state.

Until very recently [20], SDTA models do not take into account any information about the permanent faults that hardware may experience. Hence, as soon as any of the hardware components (e.g., a cache line) is detected to be faulty and disabled by the fault tolerance mechanisms in place, the WCET estimates previously derived are no longer a safe upper-bound of the WCET of the application. Similarly, WCET estimates derived with MBDTA techniques or HyDTA, remain only valid under the same processor degraded mode on which measurements were taken (typically a fault-free mode).

Generally, timing analysis must be carried out in such a way that WCET estimates are valid when the system is in its operation phase. This implies that assumptions about timing during the analysis phase match or upper-bound those at operation. In this line, the timing impact of errors at operation must be factored in during analysis so that WCET estimates are still sound in the presence of faults. This requires bounding the time for error detection/recovery, fault diagnosis/reconfiguration, and the effect on WCET estimates due to the operation on top of degraded hardware. From those, as described in Section 2, the operation under degraded hardware modes is the most challenging for WCET estimation.

3.4.1 Operation on degraded caches

It is well accepted that cache memories will experience most hardware faults due to their large area and aggressive transistor scaling to provide further cache space and reduced energy consumption. Degraded cache operation occurs since some cache lines are disabled when they are regarded as faulty. This can be simply implemented with a faulty-line bit that is set for all faulty cache lines so that they cannot report a hit on an access.

With MBDTA, that deploys time-deterministic caches (e.g., modulo placement and least recently used replacement) we must consider all fault maps, that is, all combinations of faults over N total lines in the cache. In a setup with f faulty lines in a cache with N lines, this leads to a total of C fault locations

$$C = \frac{N!}{f!(N-f)!}$$

Each fault location may have different impact on execution time and hence WCET estimates. Hence, each fault location represents a cache degraded mode. As a result, under each fault location, cache analysis should be conducted to derive a WCET bound, taking as WCET bound the highest across all fault locations. This task, however, is infeasible in the general case due to its time requirements. For instance, a 32 KB 4-way 32B/line cache would have 256 sets with four lines each. If fault location within each set is irrelevant, there would be almost 3 million different fault maps. If fault location in each set matters, then there would be more than 178 million different fault maps.

Time-randomized caches, which deploy random replacement and random placement, have the key property of removing the dependence between actual addresses accessed and their location in cache. This makes MBPTA, working on time-randomized caches, insensitive to fault location. Instead of the location MBPTA cares only about the number of faulty lines. Hence, if f faults are expected in a given cache, this leads to a single degraded mode (instead of C as shown for time-deterministic caches). This is graphically illustrated in Fig. 7, where we show that the number of faults is the only relevant parameter for MBPTA, whereas their location also matters for MBDTA, thus leading to an intractable number of fault maps to consider.

MPBTA obtains end-to-end observations by disabling f cache lines. Obtained pWCET estimates are a reliable upper-bound of the program's execution time when the system is deployed and when the cache has up to f faulty lines. That is, MBPTA takes measurements from a cache with maximum degradation (f faults). This is implemented by collecting execution times on the target processor with as many

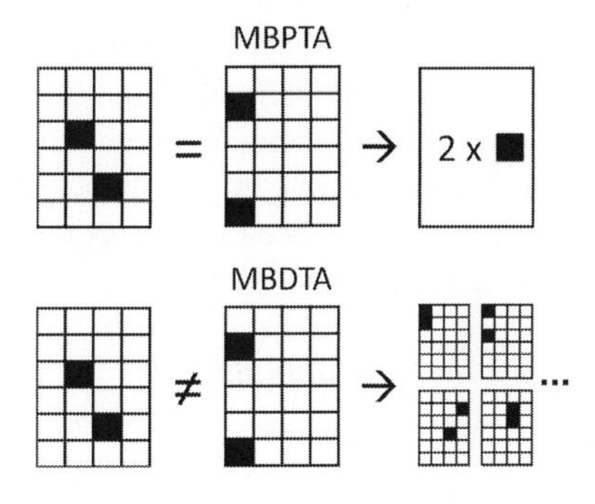

FIG. 7

Example of fault maps under MBPTA and MBDTA.

cache entries disabled as *f*. A simple hardware is required that allows disabling a number of cache entries in each cache structure.

The cache entries disabled in a fully associative cache can be either consecutive or randomly located. In a set-associative, whether several faults occur in the same cache set may impact execution time. Based on the assumption that radiation occurs in random locations in cache, the location of faulty lines can also be assumed random. Therefore, in order to collect a representative number of fault maps, each execution time measurement for MBPTA needs to occur on a random fault map. This can be achieved either by pure hardware means if hardware allows disabling a number of randomly located cache lines in a given cache memory or, with a combined hardware/software solution where hardware allows disabling specific cache lines with the appropriate instruction and software is allowed to use such hardware feature to disable some cache lines randomly chosen by software means. Note that if faults need to be tolerated, hardware support for disabling faulty cache lines needs already to be in place, as it is the case of some processors in the high-performance domain [32], and the only new requirement of this approach would be exposing this mechanism to the software.

3.4.2 Impact of anomaly detection, isolation, diagnosis, and recovery in caches

While operation on degraded caches is the most relevant concern for microcontrollers in spacecraft due to the impact of radiation, managing anomaly detection, isolation, diagnosis, and recovery in caches needs also to be accounted for in pWCET estimates. In general, these processes require a (short) fixed time to be applied. For instance, transient faults can be detected without interfering with operation by means of coding techniques (e.g., SECDED). The particular error threshold determines how many recovery actions need to be taken before isolation (i.e., reconfiguration), and so how much time to budget in the pWCET estimate to account for those recovery actions. Fault diagnosis [14] may occur in background so that if frequent errors occur in the same location the affected cache line is deemed as permanently damaged and disabled. Then, recovery, if faults are intended to lead to frequent errors, can occur immediately after the detection stage regardless of whether data were erroneous or not, thus not adding any additional delay due to faults. Alternatively, errors can be recovered only on a fault and affected instructions need to be stopped, squashed, and reexecuted, which may cost few cycles. Permanent faults may trigger a number of errors in a short period of time. Nevertheless, few recovery actions and few cycles to disable the faulty cache lead to negligible impact in the pWCET estimate, even if several transient and/or permanent faults need to be accounted for if they can occur during a single execution of the program.

3.4.3 Impact of faults in other components

Although caches are expected to account for most of the transient and permanent faults, faults can also occur in other microcontroller components with nonnegligible probability. Techniques to detect, correct, diagnose, reconfigure, and operate in degraded modes need to be devised also for those other components. Still, faults occur in random locations, so hardware mechanisms allowing disabling randomly chosen components in each run fall within the spirit of MBPTA, which naturally accounts for probabilistic events by means of measures of their impact.

Overall, we observe how MBPTA application process remains mostly unchanged to analyze deal with transient and permanent faults in caches and other components. Therefore, designing MBPTA-compliant microcontrollers is a promising approach for future spacecraft whose performance demands can only be satisfied using complex hardware implemented with smaller technology nodes.

4 CONCLUSIONS

Platform and payload software in spacecraft is subject to harsh conditions. Among those, radiation is the most distinguishing source of errors for microcontrollers in the space. The need for higher performance pushes for a shift toward more powerful microcontrollers including multiple cores, deep pipelines, and multilevel cache hierarchies. Those microcontrollers can only be implemented using smaller technology nodes. Those designs, however, are more susceptible to radiation. In particular, shifting technology toward scaled geometries increases both the probability of experiencing transient and permanent faults due to radiation from the outer space.

Frequent faults pose a serious challenge to the strict functional and timing verification processes of spacecraft systems. Therefore, adequate safety measures need to be in place to tolerate faults while still achieving the high-performance required.

We have shown how the shift toward hardware with randomization features in its timing behavior accompanied with the appropriate probabilistic and statistical analyses helps achieving the ultimate goal of deploying smaller technology nodes and high-performance features such as caches and multicores in space computing systems. Probabilistic timing analysis naturally fits the random nature of faults due to radiation and thus, it is a promising path to follow for the development of microcontrollers for future space missions.

REFERENCES

[1] M. Riera, R. Canal, J. Abella, A. Gonzalez, A detailed methodology to soft error rates in advanced technologies, in: Proceedings of the 19th Design, Automation and Test in Europe Conference, 2016.

[2] W. Ley, K. Wittmann, W. Hallmann, Handbook of Space Technology, John Wiley & Sons Ltd., Hoboken, NJ, 2009.

[3] D. Siewiorek, P. Narasimhan, Fault tolerant architectures for space and avionics, in: Joint IARP/IEEERAS/EURON & IFIP 10.4 Workshop on Dependability in Robotics and Autonomous Systems, February, 2006.

[4] K. Havelund, Monitoring the execution of spacecraft flight software, in: The Correctness, Modelling and Performance of Aerospace Systems (COMPASS) Workshop, 2009. http://compass.informatik.rwth-aachen.de/workshop.php.

[5] Gaisler Research. Processor User's Manual Version 1.0.30, XST, 2005.
[6] Cobham. GR712RC Dual-Core LEON3FT SPARC V8 Processor. 2015 User's Manual, 2015.
[7] http://microelectronics.esa.int/ngmp/.
[8] P. Roche, R. Harboe-Sorensen, Radiation Evaluation of ST Test Structures in commercial 130 nm CMOS BULK and SOI, in commercial 90 nm CMOS BULK, in commercial 65 nm CMOS BULK and SOI, Final Presentation of ESTEC, 2007. Contract No. 13528/95/NL/MV, COO-18.
[9] S. Guertin, M. White, CMOS reliability challenges—the future of commercial digital electronics and NASA, in: NEPP Electronic Technology Workshop, 2010.
[10] European Cooperation for Space Standardization, ECSS-E-ST-10-03C. Space engineering. Testing, 2012.
[11] European Cooperation for Space Standardization, ECSS-Q-ST-60-15C. Space product assurance. Radiation hardness assurance—EEE components, 2012.
[12] Cobham Gaisler, LEON3-FT SPARC V8 Processor. LEON3FT-RTAX. Data Sheet and User's Manual, v1.9, 2013.
[13] Infineon, AURIX—TriCore datasheet. Highly integrated and performance optimized 32-bit microcontrollers for automotive and industrial applications, 2012, http://www.infineon.com/.
[14] J. Abella, P. Chaparro, X. Vera, J. Carretero, A. Gonzalez, On-line failure detection and confinement in caches, in: Proceedings of the 14th International On-Line Testing Symposium (IOLTS), 2008.
[15] J. Jalle, L. Kosmidis, J. Abella, E. Quiñones, F.J. Cazorla, Bus designs for time-probabilistic multicore processors, in: Proceedings of the 17th Design, Automation and Test in Europe (DATE) Conference, 2014.
[16] G. Fernandez, J. Jalle, J. Abella, E. Quiñones, T. Vardanega, F.J. Cazorla, Resource usage templates and signatures for COTS multicore processors, in: Proceedings of the 52nd Design Automation Conference (DAC), 2015.
[17] J. Jalle, M. Fernandez, J. Abella, J. Andersson, M. Patte, L. Fossati, M. Zulianello, F.J. Cazorla, Bounding resource-contention interference in the next-generation multipurpose processor (NGMP), in: Proceedings of the 8th European Congress on Embedded Real Time Software and Systems (ERTS^2), 2016.
[18] J. Abella, J. Carretero, P. Chaparro, X. Vera, A. Gonzalez, Low Vccmin fault-tolerant cache with highly predictable performance, in: Proceedings of the 42nd International Symposium on Microarchitecture (MICRO), 2009.
[19] R. Wilhelm, J. Engblom, A. Ermedahl, N. Holsti, S. Thesing, D.B. Whalley, G. Bernat, C. Ferdinand, R. Heckmann, T. Mitra, F. Mueller, I. Puaut, P.P. Puschner, J. Staschulat, P. Stenstrom, The worst-case execution-time problem—overview of methods and survey of tools, ACM Trans. Embed. Comput. Syst. 7 (3) (2008).
[20] D. Hardy, I. Puaut, Static probabilistic worst case execution time estimation for architectures with faulty instruction caches, Real-Time Syst. J. 51 (2) (2015).
[21] J. Abella, E. Quinones, F.J. Cazorla, Y. Sazeides, M. Valero, RVC-based time-predictable faulty caches for safety-critical systems, in: 17th International On-Line Testing Symposium (IOLTS), 2011.
[22] J. Abella, E. Quinones, F.J. Cazorla, Y. Sazeides, M. Valero, RVC: a mechanism for time-analyzable real-time processors with faulty caches, in: 6th International Conference on High-Performance Embedded Architectures and Compilers, 2011.

[23] M. Slijepcevic, L. Kosmidis, J. Abella, E. Quinones, F.J. Cazorla, DTM: degraded test mode for fault-aware probabilistic timing analysis, in: 25th IEEE Euromicro Conference on Real-Time Systems, 2013.
[24] M. Slijepcevic, L. Kosmidis, J. Abella, E. Quinones, F.J. Cazorla, Timing verification of fault-tolerant chips for safety-critical applications in harsh environments, IEEE Micro 34 (6) (2014) (Special Series on Harsh Chips).
[25] E. Mezzetti, N. Holsti, A. Colin, G. Bernat, T. Vardanega, Attacking the sources of unpredictability in the instruction cache behavior, in: 16th International Conference on Real-Time and Network Systems, 2008.
[26] F.J. Cazorla, E. Quiñones, T. Vardanega, L. Cucu, B. Triquet, G. Bernat, E. Berger, J. Abella, F. Wartel, M. Houston, L. Santinelli, L. Kosmidis, C. Lo, D. Maxim, PROARTIS: probabilistically analysable real-time systems, ACM Trans. Embed. Comput. Syst. 12 (2) (2013) (Special issue on Probabilistic Embedded Computing).
[27] L. Kosmidis, J. Abella, E. Quinones, F. Cazorla, Cache design for probabilistic real-time systems, in: The Design, Automation, and Test in Europe (DATE) Conference, 2013.
[28] J. Abella, C. Hernandez, E. Quiñones, F.J. Cazorla, P. Ryan Conmy, M. Azkarate-askasua, J. Perez, E. Mezzetti, T. Vardanega, WCET analysis methods: pitfalls and challenges on their trustworthiness, in: 10th IEEE International Symposium on Industrial Embedded Systems (SIES), 2015.
[29] RVS (Rapita Verification Suite), https://www.rapitasystems.com/downloads/rvs_brochure.
[30] C. Hernandez, J. Abella, F.J. Cazorla, J. Andersson, A. Gianarro, Towards making a LEON3 multicore compatible with probabilistic timing analysis, in: 0th Data Systems in Aerospace Conference, 2015.
[31] Stuart Coles. An Introduction to Statistical Modeling of Extreme Values, Springer-Verlag London, edition 1, 2001.
[32] C. McNairy, J. Mayfield, Montecito error protection and mitigation, in: HPCRI'05: 1st Workshop on High Performance Computing Reliability Issues, in Conjunction With HPCA'05, 2005.

CHAPTER

Resilience in next-generation embedded systems*

10

E. Cheng, S. Mitra

Stanford University, Stanford, CA, United States

1 INTRODUCTION

This chapter addresses the *cross-layer resilience challenge* for designing rugged embedded systems: given a set of resilience techniques at various abstraction layers (circuit, logic, architecture, software, algorithm), how does one protect a given design from radiation-induced soft errors using (perhaps) a *combination* of these techniques, *across multiple abstraction layers*, such that overall soft error resilience targets are met at minimal costs (energy, power, execution time, area)? Specific soft error resilience targets include: *silent data corruption* (*SDC*), where an error causes the system to output an incorrect result without error indication; and *detected but uncorrected error* (*DUE*), where an error is detected (e.g., by a resilience technique or a system crash or hang) but is not recovered automatically without user intervention.

The need for *cross-layer resilience*, where multiple error resilience techniques from different layers of the system stack cooperate to achieve cost-effective error resilience, is articulated in several publications (e.g., Refs. [1–7]).

There are numerous publications on error resilience techniques, many of which span multiple abstraction layers. These publications mostly describe specific implementations. Examples include structural integrity checking [8] and its derivatives (mostly spanning architecture and software layers) or the combined use of circuit hardening, error detection (e.g., using logic parity checking and residue codes), and instruction-level retry [9–11] (spanning circuit, logic, and architecture layers). Cross-layer resilience implementations in commercial systems are often based on "designer experience" or "historical practice." There exists no comprehensive

*The material presented in this chapter has been re-used from the authors' previous article with permission. E. Cheng, et al., CLEAR: cross-layer exploration for architecting resilience—combining hardware and software techniques to tolerate soft errors in processor cores, in: Proceedings of 53rd Annual Design Automation Conference, © 2016 ACM, Inc. Reprinted with permission. http://doi.acm.org/10.1145/2897937.2897996.

Rugged Embedded Systems. http://dx.doi.org/10.1016/B978-0-12-802459-1.00010-5

framework to systematically address the cross-layer resilience challenge. Creating such a framework is difficult. It must encompass the entire design flow end-to-end, from comprehensive and thorough analysis of various combinations of error resilience techniques all the way to layout-level implementations, such that one can (automatically) determine which resilience technique or combination of techniques (either at the same abstraction layer or across different abstraction layers) should be chosen. However, such a framework is essential in order to answer important cross-layer resilience questions such as:

1. Is cross-layer resilience the best approach for achieving a given resilience target at low cost?
2. Are all cross-layer solutions equally cost-effective? If not, which cross-layer solutions are the best?
3. How do cross-layer choices change depending on application-level energy, latency, and area constraints?
4. How can one create a cross-layer resilience solution that is cost-effective across a wide variety of application workloads?
5. Are there general guidelines for new error resilience techniques to be cost-effective?

CLEAR (cross-layer exploration for architecting resilience) is a first of its kind framework, which addresses the cross-layer resilience challenge. In this chapter, the focus is on the use of CLEAR for harsh-environment-capable embedded systems that operate in the presence of radiation-induced soft errors in terrestrial settings; however, other error sources (voltage noise and circuit-aging) may be incorporated into CLEAR as well.

Although the soft error rate of an SRAM cell or a flip-flop stays roughly constant or even decreases over technology generations, the system-level soft error rate increases with increased integration [12–14]. Moreover, soft error rates can increase when lower supply voltages are used to improve energy efficiency [15,16]. The chapter focuses on *flip-flop soft errors* because design techniques to protect them are generally expensive. Coding techniques are routinely used for protecting on-chip memories. Combinational logic circuits are significantly less susceptible to soft errors and do not pose a concern [14,17]. Both single-event upsets (*SEUs*) and single-event multiple upsets (*SEMUs*) [16,18] are considered. While CLEAR can address soft errors in various digital components of a complex System-on-a-Chip (including uncore components [19] and hardware accelerators), a detailed analysis of soft errors in all these components is beyond the scope of this chapter. Hence, this chapter will focus on soft errors in processor cores.

To demonstrate the effectiveness and practicality of CLEAR, an exploration of 586 cross-layer combinations using ten representative error detection/correction techniques and four hardware error recovery techniques is conducted. These techniques span various layers of the system stack: circuit, logic, architecture, software, and algorithm (Fig. 1). This extensive cross-layer exploration encompasses over nine million flip-flop soft error injections into two diverse processor core architectures

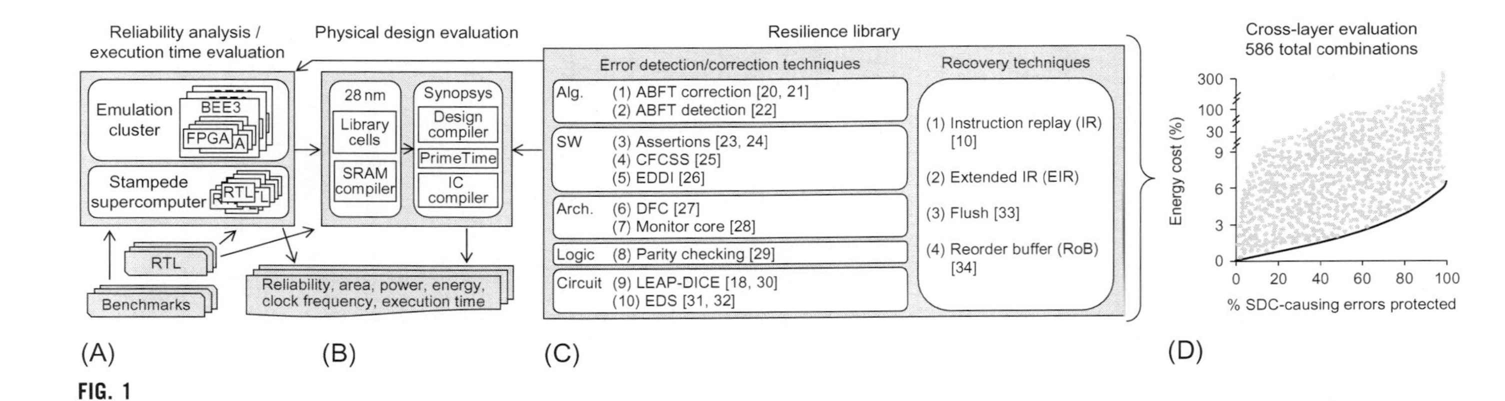

FIG. 1

CLEAR Framework: (A) BEE3 emulation cluster/Stampede supercomputer injects over nine million errors into two diverse processor architectures running 18 full-length application benchmarks. (B) Accurate physical design evaluation accounts for resilience overheads. (C)Comprehensive resilience library consisting of ten error detection/correction techniques + four hardware error recovery techniques. (D) Example illustrating thorough exploration of 586 cross-layer combinations with varying energy costs vs percentage of SDC-causing errors protected.

Table 1 Processor Designs Studied

Core	Design	Description	Clk. Freq.	Error Injections	Instructions Per Cycle
InO	Leon3 [35]	Simple, in-order (1250 flip-flops)	2.0 GHz	5.9 million	0.4
OoO	IVM [36]	Complex, superscalar, out-of-order (13,819 flip-flops)	600 MHz	3.5 million	1.3

(Table 1): a simple, in-order SPARC Leon3 core (*InO-core*) and a complex, superscalar, out-of-order Alpha IVM core (*OoO-core*), across 18 benchmarks: SPECINT2000 [37] and DARPA PERFECT [38]. Such extensive exploration enables conclusive answers to the above cross-layer resilience questions:

1. For a wide range of error resilience targets, optimized cross-layer combinations can provide low-cost solutions for soft errors.
2. Not all cross-layer solutions are cost-effective.
   a. For general-purpose processor cores, a carefully optimized combination of selective circuit-level hardening, logic-level parity checking, and microarchitectural recovery provides a highly effective cross-layer resilience solution. For example, a 50 × SDC improvement (defined in Section 2.1) is achieved at 2.1% and 6.1% energy costs for the OoO- and InO-cores, respectively. The use of selective circuit-level hardening and logic-level parity checking is guided by a thorough analysis of the effects of soft errors on application benchmarks.
   b. When the application space can be restricted to matrix operations, a cross-layer combination of algorithm-based fault tolerance (ABFT) correction, selective circuit-level hardening, logic-level parity checking, and microarchitectural recovery can be highly effective. For example, a 50 × SDC improvement is achieved at 1.9% and 3.1% energy costs for the OoO- and InO-cores, respectively. But, this approach may not be practical for general-purpose processor cores targeting general applications.
   c. Selective circuit-level hardening, guided by a thorough analysis of the effects of soft errors on application benchmarks, provides a highly effective soft error resilience approach. For example, a 50 × SDC improvement is achieved at 3.1% and 7.3% energy costs for the OoO- and InO-cores, respectively.
3. The above conclusions about cost-effective soft error resilience techniques largely hold across various application characteristics (e.g., latency constraints despite errors in soft real-time applications).
4. Selective circuit-level hardening (and logic-level parity checking) techniques are guided by the analysis of the effects of soft errors on application benchmarks. Hence, one must address the challenge of potential mismatch between

application benchmarks vs applications in the field, especially when targeting high degrees of resilience (e.g., 10 × or more SDC improvement). This challenge is overcome using various flavors of circuit-level hardening techniques (details in Section 4).

5. Cost-effective resilience approaches discussed earlier provide bounds that new soft error resilience techniques must achieve to be competitive. It is, however, crucial that the benefits and costs of new techniques are evaluated thoroughly and correctly.

2 CLEAR FRAMEWORK

Fig. 1 gives an overview of the CLEAR framework. Individual components of the framework are discussed later.

2.1 RELIABILITY ANALYSIS

CLEAR is not merely an error rate projection tool; rather, reliability analysis is a component of the overall CLEAR framework that helps enable the design of rugged systems.

Flip-flop soft error injections are used for reliability analysis with respect to radiation-induced soft errors. This is because radiation test results confirm that injection of single bit-flips into flip-flops closely models soft error behaviors in actual systems [39,40]. Furthermore, flip-flop-level error injection is crucial since naïve high-level error injections can be highly inaccurate [41]. For individual flip-flops, both SEUs and SEMUs manifest as single-bit errors. To ensure this assumption holds for both the baseline and resilient designs, it is crucial to make use of SEMU-tolerant circuit hardening and layout implementations, as is demonstrated here.

The following demonstration is based on injection of over nine million flip-flop soft errors into the register-transfer level (RTL) of the processor designs using three BEE3 FPGA emulation systems and also using mixed-mode simulations on the Stampede supercomputer (TACC at The University of Texas at Austin) (similar to Refs. [36,41–43]). This ensures that error injection results have less than a 0.1% margin of error with a 95% confidence interval per benchmark. Errors are injected uniformly into all flip-flops and application regions, to mimic real-world scenarios.

The SPECINT2000 [37] and DARPA PERFECT [38] benchmark suites are used for evaluation.[1] The PERFECT suite complements SPEC by adding applications targeting signal and image processing domains. Benchmarks are run in their entirety.

Flip-flop soft errors can result in the following outcomes [36,40,41,44,45]: *Vanished*—normal termination and output files match error-free runs, *output mismatch (OMM)*—normal termination, but output files are different from error-free runs, *unexpected termination (UT)*—program terminates abnormally, *Hang*—no

[1] 11 SPEC/7 PERFECT benchmarks for InO-cores and 8 SPEC/3 PERFECT for OoO-cores (missing benchmarks contain floating-point instructions not executable by the OoO-core RTL model).

termination or output within 2 × the nominal execution time, *error detection (ED)*—an employed resilience technique flags an error, but the error is not recovered using a hardware recovery mechanism.

Using the earlier outcomes, any error that results in OMM causes SDC (i.e., an *SDC-causing error*). Any error that results in UT, Hang, or ED causes DUE (i.e., a *DUE-causing error*). Note that, there are no ED outcomes if no error detection technique is employed. The resilience of a protected (new) design compared to an unprotected (original, baseline) design can be defined in terms of *SDC improvement* (Eq. 1a) or *DUE improvement* (Eq. 1b). The susceptibility of flip-flops to soft errors is assumed to be uniform across all flip-flops in the design (but this parameter is adjustable in our framework).

Resilience techniques that increase the execution time of an application or add additional hardware also increase the susceptibility of the design to soft errors. To accurately account for this situation, one needs to calculate, based on Ref. [46], a correction factor γ (where $\gamma \geq 1$), which is applied to ensure a fair and accurate comparison for all techniques. It is common in research literature to consider $\gamma = 1$ (which is optimistic). Although all the conclusions in this chapter would hold for $\gamma = 1$, this chapter provides a more accurate analysis by considering and reporting results using true γ values. Take, for instance, the monitor core technique; in a representative implementation, the technique increases the number of flip-flops in a resilient OoO-core by 38%. These extra flip-flops become additional locations for soft errors to occur. This results in a γ correction of 1.38 in order to account for the increased susceptibility of the design to soft errors. Techniques that increase execution time have a similar impact. For example, control flow checking by software signatures (CFCSS) incurs a 40.6% execution time impact; a corresponding γ correction of 1.41. A technique such as data flow checking (DFC), which increases flip-flop count (20%) and execution time (6.2%), would need a γ correction of 1.28 (1.2 × 1.062), since the impact is multiplicative (increased flip-flop count over an increased duration). The γ correction factor accounts for these increased susceptibilities for fair and accurate comparisons of all resilience techniques considered [46]. SDC and DUE improvements with $\gamma = 1$ can be back-calculated by multiplying the reported γ value in Table 3 and do not change the conclusions.

$$\text{SDC improvement} = \frac{(\text{original OMM count})}{(\text{new OMM count})} \times \gamma^{-1} \tag{1a}$$

$$\text{DUE improvement} = \frac{(\text{original (UT + Hang) count})}{(\text{new (UT + Hang + ED) count})} \times \gamma^{-1} \tag{1b}$$

Reporting SDC and DUE improvements allows results to be agnostic to absolute error rates. Although this chapter describes the use of error injection-driven reliability analysis, the modular nature of CLEAR allows other approaches to be swapped in as appropriate (e.g., error injection analysis could be substituted with techniques like those in Ref. [47], once they are properly validated).

Table 2 Distribution of Flip-Flops With Errors Resulting in SDC and/or DUE Over All Benchmarks Studied

Core	% FFs With SDC-Causing Errors	% FFs With DUE-Causing Errors	% FFs With Both SDC- and DUE-Causing Errors
InO	60.1	78.3	81.2
OoO	35.7	52.1	61

As shown in Table 2, across the set of applications, not all flip-flops will have errors that result in SDC or DUE (errors in 19% of flip-flops in the InO-core and 39% of flip-flops in the OoO-core always vanish regardless of the application). This phenomenon has been documented in the literature [48] and is due to the fact that errors that impact certain structures (e.g., branch predictor, trap status registers, etc.) have no effect on program execution or correctness. Additionally, this means that resilience techniques would not normally need to be applied to these flip-flops. However, for completeness, this chapter also reports design points which would achieve the maximum improvement possible, where resilience is added to every single flip-flop (including those with errors that always vanish). This maximum improvement point provides an upper bound for cost (given the possibility that for a future application, a flip-flop that currently has errors that always vanish may encounter an SDC- or DUE-causing error).

Error Detection Latency (the time elapsed from when an error occurs in the processor to when a resilience technique detects the error) is also an important aspect to consider. An end-to-end reliable system must not only detect errors, but also recover from these detected errors. Long detection latencies impact the amount of computation that needs to be recovered and can also limit the types of recovery that are capable of recovering the detected error (Section 2.4).

2.2 EXECUTION TIME

Execution time is estimated using FPGA emulation and RTL simulation. Applications are run to completion to accurately capture the execution time of an unprotected design. The error-free execution time impact associated with resilience techniques at the architecture, software, and algorithm levels are reported. For resilience techniques at the circuit and logic levels, the CLEAR design methodology maintains the same clock speed as the unprotected design.

2.3 PHYSICAL DESIGN

Synopsys design tools (Design Compiler, IC compiler, and Primetime) [49] with a commercial 28 nm technology library (with corresponding SRAM compiler) are used to perform synthesis, place-and-route, and power analysis. *Synthesis and place-and-route* (*SP&R*) was run for all configurations of the design (before and after adding resilience techniques) to ensure all constraints of the original design (e.g., timing and physical design) were met for the resilient designs. Design tools often introduce artifacts (e.g., slight variations in the final design over multiple SP&R runs) that impact the final design characteristics (e.g., area, power). These artifacts can be caused by small variations in the RTL or optimization heuristics, for example. To account for these artifacts, separate resilient designs based on error injection results for each individual application benchmark are generated. SP&R was then performed for each of these designs, and the reported design characteristics were averaged to minimize the artifacts. For example, for each of the 18 application benchmarks, a separate resilient design that achieves a $50\times$ SDC improvement using LEAP-DICE only is created. The costs to achieve this improvement are reported by averaging across the 18 designs. Relative standard deviation (i.e., standard deviation/mean) across all experiments ranges from 0.6% to 3.1%. Finally, it is important to note that all layouts created during physical design are carefully generated in order to mitigate the impact of SEMUs (as explained in Section 2.4).

2.4 RESILIENCE LIBRARY

This chapter considers ten carefully chosen error detection and correction techniques together with four hardware error recovery techniques. These techniques largely cover the space of existing soft error resilience techniques. The characteristics (e.g., costs, resilience improvement, etc.) of each resilience technique when used as a *standalone solution* (e.g., an error detection/correction technique by itself or, optionally, in conjunction with a recovery technique) are presented in Table 3.

Circuit: The *hardened flip-flops* (LEAP-DICE, light hardened LEAP (LHL), LEAP-ctrl) in Table 4 are designed to tolerate both SEUs and SEMUs at both nominal and near-threshold operating voltages [18,30]. SEMUs especially impact circuit techniques since a single strike affects multiple nodes within a flip-flop. Thus, these specially designed hardened flip-flops, which tolerate SEMUs through charge cancelation, are required. These hardened flip-flops have been experimentally validated using radiation experiments on test chips fabricated in 90, 45, 40, 32, 28, 20, and 14 nm nodes in both bulk and SOI technologies and can be incorporated into standard cell libraries (i.e., standard cell design guidelines are satisfied) [18,30,50–53]. The LEAP-ctrl flip-flop is a special design, which can operate in resilient (high resilience, high power) and economy (low resilience, low power) modes. It is useful in situations where a software or algorithm technique only provides protection when running specific applications and thus, selectively enabling low-level hardware resilience when the former techniques are unavailable may be beneficial. While *error detection sequential* (*EDS*) [31,32] was originally designed to detect timing errors, it can be used to detect flip-flop soft errors as well. While EDS incurs less overhead at the individual flip-flop level vs LEAP-DICE, for example, EDS requires delay buffers to ensure minimum hold constraints, aggregation and routing of error detection signals to an output (or recovery module), and a recovery mechanism to correct

Table 3 Individual Resilience Techniques: Costs and Improvements When Implemented as a Standalone Solution

Layer	Technique		Area Cost (%)	Power Cost (%)	Energy Cost (%)	Exec. Time Impact (%)	Avg. SDC Improve	Avg. DUE Improve	False Positive (%)	Detection Latency	γ
Circuit[2]	LEAP-DICE (no additional recovery needed)	InO	0–9.3	0–22.4	0–22.4	0	1×–5000×[3]	1×–5000×[3]	0	n/a	1
		OoO	0–6.5	0–9.4	0–9.4						
	EDS (without recovery—unconstrained)	InO	0–10.7	0–22.9	0–22.9	0	1×–100,000×[3]	0.1×[3]–1×	0	1 cycle	1
		OoO	0–12.2	0–11.5	0–11.5						
	EDS (with IR recovery)	InO	0–16.7	0–43.9	0–43.9	0	1×–100,000×[3]	1×–100,000×[3]	0	1 cycle	1.4
		OoO	0–12.3	0–11.6	0–11.6						1.06
Logic[2]	Parity (without recovery—unconstrained)	InO	0–10.9	0–23.1	0–23.1	0	1×–100,000×[3]	0.1×[5]–1×	0	1 cycle	1
		OoO	0–14.1	0–13.6	0–13.6						
	Parity (with IR recovery)	InO	0–26.9	0–44	0–44	0	1×–100,000×[3]	1×–100,000×[3]	0	1 cycle	1.4
		OoO	0–14.2	0–13.7	0–13.7						1.06
Arch.	DFC (without recovery—unconstrained)	InO	3	1	7.3	6.2	1.2×	0.5×	0	15 cycles	1.28
		OoO	0.2	0.1	7.2	7.1					1.09
	DFC (with EIR recovery)	InO	37	33	41.2	6.2	1.2×	1.4×	0	15 cycles	1.48
		OoO	0.4	0.2	7.3	7.1					1.14
	Monitor core (with RoB recovery)	OoO	9	16.3	16.3	0	19×	15×	0	128 cycles	1.38

Continued

Table 3 Individual Resilience Techniques: Costs and Improvements When Implemented as a Standalone Solution—cont'd

Layer	Technique		Area Cost (%)	Power Cost (%)	Energy Cost (%)	Exec. Time Impact (%)	Avg. SDC Improve	Avg. DUE Improve	False Positive (%)	Detection Latency	γ
Soft-ware[4]	Software assertions for general-purpose processors (without recovery—unconstrained)	InO	0	0	15.6	15.6[5]	1.5×[6]	0.6×	0.003	9.3 M cycles[7]	1.16
	CFCSS (without recovery—unconstrained)	InO	0	0	40.6	40.6	1.5×	0.5×	0	6.2 M cycles[7]	1.41
	EDDI (without recovery—unconstrained)	InO	0	0	110	110	37.8×[8]	0.3×	0	287 K cycles[7]	2.1
Alg.	ABFT correction (no additional recovery needed)	InO OoO	0	0	1.4	1.4	4.3×	1.2×	0	n/a	1.01
	ABFT detection (without recovery—unconstrained)	InO OoO	0	0	24	1–56.9[9]	3.5×	0.5×	0	9.6 M cycles[7]	1.24

Table 4 Resilient Flip-Flops

Type	SER	Area	Power	Delay	Energy
Baseline	1	1	1	1	1
LHL	2.5×10^{-1}	1.2	1.1	1.2	1.3
LEAP-DICE	2.0×10^{-4}	2.0	1.8	1	1.8
LEAP-ctrl (economy mode)	1	3.1	1.2	1	1.2
LEAP-ctrl (resilient mode)	2.0×10^{-4}	3.1	2.2	1	2.2
EDS[10]	~100% detect	1.5	1.4	1	1.4

SER, *soft error rate;* LHL, *light hardened LEAP;* EDS, *error detection sequential.*

detected errors. These factors can significantly increase the overall costs for implementing a resilient design utilizing EDS (Table 17).[2,3,4,5,6,7,8,9,10]

Logic: *Parity checking* provides error detection by checking flip-flop inputs and outputs [29]. The design heuristics presented in this chapter reduce the cost of parity while also ensuring that clock frequency is maintained as in the original design (by varying the number of flip-flops checked together, grouping flip-flops by timing slack, pipelining parity checker logic, etc.) Naïve implementations of parity checking can otherwise degrade design frequency by up to 200 MHz (20%) or increase energy cost by 80% on the InO-core. SEMUs are minimized through layouts that

[2]Circuit and logic techniques have tunable costs/resilience (e.g., for InO-cores, 5 × SDC improvement using LEAP-DICE is achieved at 4.3% energy cost while 50 × SDC improvement is achieved at 7.3% energy cost). This is achievable through selective insertion guided by error injection using application benchmarks.

[3]Maximum improvement reported is achieved by protecting every single flip-flop in the design.

[4]Software techniques are generated for InO-cores only since the LLVM compiler no longer supports the Alpha architecture.

[5]Some software assertions for general-purpose processors (e.g., Ref. [23]) suffer from false positives (i.e., an error is reported during an error-free run). The execution time impact reported discounts the impact of false positives.

[6]Improvements differ from previous publications that injected errors into architectural registers. Cho et al. [41] demonstrated such injections can be highly inaccurate; the CLEAR methodology presented in this chapter used highly accurate flip-flop-level error injections.

[7]Actual detection latency for software and algorithm techniques may be shorter in practice. On the emulation platforms, measured detection latencies include the time to trap and cleanly exit execution (on the order of a few thousand cycles).

[8]Results for EDDI with store-readback [54] are reported. Without this enhancement, EDDI provides 3.3 × SDC/0.4 × DUE improvement.

[9]Execution time impact for ABFT detection can be high since error detection checks may require computationally expensive calculations.

[10]For EDS, the costs are listed for the flip-flop only. Error signal routing and delay buffers (included in Table 3) increase overall cost [32].

ensure a minimum spacing (the size of one flip-flop) between flip-flops checked by the same parity checker. This ensures that only one flip-flop, in a group of flip-flops checked by the same parity checker, will encounter an upset due to a single strike in our 28 nm technology in terrestrial environments [55]. Although a single strike could impact multiple flip-flops, since these flip-flops are checked by different checkers, the upsets will be detected. Since this absolute minimum spacing will remain constant, the relative spacing required between flip-flops will increase at smaller technology nodes, which may exacerbate the difficulty of implementation. Minimum spacing is enforced by applying design constraints during the layout stage. This constraint is important because even in large designs, flip-flops will still tend to be placed very close to one another. Table 5 shows the distribution of distances that each flip-flop has to its next nearest neighbor in a baseline design (this does not correspond to the spacing between flip-flops checked by the same logic parity checker). As shown, the majority of flip-flops are actually placed such that they would be susceptible to an SEMU. After applying parity checking, it is evident that no flip-flop, within a group checked by the same parity checker, is placed such that it will be vulnerable to an SEMU (Table 6).

Logic parity is implemented using an XOR-tree-based predictor and checker, which detects flip-flops soft errors. This implementation differs from logic parity prediction, which also targets errors inside combinational logic [56]. XOR-tree logic parity is sufficient for detecting flip-flop soft errors (with the minimum spacing constraint applied). "Pipelining" in the predictor tree (Fig. 2) may be required to ensure 0% clock period impact. The following heuristics for forming *parity groups* (the specific flip-flops that are checked together) to minimize cost of parity (cost comparisons in Table 7) were evaluated:

1. Parity group size: flip-flops are clustered into a constant power of 2-sized group, which amortizes the parity logic cost by allowing the use of full binary trees at the predictor and checker. The last set of flip-flops will consist of modulo "group size" of flip-flops.

Table 5 Distribution of Spacing Between a Flip-Flop and Its Nearest Neighbor in a Baseline (Original, Unprotected) Design

Distance	InO-Core (%)	OoO-Core (%)
<1 flip-flop length away (i.e., flip-flops are adjacent and vulnerable to a SEMU)	65.2	42.2
1–2 flip-flop lengths away	30	30.6
2–3 flip-flop lengths away	3.7	18.4
3–4 flip-flop lengths away	0.6	3.5
>4 flip-flop lengths away	0.5	5.3

Table 6 Distribution of Spacing Between a Flip-Flop and Its Nearest Neighbor in the Same Parity Group (i.e., Minimum Distance Between Flip-Flops Checked by the Same Parity Checker)

Distance	InO-Core (%)	OoO-Core (%)
<1 flip-flop length away (i.e., flip-flops are adjacent and vulnerable to a SEMU)	0	0
1–2 flip-flop lengths away	7.8	8.8
2–3 flip-flop lengths away	5.3	10.6
3–4 flip-flop lengths away	3.4	18.3
>4 flip-flop lengths away	83.3	62.2
Average distance	4.4 flip-flops	12.8 flip-flops

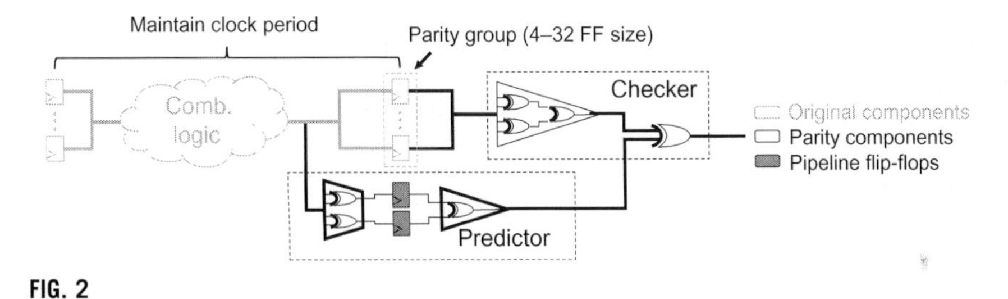

FIG. 2

"Pipelined" logic parity.

Table 7 Comparison of Heuristics for "Pipelined" Logic Parity Implementations to Protect All Flip-Flops on the InO-Core

Heuristic	Area Cost (%)	Power Cost (%)	Energy Cost (%)
Vulnerability (4-bit parity group)	15.2	42	42
Vulnerability (8-bit parity group)	13.4	29.8	29.8
Vulnerability (16-bit parity group)	13.3	27.9	27.9
Vulnerability (32-bit parity group)	14.6	35.3	35.3
Locality (16-bit parity group)	13.4	29.4	29.4
Timing (16-bit parity group)	11.5	26.8	26.8
Optimized (16-/32-bit groups)	10.9	23.1	23.1

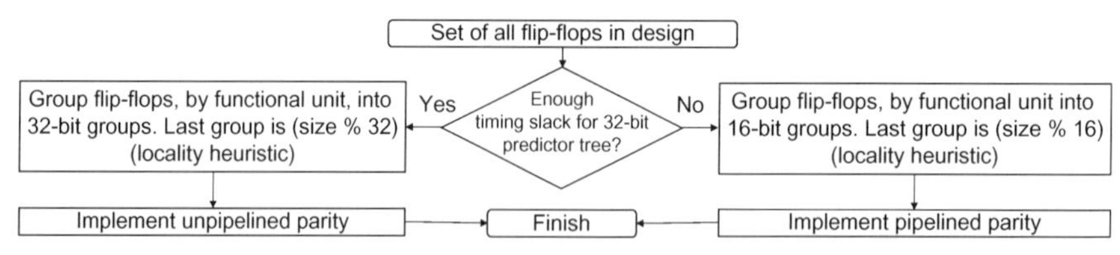

FIG. 3

Logic parity heuristic for low-cost parity implementation. 32-bit unpipelined parity and 16-bit pipelined parity were experimentally determined to be the lowest cost configurations.

2. Vulnerability: flip-flops are sorted by decreasing susceptibility to errors causing SDC or DUE and grouped into a constant power of 2-sized group. The last set of flip-flops will consist of modulo "group size" of flip-flops.
3. Locality: flip-flops are grouped by their location in the layout, in which flip-flops in the same functional unit are grouped together to help reduce wire routing for the predictor and checker logic. A constant power of 2-sized groups is formed with the last group consisting of modulo "group size" of flip-flops.
4. Timing: flip-flops are sorted based on their available timing path slack and grouped into a constant power of 2-sized group. The last set of flip-flops will consist of modulo "group size" of flip-flops.
5. Optimized: Fig. 3 describes our heuristic. This solution is the most optimized and is the configuration used to report overhead values.

When unpipelined parity can be used, it is better to use larger-sized groups (e.g., 32-bit groups) in order to amortize the additional predictor/checker logic to the number of flip-flops protected. However, when pipelined parity is required, the use of 16-bit groups was found to be a good option. This is because beyond 16-bits, additional pipeline flip-flops begin to dominate costs. These factors have driven the implementation of the previously described heuristics.

Architecture: The implementation of *DFC*, which checks static dataflow graphs, presented in this chapter includes *control flow checking*, which checks static control flow graphs. This combination checker resembles that of Ref. [27], which is also similar to the checker in Ref. [8].

Compiler optimization allows for static signatures required by the checkers to be embedded into unused delay slots in the software, thereby reducing execution time overhead by 13%.

Table 8 helps explain why DFC is unable to provide high SDC and DUE improvement. Of flip-flops that have errors that result in SDCs and DUEs (Section 2.1), DFC checkers detect SDCs and DUEs in less than 68% of these flip-flops (these 68% of flip-flops are distributed across all pipeline stages). For these 68% of flip-flops, on average, DFC detects less than 40% of the errors that result in SDCs or DUEs. This is because not all errors that result in an SDC or DUE will corrupt the dataflow or control flow signatures checked by the technique (e.g., register contents are corrupted and written out to a file, but the executed instructions remain unchanged).

Table 8 DFC Error Coverage

	InO		OoO	
	SDC	**DUE**	**SDC**	**DUE**
% flip-flops with a SDC-/DUE-causing error that are detected by DFC	57	68	65	66
% of SDC-/DUE-causing errors detected (average per FF that is protected by DFC)	30	30	29	40
Overall % of SDC-/DUE-causing errors detected (for all flip-flops in the design)	15.9	27	19.3	30
Resulting improvement (Eqs. 1a, 1b)	1.2×	1.4×	1.2×	1.4×

The combination of these factors means DFC is only detecting ~30% of SDCs or DUEs; thus, the technique provides low resilience improvement. These results are consistent with previously published data (detection of ~16% of nonvanished errors) on the effectiveness of DFC checkers in simple cores [27].

Monitor cores are checker cores that validate instructions executed by the main core (e.g., Refs. [8,28]). In this chapter, monitor cores similar to those in Ref. [28] are analyzed. For InO-cores, the size of the monitor core is of the same order as the main core, and hence, excluded from this study. For OoO-cores, the simpler monitor core can have lower throughput compared to the main core and thus stall the main core. It was confirmed (via instructions per cycle (IPC) estimation) that the monitor core implementation used does not stall the main core (Table 9).

Software: *Software assertions for general-purpose processors* check program variables to detect errors. This chapter combines assertions from Refs. [23,24] to check both data and control variables to maximize error coverage. Checks for data variables (e.g., end result) are added via compiler transformations using training inputs to determine the valid range of values for these variables (e.g., likely program invariants). Since such assertion checks are added based on training inputs, it is possible to encounter false positives, where an error is reported in an error-free run. This false positive rate was determined by training the assertions using representative inputs. However, final analysis is performed by incorporating the input data used during evaluation into the training step in order to give the technique the best possible benefit and to eliminate the occurrence of false positives. Checks for control

Table 9 Monitor Core vs Main Core

Design	Clk. Freq.	Average IPC
OoO-core	600 MHz	1.3
Monitor core	2 GHz	0.7

IPC, *instructions per cycle.*

Table 10 Comparison of Assertions Checking Data (e.g., End Result) vs Control (e.g., Loop Index) Variables

	Data Variable Check	Control Variable Check	Combined Check
Execution time impact	12.1%	3.5%	15.6%
SDC improvement	1.5 ×	1.1 ×	1.5 ×
DUE improvement	0.7 ×	0.9 ×	0.6 ×
False positive rate	0.003%	0%	0.003%

variables (e.g., loop index, stack pointer, array address) are determined using application profiling and are manually added in the assembly code.

Table 10 provides the breakdown for the contribution to cost, improvement, and false positives resulting from assertions checking data variables [23] vs those checking control variables [24]. Table 11 demonstrates the importance of evaluating resilience techniques using accurate injection [41].[11] Depending on the particular error injection model used, SDC improvement could be overestimated for one benchmark and underestimated for another. For instance, using inaccurate architecture register error injection (regU), one would be led to believe that software assertions provide 3 × the SDC improvement than they do in reality (e.g., when evaluated using flip-flop-level error injection).

In order to pinpoint the sources of inaccuracy between the actual improvement rates that were determined using accurate flip-flop-level error injection vs those published in the literature, error injection campaigns at other levels of abstraction (architecture

Table 11 Comparison of SDC Improvement and Detection for Assertions When Injecting Errors at Various Levels

App.[11]	Flip-Flop (Ground Truth)	Register Uniform (regU)	Register Write (regW)	Program Variable Uniform (varU)	Program Variable Write (varW)
bzip2	1.8 ×	1.6 ×	1.1 ×	1.9 ×	1.5 ×
Crafty	0.5 ×	0.3 ×	0.5 ×	0.7 ×	1.1 ×
gzip	2 ×	19.3 ×	1 ×	1.6 ×	1.1 ×
mcf	1.1 ×	1.3 ×	0.9 ×	1 ×	1.8 ×
Parser	2.4 ×	1.7 ×	1 ×	2.4 ×	2 ×
Avg.	1.6 ×	4.8 ×	0.9 ×	1.5 ×	1.5 ×

[11]The same SPEC applications evaluated in Ref. [23] were studied in this chapter.

register and program variable) were conducted. However, even then, there was an inability to exactly reproduce previously published improvement rates. Some additional differences in the architecture and program variable injection methodology used in this chapter compared to the published methodology may account for this discrepancy:

1. The architecture register and program variable evaluations used in this chapter were conducted on a SPARCv8 in-order design rather than a SPARCv9 out-of-order design.
2. The architecture register and program variable methodology used in this chapter injects errors uniformly into all program instructions while previous publications chose to only inject into integer instructions of floating-point benchmarks.
3. The architecture register and program variable methodology used in this chapter injects errors uniformly over the full application rather than injecting only into the core of the application during computation.
4. Since the architecture register and program variable methodology used in this chapter injects errors uniformly into all possible error candidates (e.g., all cycles and targets), the calculated improvement covers the entire design. Previous publications calculated improvement over the limited subset of error candidates (out of all possible error candidates) that were injected into and thus, only covers a subset of the design.

CFCSS checks static control flow graphs and is implemented via compiler modification similar to that in Ref. [25]. It is possible to analyze CFCSS in further detail to gain deeper understanding as to why improvement for the technique is relatively low (Table 12). Compared to DFC (a technique with a similar concept), it can be seen that CFCSS offers slightly better SDC improvement. However, since CFCSS only checks control flow signatures, many SDCs will still escape (e.g., the result of an add is corrupted and written to file). Additionally, certain DUEs, such as those which may cause a program crash, will not be detectable by CFCSS, or other software techniques, since execution may abort before a corresponding software check can be triggered. The relatively low resilience improvement using CFCSS has been corroborated in actual systems as well [57].

Error detection by duplicated instructions (*EDDI*) provides instruction redundant execution via compiler modification [26]. This chapter utilizes EDDI with store-readback [54] to maximize coverage by ensuring that values are written correctly.

Table 12 CFCSS Error Coverage

	SDC	DUE
% flip-flops with a SDC-/DUE-causing error that is detected by CFCSS	55	66
% of SDC-/DUE-causing errors that are detected per FF that is protected by CFCSS	61	14
Resulting improvement (Eqs. 1a, 1b)	1.5×	0.5×

Table 13 EDDI: Importance of Store-Readback

	SDC Improvement	% SDC Errors Detected	SDC Errors Escaped	DUE Improvement	% DUE Errors Detected	DUE Errors Escaped
Without store-readback	3.3×	86.1	49	0.4×	19	3090
With store-readback	37.8×	98.7	6	0.3×	19.8	3006

From Table 13, it is clear why store-readback is important for EDDI. In order to achieve high SDC improvements, nearly all SDC-causing errors need to be detected. By detecting an additional 12% of SDCs, store-readback increases SDC improvement of EDDI by an order of magnitude. Virtually all escaped SDCs are caught by ensuring that the values being written to the output are indeed correct (by reading back the written value). However, given that some SDC- or DUE-causing errors are still not detected by the technique, the results show that using naïve high-level injections will still yield incorrect conclusions (Table 14). Enhancements to EDDI such as Error detectors [58] and reliability-aware transforms [59], are intended to reduce the number of EDDI checks (i.e., selective insertion of checks) in order to minimize execution time impact while maintaining high overall error coverage. An evaluation of the Error detectors technique using flip-flop-level error injection found that the technique provides an SDC improvement of 2.6× improvement (a 21% reduction in SDC improvement as compared to EDDI without store-readback). However, Error detectors requires software path tracking to recalculate important variables, which introduced a 3.9× execution time impact, greater than that of the original EDDI technique. The overhead corresponding to software path tracking can be reduced by implementing path tracking in hardware (as was done in the original work), but doing so eliminates the benefits of EDDI as a software-only technique.

Algorithm: *ABFT* can detect (*ABFT detection*) or detect and correct errors (*ABFT correction*) through algorithm modifications [20,21,60,61]. Although ABFT

Table 14 Comparison of SDC Improvement and Detection for EDDI When Injecting Errors at Various Levels (Without Store-Readback)

Injection Location	SDC Improvement	% SDC Detected
Flip-flop (ground truth)	3.3×	86.1
Register Uniform (RegU)	2.0×	48.8
Register Write (RegW)	6.6×	84.8
Program Variable Uniform (VarU)	12.6×	92.1
Program Variable Write (VarU)	100,000×	100

correction algorithms can be used for detection-only (with minimally reduced execution time impact), ABFT detection algorithms cannot be used for correction. There is often a large difference in execution time impact between ABFT algorithms as well depending on the complexity of check calculation required. An ABFT correction technique for matrix inner product, for example, requires simple modular checksums (e.g., generated by adding all elements in a matrix row)—an inexpensive computation. On the other hand, ABFT detection for fast fourier transform (FFT), for example, requires expensive calculations using Parseval's theorem [62]. For the particular applications studied, the algorithms that were protected using ABFT detection often required more computationally expensive checks than algorithms that were protected using ABFT correction; therefore, the former generally had greater execution time impact (relative to each of their own original baseline execution times). An additional complication arises when an ABFT detection-only algorithm is implemented. Due to the long error detection latencies imposed by ABFT detection (9.6 million cycles, on average), hardware recovery techniques are not feasible and higher level recovery mechanisms will impose significant overheads.

Recovery: Two recovery scenarios are considered: *bounded latency*, i.e., an error must be recovered within a fixed period of time after its occurrence, and *unconstrained*, i.e., where no latency constraints exist and errors are recovered externally once detected (no hardware recovery is required). Bounded latency recovery is achieved using one of the following hardware recovery techniques (Table 15): *flush* or *reorder buffer* (*RoB*) recovery (both of which rely on flushing noncommitted instructions followed by re-execution) [33,34]; *instruction replay* (*IR*) or *extended instruction replay* (*EIR*) recovery (both of which rely on instruction checkpointing to rollback and replay instructions) [10]. EIR is an extension of IR with additional

Table 15 Hardware Error Recovery Costs

Core	Type	Area (%)	Power (%)	Energy (%)	Recovery Latency	Unrecoverable Flip-Flop Errors
InO	Instruction replay (IR) recovery	16	21	21	47 cycles	None (all pipeline FFs recoverable)
	EIR recovery	34	32	32	47 cycles	
	Flush recovery	0.6	0.9	1.8	7 cycles	FFs after memory write stage
OoO	IR recovery	0.1	0.1	0.1	104 cycles	None (all pipeline FFs recoverable)
	EIR recovery	0.2	0.1	0.1		104 cycles
	Reorder buffer (RoB) recovery	0.01	0.01	0.01	64 cycles	FFs after reorder buffer stage

buffers required by DFC for recovery. Flush and RoB are unable to recover from errors detected after the memory write stage of InO-cores or after the RoB of OoO-cores, respectively (these errors will have propagated to architecture visible states). Hence, LEAP-DICE is used to protect flip-flops in these pipeline stages when using flush/RoB recovery. IR and EIR can recover detected errors in any pipeline flip-flop. IR recovery is shown in Fig. 4 and flush recovery is shown in Fig. 5. Since recovery hardware serves as single points of failure, flip-flops in the recovery hardware itself needs to be capable of error correction (e.g., protected using hardened flip-flops when considering soft errors).

Additional techniques: Many additional resilience techniques have been published in literature; but, these techniques are closely related to our evaluated techniques. Therefore, the results presented in this chapter are believed to be representative and largely cover the cross-layer design space.

At the circuit-level, hardened flip-flops like DICE (dual interlocked storage cell) [63], BCDMR (bistable cross-coupled dual modular redundancy) [64], and BISER (built in soft error resilience) [65] are similar in cost to LEAP-DICE, the most

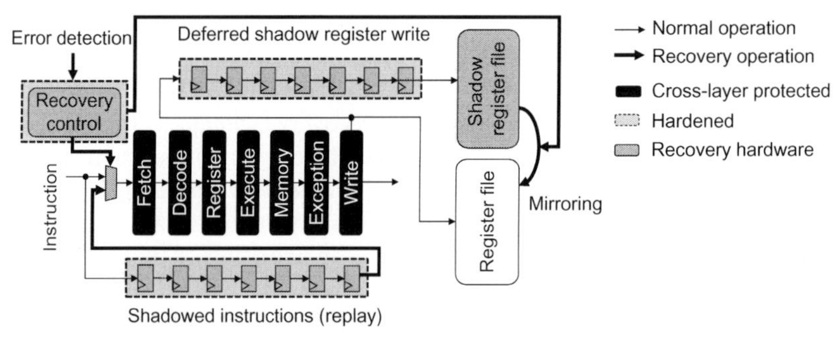

FIG. 4

Instruction replay (IR) recovery.

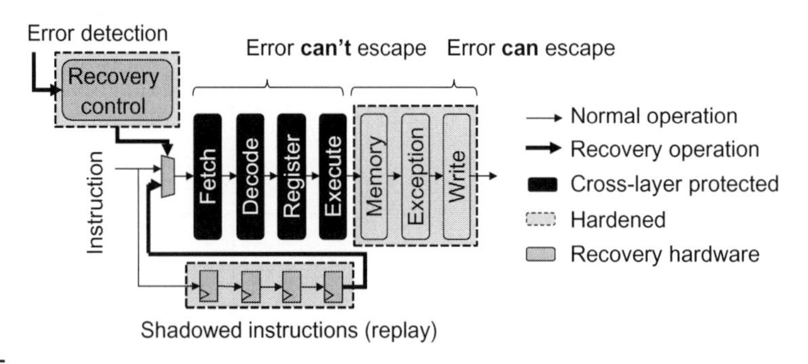

FIG. 5

Flush recovery.

resilient hardened flip-flop studied. The DICE technique suffers from an inability to tolerate SEMUs, unlike LEAP-DICE. BISER is capable of operating in both economy and resilient modes. This enhancement is provided by LEAP-ctrl. Hardened flip-flops like RCC (reinforcing charge collection) [13] offer around 5 × soft error rate improvement at around 1.2 × area, power, and energy cost. LHL provides slightly more soft error tolerance at roughly the same cost as RCC. Circuit-level detection techniques such as those in Ref. [66–68] are similar to EDS. Like EDS, these techniques can detect soft errors while offering minor differences in actual implementation. Stability checking [69] works on a similar principle of time sampling to detect errors.

Logic-level techniques like residue codes [9] can be effective for specific functional units like multipliers, but are costlier to implement than the simple XOR-trees used in logic parity. Additional logic level coding techniques like Berger codes [70] and Bose-Lin codes [71] are costlier to implement than logic parity. Like logic parity checking, residue, Berger, and Bose-Lin codes only detect errors.

Techniques like DMR (dual modular redundancy) and TMR (triple modular redundancy) at the architecture level can be easily ruled out since these techniques will incur more than 100% area, power, and energy costs. RMT (redundant multithreading) [72] has been shown to have high (>40%) energy costs (which can increase due to recovery since RMT only serves to detect errors). Additionally, RMT is highly architecture dependent, which limits its applicability.

Software techniques like Shoestring [73], Error detectors [58], Reliability-driven transforms [59], and SWIFT [74] are similar to EDDI, but offer variations to the technique by reducing the number of checks added. As a result, EDDI can be used as a bound on the maximum error detection possible. An enhancement to SWIFT, known as CRAFT [75], uses hardware acceleration to improve reliability, but doing so eliminates the benefit of EDDI as a software-only technique. Although it is difficult to faithfully compare these "selective" EDDI techniques as published (since the original authors evaluated improvements using high-level error injection at the architecture register level which are generally inaccurate), the published results for these "selective" EDDI techniques show insufficient benefit (Table 16). Enhancements

Table 16 Comparison of "Selective" EDDI Techniques as Reported in the Literature Compared to EDDI Evaluated Using Flip-Flop-level Error Injection

	Error Injection	SDC Improvement	Exec. Time Impact
EDDI with store-readback (implemented)	Flip-flop	37.8 ×	2.1 ×
Reliability-aware transforms (published)	Arch. reg.	1.8 ×	1.05 ×
Shoestring (published)	Arch. reg.	5.1 ×	1.15 ×
SWIFT (published)	Arch. reg.	13.7 ×	1.41 ×

which reduce the execution time impact provide very low SDC improvements, while those that provide moderate improvement incur high execution time (and thus, energy) impact (much higher than providing the same improvement using LEAP-DICE, for instance). Fault screening [33] is an additional software-level technique. However, this technique also checks to ensure intermediate values computed during execution fall within expected bounds, which is similar to the mechanisms behind Software assertions for general-purpose processors, and thus, is covered by the latter.

Low-level techniques: Resilience techniques at the circuit and logic layer (i.e., *low-level techniques*) are tunable as they can be selectively applied to individual flip-flops. As a result, a range of SDC/DUE improvements can be achieved for varying costs (Table 17). These techniques offer the ability to finely tune the specific flip-flops to protect in order to achieve the degree of resilience improvements required.

High-level techniques: In general, techniques at the architecture, software, and algorithm layers (i.e., *high-level techniques*) are less tunable as there is little control of the exact subset of flip-flops a high-level technique will protect. From Table 3, it can be seen that no high-level technique provides more than 38 × improvement (while most offer far less improvement). As a result, to achieve a 50 × improvement, for example, augmentation from low-level techniques at the circuit- and logic-level are required, regardless.[12,13]

3 CROSS-LAYER COMBINATIONS

CLEAR uses a top-down approach to explore the cost-effectiveness of various cross-layer combinations. For example, resilience techniques at the upper layers of the system stack (e.g., ABFT correction) are applied before incrementally moving down the stack to apply techniques from lower layers (e.g., an optimized combination of logic parity checking, circuit-level LEAP-DICE, and micro-architectural recovery). This approach (example shown in Fig. 6) ensures that resilience techniques from various layers of the stack effectively interact with one another. Resilience techniques from the algorithm, software, and architecture layers of the stack generally protect multiple flip-flops (determined using error injection); however, a designer typically has little control over the specific subset protected. Using multiple resilience techniques from these layers can lead to situations where a given flip-flop may be protected (sometimes unnecessarily) by multiple techniques. At the logic and circuit layers, fine-grained protection is available since these techniques can be applied selectively to individual flip-flops (those not sufficiently protected by higher-level techniques).

A total of 586 cross-layer combinations are explored using CLEAR (Table 18). Not all combinations of the ten resilience techniques and four recovery techniques

[12]Costs are generated per benchmark. The average cost over all benchmarks is reported. Relative standard deviation is 0.6–3.1%.

[13]DUE improvements are not possible with detection-only techniques given unconstrained recovery.

Table 17 Costs vs SDC and DUE Improvements for Tunable Resilience Techniques[12]

			Bounded Latency Recovery										Unconstrained Recovery[13]										Exec. Time Impact (%)
			SDC Improvement					DUE Improvement					SDC Improvement					DUE Improvement					
			2	5	50	500	max	2	5	50	500	max	2	5	50	500	max	2	5	50	500	max	
InO	Selective hardening using LEAP-DICE	A	0.8	1.8	2.9	3.3	9.3	0.7	1.7	3.8	5.1	9.3	0.8	1.8	2.9	3.3	9.3	0.7	1.7	3.8	5.1	9.3	0
		E	2	4.3	7.3	8.2	22.4	1.5	3.8	9.5	12.5	22.4	2	4.3	7.3	8.2	22.4	1.5	3.8	9.5	12.5	22.4	
	Logic parity only (+IR recovery)	A	17.3	18.6	20.3	20.7	26.9	16.9	18.3	21.5	22.8	23.3	1.3	2.6	4.3	4.7	10.9	–	–	–	–	–	0
		E	23.4	26	29.4	30.5	44.1	22.5	25.4	31.9	35	35.9	2.4	5	8.4	9.5	23.1						
	EDS-only (+IR recovery)	A	17.1	18.1	19.7	20.5	26.7	16.8	18	20.3	22.5	26.2	1.1	2.1	3.7	4.5	10.7	–	–	–	–	–	0
		E	23.1	25.4	28.5	29.6	43.9	22.1	25.2	31.5	39.2	43.7	2.1	4.4	7.5	8.6	22.9						
OoO	Selective hardening using LEAP-DICE	A	1.1	1.3	2.2	2.4	6.5	1.3	1.6	3.1	3.6	6.5	1.1	1.3	2.2	2.4	6.5	1.3	1.6	3.1	3.6	6.5	0
		E	1.5	1.7	3.1	3.5	9.4	2	2.3	4.2	5.1	9.4	1.5	1.7	3.1	3.5	9.4	2	2.3	4.2	5.1	9.4	
	Logic parity only (+IR recovery)	A	1.9	2.1	6.1	6.3	14.2	1.7	2.6	4.5	5	13.8	1.8	2	5.9	6.2	14.1	–	–	–	–	–	0
		E	1.6	2.4	4.1	5.1	13	2.4	3	4.4	5.4	13.6	1.5	2.3	4	5	13.6						
	EDS-only (+IR recovery)	A	1.4	1.8	3.3	4	12.3	1.3	2	3.6	4	11.8	1.3	1.7	3.2	3.9	12.2	–	–	–	–	–	0
		E	1.7	2.1	3.5	4	11.6	2.1	2.5	4.4	5.3	11.4	1.6	2	3.4	3.9	11.5						

A (area cost %), P (power cost %), E (energy cost %) (P=E for these combinations—no clock/execution time impact).

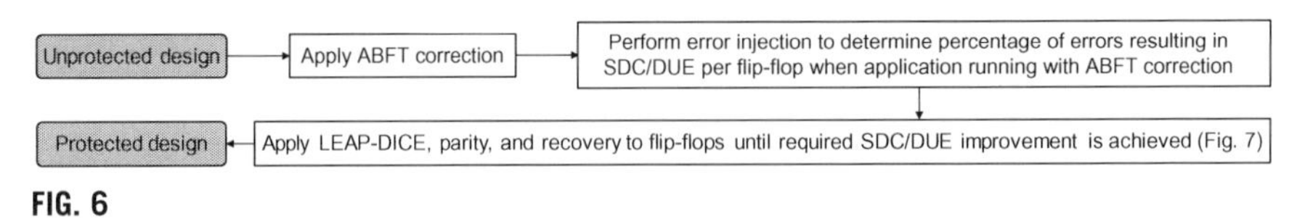

FIG. 6

Cross-layer methodology example for combining ABFT correction, LEAP-DICE, logic parity, and micro-architectural recovery.

Table 18 Creating 586 Cross-Layer Combinations

		No Recovery	Flush/RoB Recovery	IR/EIR Recovery	Total
InO	Combinations of LEAP-DICE, EDS, parity, DFC, Assertions, CFCSS, EDDI	127	3	14	144
	ABFT correction/ detection alone	2	0	0	2
	ABFT correction +previous combinations	127	3	14	144
	ABFT detection +previous combinations	127	0	0	127
	InO-core total	–	–	–	417
OoO	Combinations of LEAP-DICE, EDS, parity, DFC, monitor cores	31	7	30	68
	ABFT correction/ detection alone	2	0	0	2
	ABFT correction +previous combinations	31	7	30	68
	ABFT detection +previous combinations	31	0	0	31
	OoO-core total	–	–	–	169
		Combined total			586

are valid (e.g., it is unnecessary to combine ABFT correction and ABFT detection since the techniques are mutually exclusive or to explore combinations of monitor cores to protect an InO-core due to the high cost). Accurate flip-flop level injection and layout evaluation reveals many individual techniques provide minimal (less than 1.5 ×) SDC/DUE improvement, have high costs, or both. These conclusions are contrary to those reported in the literature (the latter having been derived using inaccurate architecture- or software-level injection). The consequence of this revelation is that most cross-layer combinations have high cost (detailed results for these costly combinations are omitted for brevity but are shown in Fig. 1).

3.1 COMBINATIONS FOR GENERAL-PURPOSE PROCESSORS

Among the 586 cross-layer combinations explored using CLEAR, a highly promising approach combines selective circuit-level hardening using LEAP-DICE, logic parity, and micro-architectural recovery (flush recovery for InO-cores, RoB recovery for OoO-cores). Thorough error injection using application benchmarks plays a critical role in selecting the flip-flops protected using these techniques. Fig. 7 and Heuristic 1 detail the methodology for creating this combination. If recovery is not needed (e.g., for unconstrained recovery), the "Harden" procedure in Heuristic 1 can be modified to always return false.

For example, to achieve a 50 × SDC improvement, the combination of LEAP-DICE, logic parity, and micro-architectural recovery provides a 1.5 × and 1.2 × energy savings for the OoO- and InO-cores, respectively, compared to selective circuit hardening using LEAP-DICE (Table 19). The relative benefits are consistent across benchmarks and over the range of SDC/DUE improvements. The overheads in Table 19 are small because only the most energy-efficient resilience solutions are reported. Most of the 586 combinations are far costlier.

The scenario where recovery hardware is not needed (e.g., unconstrained recovery) can also be considered. In this case, a minimal (<0.2% energy) savings can be achieved when targeting SDC improvement. However, without recovery hardware, DUEs increase since detected errors are now uncorrectable; thus, no DUE improvement is achievable.

Finally, one may suppose that the inclusion of EDS into cross-layer optimization may yield further savings since EDS costs ~25% less area, power, energy than LEAP-DICE. However, a significant portion of EDS overhead is not captured solely by cell overhead. In fact, the additional cost of aggregating and routing the EDS error detection signals and the cost of adding delay buffers to satisfy minimum delay

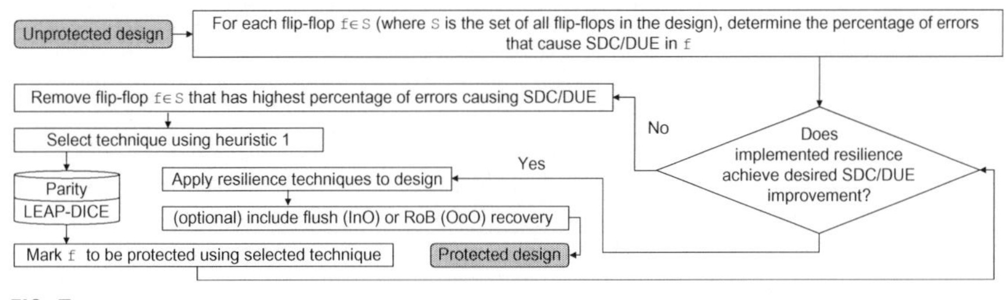

FIG. 7

Cross-layer resilience methodology for combining LEAP-DICE, parity, and micro-architectural recovery.

Table 19 Costs vs SDC and DUE Improvements for Various Combinations in General-Purpose Processors

			Bounded Latency Recovery										Unconstrained Recovery										Exec. Time Impact (%)
			SDC Improvement					DUE Improvement					SDC Improvement					DUE Improvement					
			2	5	50	500	max	2	5	50	500	max	2	5	50	500	max	2	5	50	500	max	
InO	LEAP-DICE +logic parity (+flush recovery)	A	0.7	1.7	2.5	3	8	0.6	1.5	3.6	4.4	8	0.7	1.6	2.4	2.8	7.6	–	–	–	–	–	0
		P	1.9	3.9	6.1	6.7	17.9	1.5	3.4	8.4	10.4	17.9	1.9	3.8	5.9	6.5	17.2						
		E	1.9	3.9	6.1	6.7	17.9	1.5	3.4	8.4	10.4	17.9	1.9	3.8	5.9	6.5	17.2						
	EDS +LEAP-DICE+logic parity (+flush recovery)	A	0.9	2.3	2.7	3.3	8.4	0.8	2.1	3.8	4.8	8.4	0.9	2.2	2.5	3.2	8.1	–	–	–	–	–	0
		P	1.9	4.3	6.6	7.2	19.3	1.7	3.8	8.5	11	19.3	1.9	4.2	6.3	7.1	19						
		E	1.9	4.3	6.6	7.2	19.3	1.7	3.8	8.5	11	19.3	1.9	4.2	6.3	7.1	19						
	DFC +LEAP-DICE+logic parity (+EIR recovery)	A	39.3	41.1	41.5	43.1	45	39.3	39.9	41.9	42.5	45	3.3	5.1	5.6	7.1	10.6	–	–	–	–	–	6.2
		P	32.4	35.5	38.7	41	50.9	32.5	33.9	38.4	40.7	50.9	1.4	4.8	8.1	10	18.22						
		E	44.2	56.7	60.2	62.7	60.3	45.8	48.9	58.3	63	60.3	10.6	13.9	17.4	19.9	5.5						
	Assertions +LEAP-DICE+logic parity (no recovery)	A	–	–	–	–	–	–	–	–	–	–	0.7	0	1	1.1	7.6	–	–	–	–	–	15.6
		P											1.4	1.8	2.2	2.2	17.2						
		E											17.1	17.5	18	18	24.5						
	CFCSS +LEAP-DICE+logic parity (no recovery)	A	–	–	–	–	–	–	–	–	–	–	0.3	1	1.4	1.3	7.6	–	–	–	–	–	40.6
		P											0.8	1.8	2.9	3.1	17.2						
		E											41.5	43	44.6	44.9	64.8						
	EDDI +LEAP-DICE+logic parity (no recovery)	A	–	–	–	–	–	–	–	–	–	–	0	0	0.7	0.9	7.6	–	–	–	–	–	110
		P											0	0	0.6	0.8	17.2						
		E											110	110	111	111	146						

OoO	LEAP-DICE +logic parity (+RoB recovery)	A	0.06	0.1	1.4	2.2	4.9	0.5	0.7	2.6	3	4.9	0.06	0.1	1.4	2.2	4.9	–	–	–	–	–	0
		P	0.1	0.2	2.1	2.4	7	0.1	0.1	2	1.8	7	0.1	0.2	2.1	2.4	7						
		E	0.1	0.2	2.0	2.4	7	0.1	0.1	2	1.8	7	0.1	0.2	2.1	2.4	7						
	EDS +LEAP-DICE+logic parity (+RoB recovery)	A	0.07	0.1	1.6	2.2	5.4	0.6	0.8	2.6	3	5.4	0.07	0.1	1.6	2.2	5.4	–	–	–	–	–	0
		P	0.1	0.2	2.3	2.5	8.1	0.1	0.1	2	1.8	8.1	0.1	0.2	2.3	2.5	8.1						
		E	0.1	0.2	2.3	2.5	8.1	0.1	0.1	2	1.8	8.1	0.1	0.2	2.3	2.5	8.1						
	DFC +LEAP-DICE+logic parity (+EIR recovery)	A	0.2	1	1.8	2	5.3	0.2	0.4	1.7	3.9	5.3	0.1	0.8	1.6	1.8	5.1	–	–	–	–	–	7.1
		P	1.1	1.4	2	2.8	7.2	0.2	0.2	2.6	3.3	7.2	1	1.3	1.9	2.7	7.1						
		E	21.2	21.5	22.2	23	14.8	20	20.1	22.9	23.6	14.8	10	11.4	12.1	12.9	14.7						
	Monitor core +LEAP-DICE+logic parity (+RoB rec.)	A	9	9	9.8	10.5	13.9	9	9	10.1	11.2	13.9	9	9	9.8	10.5	13.9	–	–	–	–	–	0
		P	16.3	16.3	20	20.2	23.3	16.3	16.3	20.1	21.5	22.3	16.3	16.3	20	20.2	23.3						
		E	16.3	16.3	20	20.2	23.3	16.3	16.3	20.1	21.5	22.3	16.3	16.3	20	20.2	23.3						

A (area cost %), P (power cost %), E (energy cost %).

constraints posed by EDS dominate cost and prevents cross-layer combinations using EDS from yielding benefits (Table 19). Various additional cross-layer combinations spanning circuit, logic, architecture, and software layers are presented in Table 19.

Heuristic 1: Choose LEAP-DICE or parity technique
Input: `f`: flip-flop to be protected
Output: Technique to apply to `f` (LEAP-DICE, parity)

```
1: if HARDEN(f) then return LEAP-DICE
2: if PARITY(f) then return parity
3: return LEAP-DICE

4: procedure HARDEN(f)
5:    if an error in f cannot be flushed (i.e., f is in the memory, exception, writeback stages of
        InO or after the RoB of OoO)
6:    then return TRUE; else return FALSE
7: end procedure

8: procedure PARITY(f)
9:    if f has timing path slack greater than delay imposed by 32-bit XOR-tree (this implements
        low cost parity checking as explained in Section 2.4)
10:   then return TRUE, else return FALSE
11: end procedure
```

Up to this point, SDC and DUE improvements have been considered separately. However, it may be useful to achieve a specific improvement in SDC and DUE simultaneously. When targeting SDC improvement, DUE improvement also improves (and vice-versa); however, it is unlikely that the two improvements will be the same since flip-flops with high SDC vulnerability will not necessarily be the same flip-flops that have high DUE vulnerability. A simple method for targeting joint SDC/DUE improvement is to implement resilience until SDC (DUE) improvement is reached and then continue implementing resilience to unprotected flip-flops until DUE (SDC) improvement is also achieved. This ensures that both SDC and DUE improvement meet (or exceed) the targeted minimum required improvement.

Table 20 Cost to Achieve Joint SDC/DUE Improvement With a Combination of LEAP-DICE, Parity, and Flush/RoB Recovery

Joint SDC/DUE Improvement	InO			OoO		
	Area (%)	Power (%)	Energy (%)	Area (%)	Power (%)	Energy (%)
2×	0.7	2	2	0.6	0.1	0.1
5×	1.9	4.2	4.2	0.9	0.4	0.4
50×	4.1	9	9	2.8	2.2	2.2
500×	4.6	10.8	10.8	3.1	2.8	2.8
max	8	17.9	17.9	4.9	7	7

Table 20 details the costs required to achieve joint SDC/DUE improvement using this methodology when considering a combination of LEAP-DICE, parity, and flush/RoB recovery.

3.2 TARGETING SPECIFIC APPLICATIONS

When the application space targets specific algorithms (e.g., matrix operations), a cross-layer combination of LEAP-DICE, parity, ABFT correction, and micro-architectural error recovery (flush/RoB) provides additional energy savings (compared to the general-purpose cross-layer combinations presented in Section 3.1). Since ABFT correction performs in-place error correction, no separate recovery mechanism is required for ABFT correction. For this study, it is possible to apply ABFT correction to three of the PERFECT benchmarks: `2d_convolution`, `debayer_filter`, and `inner_product` (the rest were protected using ABFT detection).

The results in Table 21 confirm that combinations of ABFT correction, LEAP-DICE, parity, and micro-architectural recovery provide up to $1.1\times$ and $2\times$ energy savings over the previously presented combination of LEAP-DICE, parity, and recovery when targeting SDC improvement for the OoO- and InO-cores, respectively. However, as will be discussed in Section 3.2.1, the practicality of ABFT is limited when considering general-purpose processors.

When targeting DUE improvement, including ABFT correction provides no energy savings for the OoO-core. This is because ABFT correction (along with most architecture and software techniques like DFC, CFCSS, and Assertions) performs checks at set locations in the program. For example, a DUE resulting from an invalid pointer access can cause an immediate program termination before a check is invoked. As a result, this DUE would not be detected by the resilience technique.

Although ABFT correction is useful for general-purpose processors limited to specific applications, the same cannot be said for ABFT detection (Table 21). Fig. 8 shows that, since ABFT detection cannot perform in-place correction, ABFT detection benchmarks cannot provide DUE improvement (any detected error necessarily increases the number of DUEs). Additionally, given the lower average SDC improvement and generally higher execution time impact for ABFT detection algorithms, combinations with ABFT detection do not yield low-cost solutions.

3.2.1 Additional considerations for ABFT

Since most applications are not amenable to ABFT correction, the flip-flops protected by ABFT correction must also be protected by techniques such as LEAP-DICE or parity (or combinations thereof) for processors targeting general-purpose applications. This requires circuit hardening techniques (e.g., Refs. [65,76]) with the ability to selectively operate in an error-resilient mode (high resilience, high energy) when ABFT is unavailable, or in an economy mode (low resilience, low-power mode) when ABFT is available. The LEAP-ctrl flip-flop accomplishes this task. The addition of LEAP-ctrl can incur an additional ~1% energy cost and ~3% area cost (Table 21).

Table 21 Costs vs SDC and DUE Improvement for Various Cross-Layer Combinations Involving ABFT

			Bounded Latency Recovery										Unconstrained Recovery										Exec. Time Impact (%)
			SDC Improvement					DUE Improvement					SDC Improvement					DUE Improvement					
			2	5	50	500	max	2	5	50	500	max	2	5	50	500	max	2	5	50	500	max	
InO	ABFT correction +LEAP-DICE +logic parity (+flush recovery)	A	0	0.4	1.0	1.2	8	0.3	0.4	1.5	2.7	8	0	0.4	0.9	1.1	7.6	–	–	–	–	–	1.4
		P	0	0.7	1.7	1.8	17.9	1	1	3.3	5.7	17.9	0	0.7	1.6	1.8	17.2						
		E	1.4	2.2	3.1	3.2	19.6	2.4	2.4	4.8	7.2	19.6	1.4	2.2	3	3.2	18.8						
	ABFT detection +LEAP-DICE +logic parity (no recovery)	A	–	–	–	–	–	–	–	–	–	–	0	1.2	2	2.5	7.6	–	–	–	–	–	24
		P											0	2.4	4.8	5.7	17.2						
		E											1.4	27	30	31.1	45.3						
	ABFT correction +LEAP-ctrl +LEAP-DICE +logic parity (+flush recovery)	A	1.5	2.5	3.8	4.1	8	1	1	4.1	5	8	1.5	2.3	3.4	4	7.6	–	–	–	–	–	1.4
		P	0.6	1.3	2.6	2.8	17.9	1.3	1.3	4.6	7	17.9	0.6	1.2	2.6	2.7	17.2						
		E	1.9	2.7	4.0	4.2	19.6	2.8	2.8	6.1	8.5	19.6	1.9	2.6	4	4.1	18.8						
OoO	ABFT correction +LEAP-DICE +logic parity (+RoB recovery)	A	0	0.01	0.3	0.5	4.9	0.4	0.6	2.1	3	4.9	0	0.01	0.3	0.5	4.8	–	–	–	–	–	1.4
		P	0	0.01	0.5	0.8	7	0.1	0.1	3	1.6	7	0	0.01	0.5	0.8	6.9						
		E	1.4	1.5	1.9	2.2	8.5	1.5	1.5	4.2	3	8.5	1.4	1.5	1.9	2.2	8.4						
	ABFT detection +LEAP-DICE +logic parity (no recovery)	A	–	–	–	–	–	–	–	–	–	–	0	0.1	0.7	1.2	4.8	–	–	–	–	–	24
		P											0	0.2	1.2	1.6	6.9						
		E											24	24.2	25.5	26	32.6						
	ABFT correction +LEAP-ctrl +LEAP-DICE +logic parity (+RoB recovery)	A	1.5	1.8	2.9	3.2	4.9	0.6	0.9	2.8	3.6	4.9	1.5	1.8	2.9	3.2	4.9	–	–	–	–	–	1.4
		P	0.3	0.3	1.0	1.3	7	0.1	0.1	3	1.6	7	0.3	0.3	1	1.3	6.9						
		E	1.7	1.7	2.5	2.7	8.3	1.5	1.5	4.3	3.1	8.5	1.7	1.7	2.5	2.7	8.4						

A (area cost %), P (power cost %), E (energy cost %).

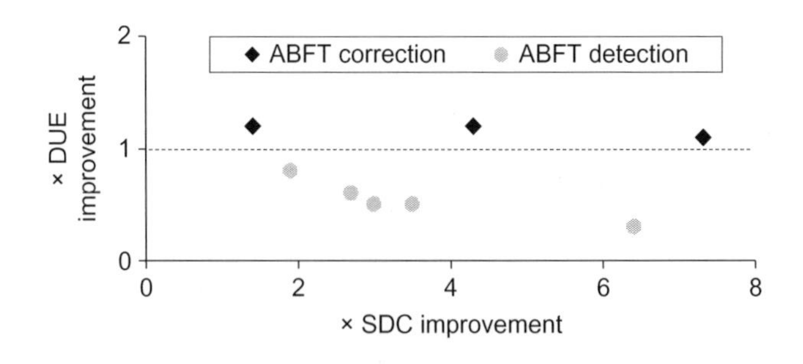

FIG. 8

ABFT correction and ABFT detection benchmark comparison.

Table 22 Impact of ABFT Correction on Flip-Flops

Core	% FFs With an Error Corrected by Any ABFT Algorithm (∪)	% FFs With an Error Corrected by Every ABFT Algorithm (∩)
InO	44	5
OoO	22	2

Although 44% (22% for OoO-cores) of flip-flops would need to be implemented using LEAP-ctrl, only 5% (2% for OoO-cores) would be operating in economy mode at any given time (Table 22). Unfortunately, this requirement of fine-grained operating mode control is difficult to implement in practice since it would require some firmware or software control to determine and pass information to a hardware controller indicating whether or not an ABFT application was running and which flip-flops to place in resilient mode and which to place in economy mode (rather than a simple switch setting all such flip-flops into the same operating mode). Therefore, cross-layer combinations using ABFT correction may not be practical or useful in general-purpose processors targeting general applications.

4 APPLICATION BENCHMARK DEPENDENCE

The most cost-effective resilience techniques rely on selective circuit hardening/parity checking guided by error injection using application benchmarks. This raises the question: what happens when the applications in the field do not match application benchmarks? This situation is referred to as *application benchmark dependence*.

To quantify this dependence, one can randomly select 4 (of 11) SPEC benchmarks as a *training set*, and use the remaining 7 as a *validation set*. Resilience is implemented using the training set and the resulting design's resilience is determined using the validation set. Therefore, the training set tells which flip-flops to protect and the validation set allows for the determination of what the actual improvement would be when this same set of flip-flops is protected. This chapter evaluated 50 training/validation pairs.

Since high-level techniques cannot be tuned to achieve a given resilience improvement, each technique is analyzed as a standalone technique to better understand how they perform individually. For standalone resilience techniques, the average inaccuracy between the results of trained and validated resilience is generally very low (Table 23) and is likely due to the fact that the improvements that the techniques themselves provide are already very low. *p*-Values [77], which provide a measure of how likely the validated improvement and trained improvement would match, are also reported.

Table 24 indicates that validated SDC and DUE improvements are generally underestimated. Fortunately, when targeting $<10\times$ SDC improvement, the underestimation is minimal. This is due to the fact that the most *vulnerable* 10% of flip-flops (i.e., the flip-flops that result in the most SDCs or DUEs) are consistent across benchmarks. Since the number of errors resulting in SDC or DUE is not uniformly distributed among flip-flops, protecting these top 10% of flip-flops will result in the $\sim 10\times$ SDC improvement regardless of the benchmark considered. The vulnerabilities of the remaining 90% of flip-flops are more benchmark-dependent. Concretely, it is possible to analyze benchmark similarity by examining the vulnerable flip-flops indicated by each application benchmark. Per benchmark, one can group the most vulnerable 10% of flip-flops into a subset (e.g., subset 1). The next 10% of vulnerable flip-flops (e.g., 10–20%) are grouped into subset 2 (and so on up to subset 10). Therefore, given the 18 benchmarks used in this chapter, create 18 distinct subset 1's, 18 distinct subset 2's, and so on, are created. Each group of 18 subsets (e.g., all subset 1's) can then be assigned a similarity as given in Eq. (2). The similarity of subset "*x*" is the number of flip-flops that exist in all subset "*x*'s" (e.g., subset intersection) divided by the number of unique flip-flops in every subset "*x*'s" (e.g., subset union). From Table 25, it is clear that only the top 10% most vulnerable flip-flops have very high commonality across all benchmarks (the last 2 subsets have high similarity because these are the flip-flops that have errors that always vanish). All other flip-flops are relatively distributed across the spectrum depending on the specific benchmark being run.

$$\text{Similarity}(\text{subset``}x\text{''}) = \frac{|\cap(\text{all flip}-\text{flops in every subset ``}x\text{''})|}{|\cup(\text{all flip}-\text{flops in every subset ``}x\text{''})|} \tag{2}$$

It is clear that for highly resilient designs, one must develop methods to combat this sensitivity to benchmarks. Benchmark sensitivity may be minimized by training using additional benchmarks or through better benchmarks (e.g., Ref. [78]). An alternative approach is to apply the CLEAR framework using available benchmarks, and then replace all remaining unprotected flip-flops using LHL (Table 4). This enables the resilient designs to meet (or exceed) resilience targets at $\sim 1\%$ additional cost for SDC and DUE improvements $>10\times$.

Table 23 Trained vs Validated Improvement for High-Level Techniques. Underestimation Low Because Improvements are Already Low

		SDC				DUE			
Core	**Technique**	**Train**	**Validate**	**Underestimate (%)**	***p*-Value**	**Train**	**Validate**	**Underestimate (%)**	***p*-Value**
InO	DFC	1.3×	1.2×	−7.7	3.8×10^{-9}	1.4×	1.3×	−7.1	3.9×10^{-17}
	Assertions	1.5×	1.4×	−6.7	2.4×10^{-1}	0.6×	0.6×	0	8×10^{-2}
	CFCSS	1.6×	1.5×	−6.3	5.7×10^{-1}	0.6×	0.6×	0	9.2×10^{-1}
	EDDI	37.8×	30.4×	−19.6	6.9×10^{-1}	0.4×	0.4×	0	2.2×10^{-1}
	ABFT correction	4.3×	3.9×	−9.3	6.7×10^{-1}	1.2×	1.2×	0	1.8×10^{-1}
OoO	DFC	1.3×	1.2×	−7.7	1.9×10^{-5}	1.4×	1.3×	−7.1	1.4×10^{-10}
	Monitor core	19.6×	17.5×	−5.6	8.3×10^{-3}	15.2×	13.9×	−8.6	3.5×10^{-7}
	ABFT correction	4.3×	3.7×	−14	7.2×10^{-1}	1.1×	1.1×	0	1.5×10^{-1}

Table 24 Improvement, Cost Before and After Applying LHL to Otherwise Unprotected Flip-Flops

	SDC Improvement			Cost Before LHL Insertion		Cost After LHL Insertion		DUE Improvement			Cost Before LHL Insertion		Cost After LHL Insertion	
Core	Train	Validate	After LHL	Area (%)	Power/ Energy (%)	Area (%)	Power/ Energy (%)	Train	Validate	After LHL	Area (%)	Power/ Energy (%)	Area (%)	Power/ Energy (%)
InO	5×	4.8×	19.3×	1.6	3.6	3.1	5.7	5×	4.7×	18.7×	1.5	3.4	3.3	5.9
	10×	9.6×	38.2×	1.7	3.9	3.1	5.7	10×	8.7×	34.6×	1.9	4.2	3.5	6.5
	20×	19.1×	75.8×	1.9	4.4	3.2	6.1	20×	16.3×	64.5×	2.4	5.3	3.7	7
	30×	26.8×	105.6×	2.2	4.8	3.2	6.3	30×	23.5×	92.7×	2.8	6.6	3.7	8.1
	40×	32.9×	129.4×	2.3	5.3	3.3	6.7	40×	29.9×	117.6×	3.3	7.5	4.1	8.7
	50×	38.9×	152.3×	2.4	5.7	3.3	6.9	50×	35.9×	140.6×	3.6	8.4	4.2	9.4
	500×	433.1×	1326.1×	2.9	6.3	3.4	7.1	500×	243.5×	840.3×	4.4	10.4	4.8	10.9
	Max	5568.9×	5568.9×	8	17.9	8	17.9	Max	5524.7×	5524.7×	8	17.9	8	17.9
OoO	5×	4.8×	35.1×	0.1	0.2	0.9	1.8	5×	4.4×	28.7×	0.7	0.1	1.8	1.7
	10×	8.8×	40.7×	0.4	0.6	1.1	2.1	10×	8.7×	36.6×	1.1	0.5	2.1	2
	20×	18.8×	65.6×	0.7	1	1.3	2.3	20×	17.3×	70.2×	1.5	0.9	2.5	2
	30×	21.3×	82.3×	0.9	1.4	1.4	2.4	30×	22.2×	81.5×	1.8	1.3	2.6	2.1
	40×	26.4×	130.2×	1.2	1.7	1.7	2.5	40×	26.1×	115.1×	2.1	1.6	2.8	2.4
	50×	32.1×	204.3×	1.4	2.1	1.9	2.7	50×	29.8×	121.3×	2.5	2	3.1	2.6
	500×	301.4×	1084.1×	2.2	2.4	2.4	2.8	500×	153.2×	625.1×	2.9	1.9	3.4	2.7
	Max	6625.8×	6625.8×	4.9	7	4.9	7	Max	6802.6×	6802.6×	4.9	7	4.9	7

Table 25 Subset Similarity Across All 18 Benchmarks for the InO-Core (Subsets Broken into Groups Consisting of 10% of All Flip-Flops)

Subset (Ranked by Decreasing SDC + DUE Vulnerability) (%)	Similarity (Eq. 2)
1: 0–10	0.83
2: 10–20	0.05
3: 20–30	0
4: 30–40	0
5: 40–50	0
6: 50–60	0
7: 60–70	0
8: 70–80	0
9: 80–90	0.71
10: 90–100	1

5 THE DESIGN OF NEW RESILIENCE TECHNIQUES

CLEAR has been used to comprehensively analyze the design space of existing resilience techniques (and their combinations). As new resilience techniques are proposed, CLEAR can incorporate and analyze these techniques as well. However, CLEAR can also be used today to guide the design of new resilience techniques.

All resilience techniques will lie on a two-dimensional plane of energy cost vs SDC improvement (Fig. 9). The range of designs formed using combinations of LEAP-DICE, parity, and micro-architectural recovery forms the lowest-cost cross-layer combination available using today's resilience techniques. In order for new resilience techniques to be able to create competitive cross-layer combinations,

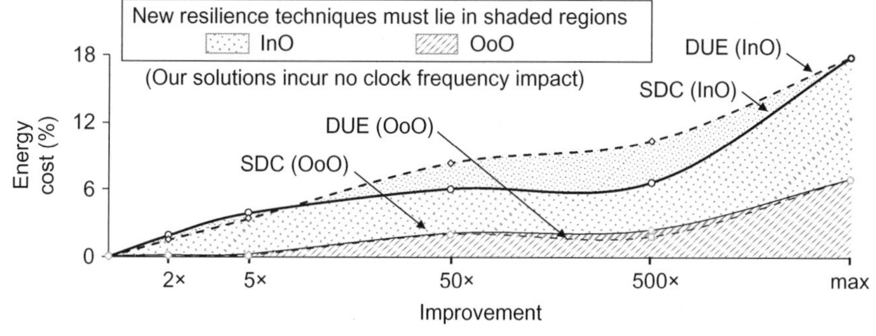

FIG. 9

New resilience techniques must have cost and improvement tradeoffs that lie within the shaded regions bounded by LEAP-DICE + parity + micro-architectural recovery.

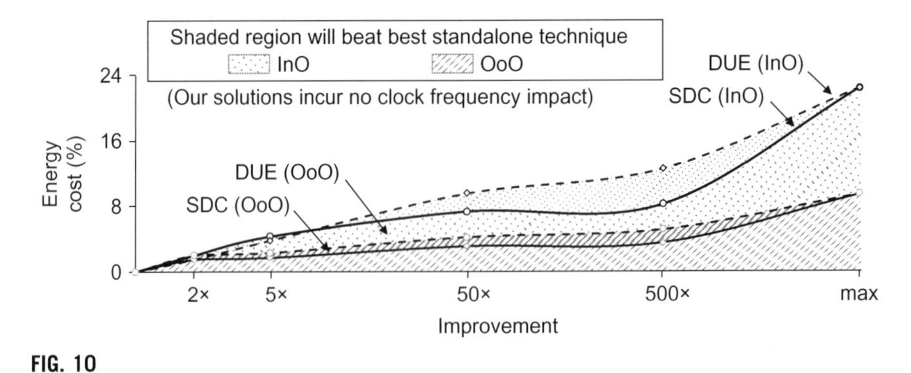

FIG. 10

To be competitive standalone solutions, new resilience techniques must have cost and improvement tradeoffs that lie within the shaded regions bounded by LEAP-DICE.

they must have energy and improvement tradeoffs that place the technique under the region bounded by our LEAP-DICE, parity, and micro-architectural recovery solution. Since certain standalone techniques, like LEAP-DICE, can also provide highly competitive solutions, it is useful to understand the cost vs improvement tradeoffs for new techniques in relation to this best standalone technique as well (Fig. 10).

6 CONCLUSIONS

CLEAR is a first of its kind cross-layer resilience framework that enables effective exploration of a wide variety of resilience techniques and their combinations across several layers of the system stack. Extensive cross-layer resilience studies using CLEAR demonstrate:

1. A carefully optimized combination of selective circuit-level hardening, logic-level parity checking, and micro-architectural recovery provides a highly cost-effective soft error resilience solution for general-purpose processors.
2. Selective circuit-level hardening alone, guided by thorough analysis of the effects of soft errors on application benchmarks, also provides a cost-effective soft error resilience solution (with ~1% additional energy cost for a 50× SDC improvement compared to the earlier approach).
3. ABFT correction combined with selective circuit-level hardening (and logic-level parity checking and micro-architectural recovery) can further improve soft error resilience costs. However, existing ABFT correction techniques can only be used for a few applications; this limits the applicability of this approach in the context of general-purpose processors.
4. Based on this analysis, it is possible to derive bounds on energy costs vs degree of resilience (SDC or DUE improvements) that new soft error resilience techniques must achieve to be competitive.

5. It is crucial that the benefits and costs of new resilience techniques are evaluated thoroughly and correctly. Detailed analysis (e.g., flip-flop-level error injection or layout-level cost quantification) identifies hidden weaknesses that are often overlooked.

While this chapter focuses on soft errors in processor cores, cross-layer resilience solutions for accelerators and uncore components as well as other error sources (e.g., voltage noise) may have different tradeoffs and may require additional modeling and analysis capabilities. However, this CLEAR methodology and framework can be applied to and enables the design of both current and next-generation rugged embedded systems that are harsh environment capable.

ACKNOWLEDGMENTS

This work is supported in part by DARPA MTO (contract no. HR0011-13-C-0022), DTRA, NSF, and SRC. The views expressed are those of the authors and do not reflect the official policy or position of the Department of Defense or the U.S. Government. Distribution Statement "A" (Approved for Public Release, Distribution Unlimited).

REFERENCES

[1] S. Borkar, Designing reliable systems from unreliable components: the challenges of transistor variability and degradation, IEEE Micro 25 (6) (2005) 10–16.
[2] F. Cappello, et al., Toward exascale resilience: 2014 update, Supercomput. Front. Innov. 1 (1) (2014) 5–28.
[3] N.P. Carter, H. Naeimi, D.S. Gardner, Design techniques for cross-layer resilience, in: Design, Automation and Test in Europe, 2010.
[4] A. DeHon, H.M. Quinn, N.P. Carter, Vision for cross-layer optimization to address the dual challenges of energy and reliability, in: Design, Automation, & Test in Europe Conference, 2010.
[5] M.S. Gupta, et al., Cross-layer system resilience at affordable power, in: IEEE International Reliability Physics Symposium, 2014.
[6] J. Henkel, et al., Multi-layer dependability: from microarchitecture to application level, in: ACM/EDAC/IEEE Design Automation Conference, 2014.
[7] M. Pedram, et al., Report for the NSF workshop on cross-layer power optimization and management, in: NSF, 2012.
[8] D.J. Lu, Watchdog processor and structural integrity checking, IEEE Trans. Comput. C-31 (7) (1982) 681–685.
[9] H. Ando, et al., A 1.3-GHz fifth-generation SPARC64 microprocessor, IEEE J. Solid-State Circuits 38 (11) (2003) 1896–1905.
[10] P.J. Meaney, et al., IBM z990 soft error detection and recovery, IEEE Trans. Device Mater. Reliab. 5 (3) (2005) 419–427.
[11] B. Sinharoy, et al., IBM POWER7 multicore server processor, IBM J. Res. Dev. 55 (3) (2011) 1:1–1:29.

[12] S. Mitra, et al., The resilience wall: cross-layer solution strategies, in: International Symposium on VLSI Technology, Systems and Applications, 2014.
[13] N. Seifert, et al., On the radiation-induced soft error performance of hardened sequential element in advanced bulk CMOS technologies, in: IEEE International Reliability Physics Symposium, 2010.
[14] N. Seifert, et al., Soft error susceptibilities of 22 nm tri-gate devices, IEEE Trans. Nucl. Sci. 59 (6) (2012) 2666–2673.
[15] N.N. Mahatme, et al., Impact of supply voltage and frequency on the soft error rates of logic circuits, IEEE Trans. Nucl. Sci. 60 (6) (2013) 4200–4206.
[16] R. Pawlowski, et al., Characterization of radiation-induced SRAM and logic soft errors from 0.33 V to 1.0 V in 65 nm CMOS, in: IEEE Custom Integrated Circuits Conference, 2014.
[17] B. Gill, N. Seifert, V. Zia, Comparison of alpha-particle and neutron-induced combinational and sequential logic error rates at the 32 nm technology node, in: IEEE International Reliability Physics Symposium, 2009.
[18] H.-H.K. Lee, et al., LEAP: layout design through error-aware transistor positioning for soft-error resilient sequential cell design, in: IEEE International Reliability Physics Symposium, 2010.
[19] H. Cho, et al., Understanding soft errors in uncore components, in: ACM/EDAC/IEEE Design Automation Conference, 2015.
[20] Z. Chen, J. Dongarra, Numerically stable real number codes based on random matrices, in: Lecture Notes in Computer Science, 2005.
[21] K.-H. Huang, J.A. Abraham, Algorithm-based fault tolerance for matrix operations, IEEE Trans. Comput. C-33 (6) (1984) 518–528.
[22] D.L. Tao, C.R.P. Hartmann, A novel concurrent error detection scheme for FFT networks, IEEE Trans. Parallel Distrib. Syst. 4 (2) (1993) 198–221.
[23] S.K. Sahoo, et al., Using likely program invariants to detect hardware errors, in: IEEE/IFIP International Conference on Dependable Systems & Networks, 2008.
[24] S.K.S. Hari, S.V. Adve, H. Naeimi, Low-cost program-level detectors for reducing silent data corruptions, in: IEEE/IFIP International Conference on Dependable Systems & Networks, 2012.
[25] N. Oh, P.P. Shirvani, E.J. McCluskey, Control flow checking by software signatures, IEEE Trans. Reliab. 51 (1) (2002) 111–122.
[26] N. Oh, P.P. Shirvani, E.J. McCluskey, Error detection by duplicated instructions in superscalar processors, IEEE Trans. Reliab. 51 (1) (2002) 63–75.
[27] A. Meixner, M.E. Bauer, D.J. Sorin, Argus: low-cost comprehensive error detection in simple cores, in: IEEE/ACM International Symposium on Microarchitecture, 2007.
[28] T.M. Austin, DIVA: a reliable substrate for deep submicron microarchitecture design, in: IEEE/ACM International Symposium on Microarchitecture, 1999.
[29] L. Spainhower, T.A. Gregg, IBM S/390 parallel enterprise server G5 fault tolerance: a historical perspective, IBM J. Res. Dev. 43 (5.6) (1999) 863–873.
[30] K. Lilja, et al., Single-event performance and layout optimization of flip-flops in a 28-nm bulk technology, IEEE Trans. Nucl. Sci. 60 (4) (2013) 2782–2788.
[31] K.A. Bowman, et al., Energy-efficient and metastability-immune resilient circuits for dynamic variation tolerance, IEEE J. Solid-State Circuits 44 (1) (2009) 49–63.
[32] K.A. Bowman, et al., A 45 nm resilient microprocessor core for dynamic variation tolerance, IEEE J. Solid-State Circuits 46 (1) (2011) 194–208.
[33] P. Racunas, et al., Perturbation-based fault screening, in: IEEE International Symposium on High Performance Computer Architecture, 2007.

[34] N.J. Wang, S.J. Patel, ReStore: symptom based soft error detection in microprocessors, in: IEEE/IFIP International Conference on Dependable Systems & Networks, 2005.
[35] Aeroflex G., Leon3 processor, http://www.gaisler.com.
[36] N.J. Wang, et al., Characterizing the effects of transient faults on a high-performance processor pipeline, in: IEEE/IFIP International Conference on Dependable Systems & Networks, 2004.
[37] J.L. Henning, SPEC CPU2000: measuring CPU performance in the new millennium, Computer 33 (7) (2000) 28–35.
[38] DARPA PERFECT benchmark suite, http://hpc.pnl.gov/PERFECT.
[39] C. Bottoni, et al., Heavy ions test result on a 65 nm sparc-v8 radiation-hard microprocessor, in: IEEE International Reliability Physics Symposium, 2014.
[40] P.N. Sanda, et al., Soft-error resilience of the IBM POWER6 processor, IBM J. Res. Dev. 52 (3) (2008) 275–284.
[41] H. Cho, et al., Quantitative evaluation of soft error injection techniques for robust system design, in: ACM/EDAC/IEEE Design Automation Conference, 2013.
[42] J.D. Davis, C.P. Thacker, C. Chang, BEE3: revitalizing computer architecture research, Microsoft Research Technical Report, 2009.
[43] P. Ramachandran, et al., Statistical fault injection, in: IEEE/IFIP International Conference on Dependable Systems & Networks, 2008.
[44] S.E. Michalak, et al., Assessment of the impact of cosmic-ray-induced neutrons on hardware in the roadrunner supercomputer, IEEE Trans. Device Mater. Reliab. 12 (2) (2012) 445–454.
[45] N.J. Wang, A. Mahesri, S.J. Patel, Examining ACE analysis reliability estimates using fault-injection, in: ACM/IEEE International Symposium on Computer Architecture, 2007.
[46] H. Schirmeier, C. Borchert, O. Spinczyk, Avoiding pitfalls in fault-injection based comparison of program susceptibility to soft errors, in: IEEE/IFIP International Conference on Dependable Systems & Networks, 2015.
[47] S. Mirkhani, et al., Efficient soft error vulnerability estimation of complex designs, in: Design, Automation and Test in Europe, 2015.
[48] M. Sullivan, et al., An analytical model for hardened latch selection and exploration, in: Silicon Errors in Logic—System Effects, 2016.
[49] Synopsys, Inc., Synopsys design suite, http://www.synopsys.com.
[50] K. Lilja, et al., SER prediction in advanced finFET and SOI finFET technologies; challenges and comparisons to measurements, in: Single Event Effects Symposium, 2016.
[51] R.C. Quinn, et al., Frequency trends observed in 32 nm SOI flip-flops and combinational logic, in: IEEE Nuclear and Space Radiation Effects Conference, 2015.
[52] R.C. Quinn, et al., Heavy ion SEU test data for 32 nm SOI flip-flops, in: IEEE Radiation Effects Data Workshop, 2015.
[53] M. Turowski, et al., 32 nm SOI SRAM and latch SEU cross-sections measured (heavy ion data) and determined with simulations, in: Single Event Effects Symposium, 2015.
[54] D. Lin, et al., Effective post-silicon validation of system-on-chips using quick error detection, IEEE Trans. Comput. Aided Des. Integr. Circuits Syst. 33 (10) (2014) 1573–1590.
[55] O.A. Amusan, Effects of single-event-induced charge sharing in sub-100 nm bulk CMOS technologies, Vanderbilt University Dissertation, Vanderbilt, Nashville, TN, 2009.
[56] S. Mitra, E.J. McCluskey, Which concurrent error detection scheme to choose? in: IEEE International Test Conference, 2000.
[57] M.N. Lovellette, et al., Strategies for fault-tolerant, space-based computing: lessons learned from the ARGOS testbed, in: IEEE Aerospace Conference, 2002.

[58] K. Pattabiraman, Z. Kalbarczyk, R.K. Iyer, Automated derivation of application-aware error detectors using static analysis: the trusted illiac approach, IEEE Trans. Dependable Secure Comput. 8 (1) (2009) 44–57.
[59] S. Rehman, et al., Reliability-driven software transformations for unreliable hardware, IEEE Trans. Comput. Aided Des. Integr. Circuits Syst. 33 (11) (2014) 1597–1610.
[60] G. Bosilca, et al., Algorithmic based fault tolerance applied to high performance computing, J. Parallel Distrib. Comput. 69 (4) (2009) 410–416.
[61] V.S.S. Nair, J.A. Abraham, Real-number codes for fault-tolerant matrix operations on processor arrays, IEEE Trans. Comput. 39 (4) (1990) 426–435.
[62] A.L.N. Reddy, P. Banerjee, Algorithm-based fault detection for signal processing applications, IEEE Trans. Comput. 39 (10) (1990) 1304–1308.
[63] T. Calin, M. Nicolaidis, R. Velazco, Upset hardened memory design for submicron CMOS technology, IEEE Trans. Nucl. Sci. 43 (6) (1996) 2874–2878.
[64] J. Furuta, et al., A 65 nm bistable cross-coupled dual modular redundancy flip-flop capable of protecting soft errors on the C-element, in: IEEE Symposium on VLSI Circuits, 2010.
[65] S. Mitra, et al., Robust system design with built-in soft error resilience, Computer 38 (2) (2005) 43–52.
[66] D. Blaauw, et al., Razor II: in situ error detection and correction for PVT and SER tolerance, in: IEEE International Solid-State Circuits Conference, 2008.
[67] M. Fojtik, et al., Bubble razor: eliminating timing margins in an ARM cortex-M3 processor in 45 nm CMOS using architecturally independent error detection and correction, IEEE J. Solid-State Circuits 37 (11) (2013) 1545–1554.
[68] M. Nicolaidis, Time redundancy based soft-error tolerance to rescue nanometer technologies, in: IEEE VLSI Test Symposium, 1999.
[69] P. Franco, E.J. McCluskey, On-line delay testing of digital circuits, in: IEEE VLSI Test Symposium, 1994.
[70] J.M. Berger, A note on error detection codes for asymmetric channels, Inf. Control. 4 (1961) 68–73.
[71] B. Bose, D.J. Lin, Systematic unidirectional error-detecting codes, IEEE Trans. Comput. 34 (11) (1985) 63–69.
[72] S.S. Mukherjee, M. Kontz, S.K. Reinhardt, Detailed design and evaluation of redundant multi-threading alternatives, in: ACM/IEEE International Symposium on Computer Architecture, 2002.
[73] S. Feng, et al., Shoestring: probabilistic soft error reliability on the cheap, in: ACM International Conference on Architectural Support for Programming Languages and Operating Systems, 2010.
[74] G. Reis, et al., SWIFT: software implemented fault tolerance, in: IEEE/ACM International Symposium on Code Generation and Optimization, 2005.
[75] G.A. Reis, et al., Design and evaluation of hybrid fault-detection systems, in: ACM/IEEE International Symposium on Computer Architecture, 2005.
[76] M. Zhang, et al., Sequential element design with built-in soft error resilience, in: IEEE Transactions on Very Large Scale Integration (VLSI) Systems, 2006.
[77] R.L. Wasserstein, N.A. Lazar, The ASA's statement on *p*-values: context, process, and purpose, Am. Stat. 70 (2) (2016) 129–133.
[78] S. Mirkhani, B. Samynathan, J.A. Abraham, Detailed soft error vulnerability analysis using synthetic benchmarks, in: IEEE VLSI Test Symposium, 2015.

Index

Note: Page numbers followed by *f* indicate figures, *t* indicate tables, *b* indicate boxes, and *np* indicate footnotes.

D

E

F

G

H

I

N

O

P

S

T

U

V

W

X

Z

Printed in the United States
By Bookmasters